recent advances in phytochemistry

volume 1

Proceedings of the Sixth Annual Symposium of the Phytochemical Society of North America

APPLETON-CENTURY-CROFTS
Division of Meredith Corporation
NEW YORK 1968

618-1

Library of Congress Card Number: 67-26242

PRINTED IN THE UNITED STATES OF AMERICA
F-58420

Dedicated to

Ralph Eugene Alston (1925-1967)
late Professor of Botany, The University of Texas at Austin

The zeal and integrity with which he approached, not only research and teaching, but life itself, represent a continuing inspiration.

Contributors

R. E. ALSTON†, The Department of Botany and the Cell Research Institute, The University of Texas at Austin, Austin, Texas

A. S. DREIDING, Organisch-Chemisches Institut der Universitat Zurich, Zürich, Switzerland

H. ERDTMAN, Institutionen för Organisk Kemi, Kungl. Tekniska Högskolan, Stockholm, Sweden

M. G. ETTLINGER, Organic Chemistry Laboratory, Royal Veterinary and Agricultural College, Copenhagen, Denmark

H. GRISEBACH, Chemisches Laboratorium der Universität Freiburg, Freiburg i. Br., Germany

W. HERZ, Department of Chemistry, Florida State University, Tallahassee, Florida

A. KJAER, Organic Chemical Laboratory, Royal Veterinary and Agricultural College, Copenhagen, Denmark

T. J. MABRY, The Cell Research Institute and Department of Botany, The University of Texas at Austin, Austin, Texas

A. C. OEHLSCHLAGER, Université de Strasbourg, Faculté des Sciences, Institut de Chimie, Strasbourg, France

W. D. OLLIS, Department of Chemistry, The University, Sheffield, England

G. OURISSON, Université de Strasbourg, Faculté des Sciences, Strasbourg, France

G. PONSINET, Université de Strasbourg, Faculté des Sciences, Strasbourg, France

N. A. SØRENSEN, Institut for Organisk Kjemi, Norges Tekniske Høgskole, Trondheim, Norway

F. R. STERMITZ, Department of Chemistry, Colorado State University, Fort Collins, Colorado

W. G. WHALEY, The Cell Research Institute and Department of Botany, The University of Texas at Austin, Austin, Texas

†Deceased

Preface

This volume contains a series of invited papers which were presented at the sixth annual meeting of the Plant Phenolics Group of North America. The meeting, held in Austin, Texas, April 6-8, 1966, was organized as a symposium emphasizing recent advances in phytochemistry and preceded by only a few months the reorganization of the Plant Phenolics Group into the Phytochemical Society of North America. Through a grant from the Graduate School of the University of Texas, funds were provided for the participation in the symposium of a number of invited lecturers from other countries.

It was the intent of the organizers of the symposium that the concept of "recent advances" was definitive enough so that no secondary theme was needed. As expected, the individual speakers dealt variously with the specific groups of compounds representing their special interests. However, since there is considerable interest at this time in the use of natural products in systematic botany, a number of speakers emphasized this interest in chemotaxonomy in the course of their presentations. Furthermore, advances in our knowledge of the chemistry of plant constituents are increasingly connected with information and speculation regarding their biosynthesis. These aspects, too, were emphasized in several presentations.

In addition to the coverage of particular groups of natural products, two special contributions were solicited to provide an increased biological perspective for the audience. Dr. W. G. Whaley, Director of the Cell Research Institute, The University of Texas, began the series of special lectures with a discussion of Chemistry—The Frontier of Biological Science. Dr. Holger Erdtman, of the Royal Institute of Technology, Stockholm, discussed Chemical Principles in Chemosystematics in a special evening lecture. These two papers form the opening chapters of this volume.

An incidental but nonetheless important derivative of the symposium was the repeated emphasis upon the newer analytical techniques available to the chemist, particularly in structure determination, when only small quantities of material are available, and in techniques of separation. These rapid advances in methodology will have a profound effect upon the research goals and general trends in the field of natural product chemistry in the future. It is inevitable that comparative aspects of plant chemistry will be emphasized either with or without a direct focus upon systematics. Substances formerly regarded as trace components, and metabolic intermediates, are being detected and identified with ever-increasing ease, and major advances in our knowledge of the biogenesis of secondary products will surely be forthcoming. Harbingers of these advances are encountered liberally among the papers included in this volume. Literature covered either in the text or as addenda goes through the period to the end of 1966.

The editors are grateful to the contributors who have cooperated fully in this endeavor. We also wish to thank Mrs. Virginia Findeisen, Dr. Madeline Wu and Mrs. Gail Stotland for their many services in the preparation of the manuscripts and in the performance of many other duties associated with the completion of this volume.

T. J. MABRY

R. E. ALSTON

V. C. RUNECKLES

We feel that it is appropriate to dedicate this volume to Dr. Ralph E. Alston. Though only 41 years old at his death, Dr. Alston leaves a record of accomplishments which justly deserve recognition. After completing a B.S. at the College of William and Mary and a Ph.D. at Indiana University in 1955, Dr. Alston joined the Botany Department, The University of Texas, becoming Professor in 1963. Early in his academic life, he set his course, knowing what he wanted to do and how to go about it. His research spanned all aspects of the genetics and chemistry of plant secondary compounds. His publications (more than 50 in number including two books: *Biochemical Systematics and Cellular Continuity and Development* early focused attention on biochemical approaches to biological problems. It is hoped that this volume, to which he contributed both as author and editor, will stand as a small testament to his goals and accomplishments.
to his goals and accomplishments.

T. J. MABRY

V. C. RUNECKLES

Contents

PART III ACETATE AND MEVALONATE-DERIVED COMPOUNDS

PART IV FLAVONOIDS

part I

THE ROLE OF CHEMISTRY IN MODERN BIOLOGY

1

CHEMISTRY–THE FRONTIER OF BIOLOGICAL SCIENCE

W. GORDON WHALEY
The Cell Research Institute
The University of Texas
at Austin

When I was a graduate student, the group in which I was working was attacking the most important biological problem of all—the relation between the gene and the myriad characteristics of the organism. We couldn't take too seriously the dated efforts of our fellow students in physiology or morphology or any of the other classical areas. We had the newly developed concept of the gene well in hand and we were fully confident we could relate all the features of the organism to it.

The attitude we had, with almost a touch of arrogance, is probably good for graduate students. Any contributing scientist gains something from it, but, in time, he learns he must develop techniques for approaching specific problems. I must agree with Theodosius Dobzhansky's remark that, if the biologist is to solve problems, he must be both a reductionist and a compositionist, for the biologist is dealing with populations, species, individuals, organs, tissues, cells, organelles, various macromolecular aggregates and, ultimately, in his reductionist role, with the characters of the molecules and the atoms involved. I want to make a strong case for the biologist's responsibilities in interpreting systems that are far more complex, and integrated at many more levels, than are those systems which the physical scientist ordinarily encounters. But I want to make an equally strong case for the really exciting development that has come with the adaptation of methods of the physical sciences to biological materials. From these advances has emerged something called molecular biology, cellular chemistry, or any of a half dozen other names. Call it what you will, it represents the application of techniques from the physical sciences to problems centered on the heredity, development, and behavior of organisms.

I should like to leave aside the problems of compositionist biology, or, if you want, organismic biology and deal with a few straightforward examples in which experimental designs, techniques, and patterns of interpretation from chemistry have provided at least partial answers to some of the questions raised above.

My first example bears on genetics. Since Correns' (1909) studies in the first decade of the twentieth century, of the inheritance of some unusual instances of leaf coloration, the existence of some sort of units of inheritance in the cytoplasm has seemed likely. There followed, at intervals, convincing demonstrations that this must be so.

In 1946 Rhoades published a now classic paper on the so-called "iojap" mutant in maize, setting forth the distinct possibility that genetic mutation, by then established for chromosomal genes, might also take place in plastids, thus implying the existence of entities analogous to chromosomal genes in these organelles of the cytoplasm. In the meantime there had been accumulating evidence that DNA is the essential unit of inheritance. Subsequently, there was development of the now famous Watson–Crick model of the DNA molecule, which led to plausible ideas about its replication. Then came the knowledge that DNA codes cellular activity through RNA and proteins (for a review, see Muller, 1962). Linked to the accumulation of data about Mendelian inheritance, these developments laid the ground for a fundamental understanding of how the gene might work in inheritance and development, but it left unanswered the question of this so-called cytoplasmic inheritance, which 60-years ago had been shown not to follow a Mendelian pattern. Rhoades' finding of a possible plastid mutation posed a very clear question for the botanist in particular. Is there DNA in the plastid? Or is there some other sort of genetic mechanism involved? The past five

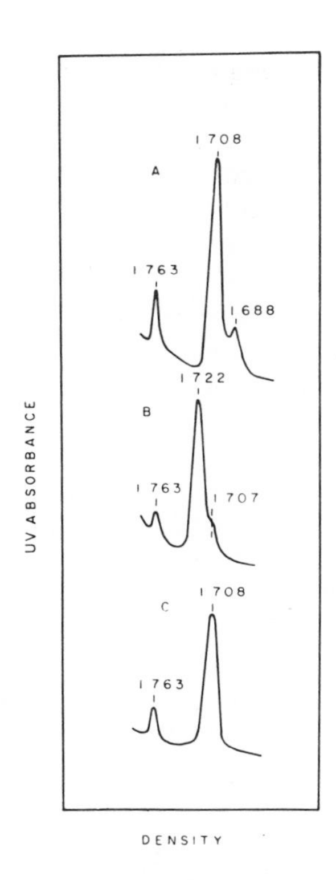

Fig. 1. Microdensitometer tracings of ultraviolet photographs of DNA from *Euglena gracilis* var. *bacillaris* banded in cesium chloride density gradients. A. Native DNA of light-grown wild-type cells. B. Heat-denatured DNA of light-grown wild-type cells. C. Native DNA of cells of a mutant with chloroplasts destroyed by ultraviolet irradiation. For a full description see Schiff and Epstein (1965).

years have brought the answer. There is indeed DNA in the plastids. It is in the form of strands visible in electron micrographs; it is reflected in the differences in base ratios of the RNA in plastids and that in the cytoplasm generally; and in differences in protein synthesis within the plastids and in the cytoplasmic matrix. All the evidence has been accumulated by applying techniques from the physical sciences.

Let me introduce one example that I think makes the point of what chemistry has contributed to the resolution of the problem of cytoplasmic DNA. Figure 1 presents three microdensitometer tracings of ultraviolet photographs of DNA from the green alga *Euglena gracilis* var. *bacillaris*, banded in cesium chloride density gradients. In all three tracings, the 1.73 density band is an added DNA of known density included as a marker. In tracing A, 1.708 is the major DNA band, taken to represent the nuclear DNA. In light-grown cultures there is, however, a satellite band at 1.688. We need not be concerned with tracing B for it simply shows a shift resulting from heat denaturation. Tracing C shows no satellite band. It is possible to destroy the chloroplasts in *Euglena* by exposure to ultraviolet radiation and if the nucleus is protected, the cell may divide to produce a mutant lacking chloroplasts. It was such a mutant that gave tracing C and thus provided proof for the presence of DNA in the plastids (Schiff and Epstein, 1965). One can also remove DNA by treatment with DNAse. Where this has been done one no longer sees those strands that previously had been interpreted as DNA in the chloroplasts. Similar treatment has now also been found to eliminate seemingly comparable strands of DNA from mitochondria. There is no time to detail the story here but the biologist now finds himself faced with the possibility that part of the cell's genome is in the plastids and that, probably, part of it is in the mitochondria. This raises some fundamental questions—not only with respect to cell structure—but also with respect to evolution.

Let's take a quite different example of how chemistry has provided insights into difficult biological problems. The biologist has built up, through essentially mechanical interpretations at first, then through stages of knowledge, models of the form of contractile protein units, and he has a not-yet-complete but increasingly satisfactory concept of how higher organisms carry out the movements that are part of their activities. But he has remained puzzled about the capacity for movement exhibited by the so-called lower organisms. Recently Wohlfarth-Bottermann (1964) at Bonn and then others (McManus and Roth, 1965), including members of our own laboratory, began to demonstrate fibrillar components in the amoeboid plasmodia of slime molds—lowly and, I should add, strange forms of life. Fibrillar layers are found just inside the plasma membrane and around certain of the vacuoles that are more or less constantly changing shape with

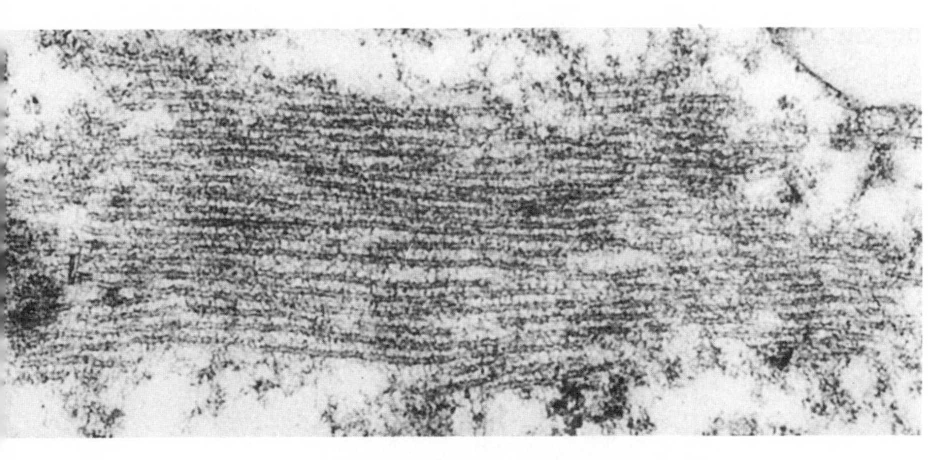

Fig. 2. Filament bundle in the cytoplasm of *Clostoderma debaryanum*. For a full description see McManus and Roth (1965).

the activities of the organism (Fig. 2). Wohlfarth-Bottermann has demonstrated that the alignment of these fibrils is associated with the shape or movement characteristics of the organism. He has also extracted the fibrils by the same procedures used for extraction of certain muscle proteins and found them to share the chemical characteristics of actomyosin or myosin-B (Fig. 3). The fibrils have also been shown to be ATP sensitive in a manner similar to muscle proteins. Thus, a combination of electron microscopy and chemical analysis of these fibrils has exposed what may represent a primitive form of one of the most highly organized and efficiently operative components of higher animal muscle.

Let me turn now to an entirely different sort of problem. For many years it has been known that the celluose microfibrils that are a conspicuous feature of plant cell walls are components of some very distinct architecture (Fig. 4). The hypotheses that have been advanced to explain the design and orientation of these supermolecular celluose units usually implicate inexplicable influences from within the cell.

A few bacteria, notably *Acetobacter xylinum* form cellulose microfibrils that can be studied more or less isolated in a growth medium. Their study has added substantially to our knowledge of the development and structure of the microfibril. Ben-Hayyim

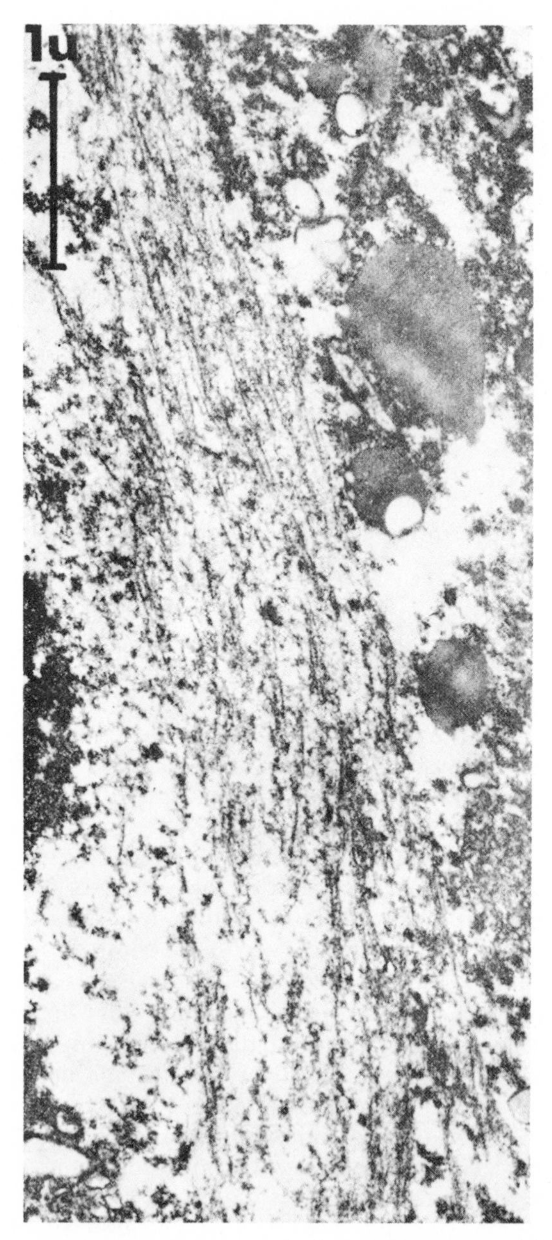

Fig. 3. A ultrathin section of glycerine-extracted fibrils from *Physarum polycephalum*. For a full description see Wohlfarth-Bottermann (1964).

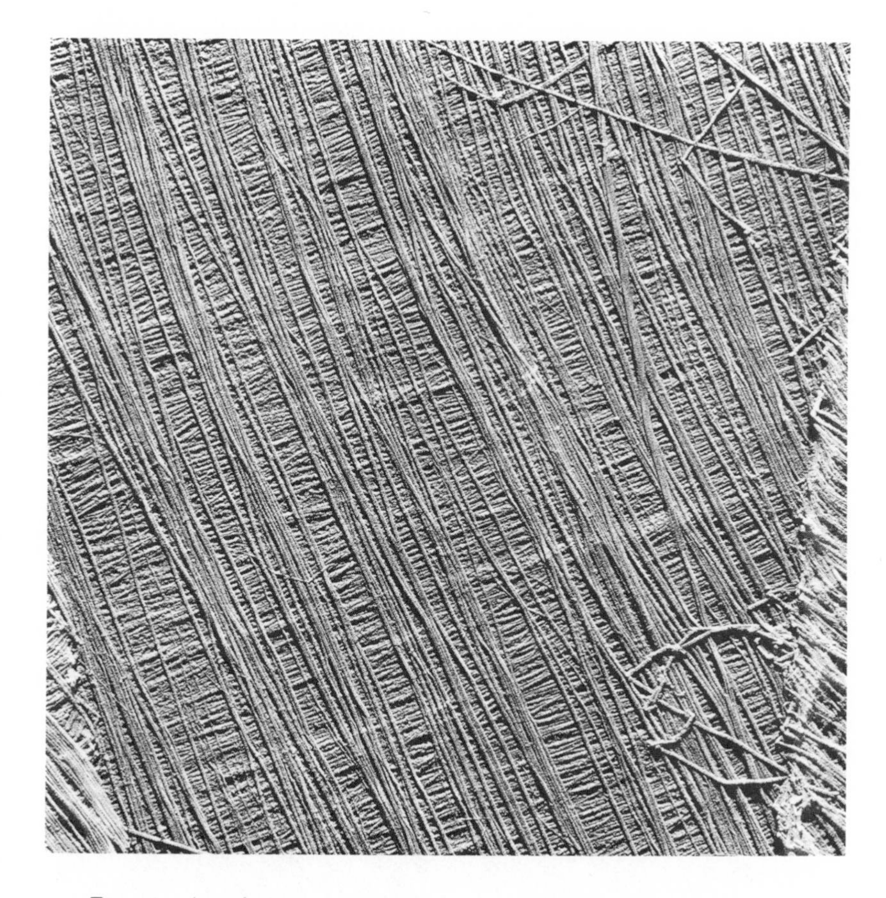

Fig. 4. An electron micrograph of two layers of fibrillar structure in the wall of *Cladoprora rupestris*. From Preston (1961).

and Ohad (1965) have demonstrated the transfer of the glucosyl moiety from uridine diphosphate glucose in the presence of extracts from *Acetobacter xylinum* (which are free of cells) to form a mixture of alkali-soluble and alkali-insoluble cellodextrins. There is, at this stage, no formation of cellulose fibrils. Neither are fibrils present in what are described as a "pre-fibrous" form of cellulose released into the medium from whole cells. These workers therefore envision formation of the fibrils by crystallization, probably without the involvement of any extra-protoplastic enzymes. For instance, it is known that when soluble Na-carboxymethylcellulose (CMC) is present in such a crystallizing system, it is incorporated directly into the network of form-

Fig. 5. Orientation of cellulose fibrils formed in the pellicle of *Acetobacter xylinum* in the presence of sodium carboxymethyl cellulose. B–attached bacterial cell wall. For a full description see Ben-Hayyim and Ohad (1965).

ing fibrils. Although the presence of CMC introduces some variables into the process, in experiments with *Acetobacter xylinum* the ultimate structure is a system of oriented cellulose fibers in the pellicle (Fig. 5). When phosphomannan is substituted for CMC the same sort of ultimate structure is derived. Let me quote Ben-Hayyim and Ohad (1965) as to their conclusion:

> The biogenesis of oriented cellulose fibrils is envisaged as a process comprising the following steps: Polymerization of the monomeric precursor, diffusion of the molecule to crystallization sites, crystallization and orientation. It is proposed that charged polysaccharides play a role similar to that of CMC in affecting the orientation of cellulose fibrils in the plant cell wall (p. 191).

The main point I wish to make is that here is a problem that any plant anatomist would have defended fiercely as being anatomical or morphological in character, which, when we really begin to analyze it, is subject to consideration in terms of some of the fundamental forces acting between molecular units.

I shall conclude with still another area. One of the most important recently developed techniques of value to organic chemists is that of nuclear magnetic resonance spectroscopy. It has opened new avenues of approach to a whole world of problems involving small molecules, including the phenolics and other secondary metabolic substances. The principles and methods involved make its use applicable to the small amounts of substances available in many types of biological experiments. One group in our laboratory here in Austin started out with an interest in certain systematic problems where relationships might be reflected in the biosynthesis of flavonoids, and they and others have already dealt with the applicability of nuclear magnetic resonance spectroscopy to problems of consequence in this area. But there is no separating of the problems of genetics, of development, or of evolutionary relationships, except for purposes of supplying the approach. Genetics, ontogeny, phylogeny, the characteristics and behavior of organisms are all simply phrases to describe portions of the spectrum that is biological science. A paper by Mandel in the Journal of Biological Chemistry just a year ago (1965) dealt with the nuclear magnetic resonance spectra of ribonuclease, lysozyme, and cytochrome *c*. Mandel made very informative analyses of the substances he chose to work with.

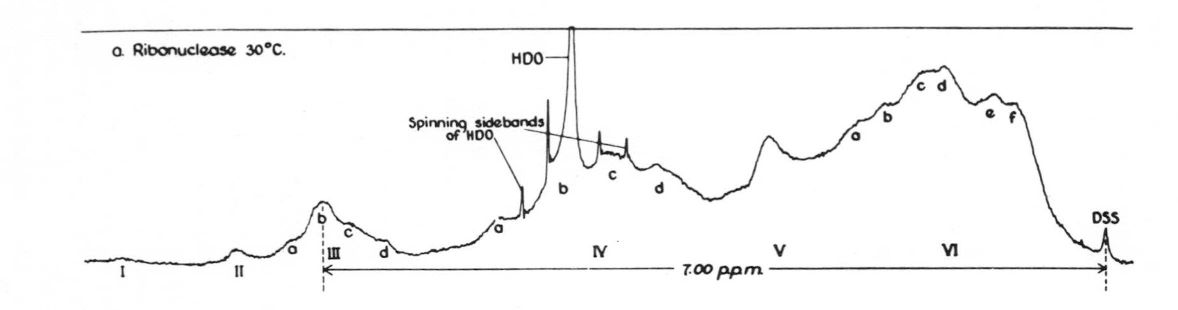

Fig. 6. Spectrum of ribonuclease obtained by nuclear magnetic resonance techniques. For a full description see Mandel (1965).

But he also made an important observation that is best stated in his own words:

> The real usefulness of high resolution nuclear magnetic resonance, however, lies in the fact that because of the chemical shift, specific amino acids can be identified and isolated in the protein spectra. In ribonuclease, Peak II arises solely from the C-2 imidazole proton of histidine (Fig. 6). Although McDonald and Phillips have shown the chemical shift of this proton to be pH dependent, the above assignment was made on the basis of relative intensities and on the fact that the protein spectra is [*sic*] essentially identical at pH 2 and pH 9. Histidine residues are thought to be involved in the active site. This provides us with a window to observe the active site under various conditions (p. 1590).

It may be apparent that I am motivated by an ever-increasing excitement that relates to the problems that became part of my thinking in the late 1930's. We now have knowledge of the structure of genetic DNA. We know that it codes RNA and that somehow this coded material provides for the formation of specific proteins in the changing patterns that relate directly to differentiation in the organism. This differentiation may be changed by mutation and become the basis of divergence in evolution. The problems still have the same broad outlines but it is now clear that resolution of them depends on the application of techniques and principles of interpretation developed by chemistry. With these techniques and these principles we are in an increasingly better position to discover more and more about living things. This part of the matter I now view quite differently than I did 30 years ago, for I now realize that success poses tremendous problems of integrating this knowledge and using it wisely. The cliché is "using it wisely in the best interests of man." But man, for all the level of his activities, is only part of a vast, intricate array of organisms whose genetic bases, developmental patterns, mature characters, and behavior interrelate among themselves and to the so-called inorganic world in many ways, some of them far beyond our understanding. The biologist does well in applying the chemist's knowledge to every possible part of this great web, but both the biologist and the chemist would be well advised to listen carefully to every other branch of knowledge, old and new, for every time they solve a problem there looms a larger, more complex one ahead.

REFERENCES

Ben-Hayyim, G., and I. Ohad. 1965. Synthesis of cellulose by *Acetobacter xylinum*. VIII. On the formation and orientation of bacterial cellulose fibrils in the presence of acidic polysaccharides. J. Cell Biol., 25: 191-207.

Correns, C. 1909. Verebungsversuche mit blas (gelb) grünen und bunblättrigen Sippen bei *Mirabilis*, *Urtica*, und *Lunaria*. Z. Vererbungslehre, 1: 291-329.

Mandel, M. 1965. Proton magnetic resonance spectra of some proteins. I. Ribonuclease, oxidized ribonuclease, lysozyme and cytochrome *c*. J. Biol. Chem., 240: 1586-1592.

McManus, Sister M. A., and L. E. Roth. 1965. Fibrillar differentiation in myxomycete plasmodia. J. Cell Biol., 25: 305-318.

Muller, H. J. 1962. Genetic nucleic acid. The Graduate Journal, 5: 133-159.

Preston, R D. 1961. Cell wall organization and wall growth in the filamentous green algae, *Cladophora* and *Chaetomorpha*. I. The basic structure and its formation. Proc. Roy. Soc. London, series B, 154: 70-94.

Rhoades, M. M. 1946. Plastid mutations. Sympos. Quant. Biol., 11: 202-207.

Schiff, J. A., and H. T. Epstein. 1965. The continuity of the chloroplast in *Euglena*. *In* Locke, M., ed., Reproduction: Molecular, Subcellular and Cellular, 131-189, New York, Academic Press.

Wohlfarth-Bottermann, K. E. 1964. Differentiation of the ground cytoplasm and their significance for the generation of the motive force of ameboid movement. *In* Allen, R. D., and Kamiya, N., eds., Primitive Motile Systems in Cell Biology, 79-109, New York, Academic Press.

2

CHEMICAL PRINCIPLES IN CHEMOSYSTEMATICS

HOLGER G. H. ERDTMAN
Department of Organic Chemistry
Royal Institute of Technology
Stockholm, Sweden

I. INTRODUCTION

Botanical classifications rest largely on comparative studies of morphological and anatomical characters. Roads to phylogenetic systems were opened by Darwin, and a deeper understanding of the mechanism of heredity resulted from the fundamental experimental studies of Mendel and his successors. Their work made it abundantly clear that chemical as well as morphological characters of plants are determined by genetic factors by mechanisms the more precise natures of which are now gradually being unveiled.

The classical morphological approach to plant taxonomy has many great advantages over the chemical approach. This is so particularly because of the ease with which most nonchemical characters can be observed. Certain botanical characters appear to be highly conservative and are, for that reason, very useful for the delineation of higher systematic categories. Others serve the same purpose on the family or generic level. No wonder that the botanical taxonomists have made great strides in this field during the two centuries following Linnaeus' attempt to bring some order out of chaos.

No botanists, however, would claim that any of the many, present systems of classification that have been put forward represents an ideal or final solution. There are still many problems existent at all taxonomic levels and, in particular, the interrelations between higher categories such as orders are very unclear. Many difficult taxonomic problems arise from evolutionary pheno mena such as parallel development, convergence, or diversification, which may cause botanists to assume closer or more distant genetic relations between plants or groups of plants than those actually existing.

It is evident that systematic investigations of the chemical characters of plants are likely to become of great supplementary value to classical plant taxonomy. Chemical characters of plants have the great advantage that they can be exactly defined.

Chemistry is concerned with molecules, their structures, and their reactions, matters that escape direct visual inspection, and these facts explain why progress in the chemical characterization of plants has been relatively slow, and why chemistry has played and still plays such a humble role in plant taxonomy. Also the differentiation of science into separate compartments has led its respective devotees to speak almost completely different scientific languages. Bridging the gaps between the sciences is a slow process and not always a gratifying task for the bridge builders. Not long ago biochemists were rated as chemists by the biologists and as biologists among the chemists. Nevertheless experience shows that even such hybrids may grow luxuriantly and develop vigorously. "Chemical plant taxonomy," "biochemical systematics," or "chemotaxonomy" is such a hybrid science.

II. CHANGING HORIZONS IN NATURAL PRODUCT CHEMISTRY

The basic ideas of chemotaxonomy involving the use of chemical characters for classification purposes are by no means new. They are, in fact, pre-Darwinian, and it is not surprising that they occurred to many of the early pharmaceutical chemists who had a professional interest in drugs and their active constituents, and combined a training in chemistry with a considerable knowledge in systematic botany. To some extent pharmacognosy has acted as a forerunner to chemotaxonomy.

A number of our best-known natural products were isolated many years ago and their elementary composition correctly determined. The elucidation of their structures and configurations, however, belongs to a much later period. Even in 1958, when Karrer (1958) published his valuable compilation of low-molecular-weight plant products that were then more or less completely elucidated only a little over 2500 compounds, other than alkaloids, could be listed. The number of known alkaloids alone is presently larger than this figure.

The enormous progress in plant chemistry that we are witnessing today is partly the result of the efforts of the much larger number of people who are able to devote themselves to studies on natural products, but even more so to the great arsenal of physical methods and instruments now available. Most of these methods were unheard of only 30 years ago. Many of the instruments have been developed or constructed largely to meet the needs of the natural product chemists.

Mass spectrometry is increasingly replacing the classical elementary analyses and determinations of molecular weights. The combination of gas chromatography and mass spectrometry makes it possible to investigate products available only in trace amounts and to identify their components by comparing their mass spectra with those of known substances. The fragmentation reactions of organic compounds are being intensely studied and, no doubt, in the future it will become possible to elucidate the structure of many substances by mass spectrometry alone. The UV, IR, and NMR spectra of many substances give valuable information about their structures, configurations, and conformations. Many important but difficult problems concerning the absolute configuration of natural products are now easily solved by examination of their ORD curves. It is, indeed, only the amorphous, irregularly constructed polymeric substances or mixtures of substances that still resist exact structural elucidation. Here even X-ray analysis fails. Natural product chemists fully appreciate the advantages of these new methods, which discharge most of the exhausing labors of the chemist and now enable them to attack problems that only a decade ago appeared unsurmountable.

As a result of these spectacular developments our knowledge concerning the structure of natural compounds is growing at a tremendous rate. This is all very pleasing, but nevertheless it creates a serious dilemma. Our knowledge of the chemical and physiological properties of the multitude of new substances is by no means increasing at a comparable rate. During the laborious classical structural elucidations of complex molecules, many observations were made that greatly contributed to the development of organic chemistry. The X-ray analysis of a new compound tells us almost nothing about its properties. Many compounds of known structure and of potential physiological interest are available only in minute quantities, and one can foresee that much of the time and labor formerly expended on degradative work will in the future be devoted to the synthesis of such compounds as well as of their analogues.

It is clear that these dramatic changes and advances will influence profoundly the whole philosophy of natural product research. Formerly the goal was essentially structure elucidation. In the future, however, this will certainly be only the beginning rather than the end of the story. For example, the complete elucidation of the structure and configuration of natural products is a necessary prerequisite for successful attacks against such challenging and important problems as their biosynthesis. Similarly, structural investigations and studies on the distribution of natural products in the plant kingdom open roads to fields such as taxonomy and, perhaps, plant evolution.

III. IMPORTANCE OF BIOSYNTHESIS IN CHEMOTAXONOMY

For purposes of classification, knowledge about the genesis of plant constituents is just as important as knowledge of their struc tures. The reason is simply that identical compounds may be synthesized in different ways in different plants, being therefore analogous but not homologous in the biological sense of the terms.

Our knowledge concerning the exact course of the biological synthesis of natural products remains quite limited. Biochemical studies of fundamental importance have revealed the existence of several rather general biosynthetic pathways such as the well-known acetate—malonate—fatty acid, the shikimic–prephenic acid, and the mevalonic acid pathways. According to Lynen (1961), fatty acids are synthesized by means of highly organized multi-enzyme complexes which effect the complete synthesis of a fatty acid. Thus, the acid is released from the enzyme system only when it has reached a certain chain length. The shikimic–prephenic acid pathway opens the road to a large variety of aromatic and heterocyclic compounds; and a multitude of isoprenoids are formed along the mevalonic acid pathway. There are, of course, many other routes that have been discovered by the biochemists and, no doubt, many new ones will be found in the future.

There is, however, one important and obviously very general route that has been inferred by the organic chemists almost exclusively on the basis of comparative studies of the structure of natural products. This route is often called the acetate or polyketide route. However, very little is known about the enzymatic requirements for the assembling of the hypothetical polyketide chains and their further transformations. It is recognized that coenzyme A must play an important role in these reactions, while one terminal acetate group can be substituted for by a number of other carboxylic acids; examples include cinnamic acid and nicotinic acid. The acetate hypothesis is founded essentially on the observation that numerous natural products possess a characteristic hydroxylation pattern which indicates that they have been formed from acetic acid molecules arranged head to tail. This hypothesis is old and goes back to Collie and to Robinson, but it has been elaborated in a brilliant manner by Birch (Birch, 1957; Birch and Donovan, 1953). On the basis of the acetate hypothesis it has been possible to predict structural details that were later confirmed experimentally, and to correct structures that were only partly in accordance with the "acetate rule" and therefore appeared suspect.

It is not necessary here to discuss and exemplify this well-known and very useful hypothesis in detail. By intelligent consideration of the many possibilities of folding the hypothetical polyketide chain, followed by condensation, and, sometimes, decarboxylation reactions, the structure of an astonishing number of complex natural compounds may be interpreted. The acetate hypothesis has been used so successfully that one is almost tempted to think it is too good to be true (cf. Geisman 1963).

On the basis of laboratory experiments, chemists have a tendency to look upon biosynthesis as proceeding stepwise, one intermediate being handed over from one catalytic site to another where it will undergo further changes. Such transport from one site of synthesis to another occurs, but many biosynthetic reactions, perhaps most, seem to be carried out concertedly in highly organized, multienzyme systems or in intimate collaborations of several such enzymatic units.

It is natural that organic chemists as well as biochemists are interested in investigating the fates of molecules introduced into such enzyme systems in order to see whether they can be utilized as precursors and become incorporated into the final product. As long as one is using simple molecules, such as those produced in basic metabolism, the compounds are readily accepted and transformed into more complex molecules. The structures and configurations of the final products clearly depend on the nature of the special enzyme system into which acetate, mevalonic acid, or other simple metabolites are entering.

It is also natural that organic chemists are intrigued by the possibility of offering to living cells even more complex molecules which, on biogenetic grounds, may be actual precursors or may be able to act as prefabricated natural intermediates. This is perhaps more risky because it is difficult to know whether the enzyme systems are able to accept such products from their surroundings. They may prefer going ahead with their own "intermediates" produced in situ. In many cases, however, results of feeding experiments using the more sophisticated precursors have given very interesting results; while in others the results have been less convincing or have been disputed. There are many pitfalls in this type of biosynthetic experiment. For example, a precursor may be introduced during a period when the biosynthesis that is being investigated is not actively proceeding in the organ.

As the result of biosynthetic experiments using the feeding technique it has been suggested that the pyrrolidine ring in nicotine is formed from a symmetrical intermediate, perhaps putrescine. However, in a recent paper by Liebman et al. (1965), it has been shown that this may not be the case during the natural synthesis of the alkaloid. They wrote, "The most significant conclusion that may be drawn from these results is that the labeling pattern of the pyrrolidine ring found from $^{14}CO_2$ exposure is greatly different from that produced from precursor feedings.'' ''The labeling pattern produced by short term $^{14}CO_2$ exposure must result predominantly from an unsymmetrical intermediate, and the question of pyrrolidine ring biosynthesis must now be considered anew.''

Spenser and Tiwari (1966) reported the incorporation of various C_6C_3 ''precursors'' into the ''modified aporphine alkaloid'' aristolochic acid (V). They found that D,L-tyrosine (I) labeled at C-3 in the side chain was incorporated in both C_6 C_3-derived parts of the molecule; C-2-labeled D,L-dopa (II) gave labeled acid which on decarboxylation did not lose any activity. C-2-Labeled dihydroxyphenylethylamine (III) was also incorporated but only in the upper part of the aristolochic acid (as written in formula V). Finally, C-2-labeled D,L-noradrenaline (IV) was incorporated and in the same manner as the phenylethylamine derivative.

Hence, one must conclude that only the simple amino acid tyrosine can be used for the synthesis of the whole molecule of aristolochic acid, but that other related compounds can be accepted under experimental conditions for the synthesis of one part of the molecule. Presumably tyrosine is the normal precursor of aristolochic acid.

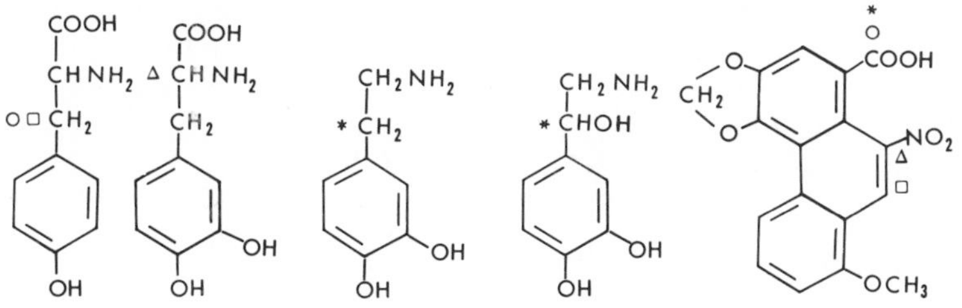

I II III IV V

Aristolochic acid

These and several other conflicting or puzzling results obtained from incorporation experiments call for great caution in the interpretation of biosynthetic experiments.

IV. STRUCTURAL COMPLEXITY AND TAXONOMIC WEIGHT

It has long been realized that there is often a relationship between the chemical complexity of natural products and the extent of their distribution in nature. Fatty acids are almost omnipresent, and relatively simple enzyme systems of wide occurrence and perhaps low specificity must be responsible for their synthesis. For the formation of hydroxylated or unsaturated fatty acids, additional enzymes are required, and it is not surprising that many such acids are restricted to just one genus or family, or, as in the case of petroselinic acid, to some presumably related plant families such as Umbelliferae and Araliaceae. Further elaborations of the fatty acid theme are the many acetylenic compounds from fungi as well as angiosperms, such as those from Compositae, Umbelliferae, Araliaceae, Oleacae, and Santalaceae. Still more complicated substances, which are undoubtedly derived from the fatty acid route, are aromatic compounds such as capillene (VI), in which the phenyl group certainly arises through cyclization of an unsaturated C_6 chain, as well as furanes, spiroketals, and thiophenes; for example, terthienyl (VII). The Compositae appear to be "experts" in such sulfur chemical extravagances, but junipal (VIII) shows that such structures are not entirely unfamiliar to fungi.

$C_6H_5-CH_2-CH=CH-CH=CH-CH_3$

VI. Capillene

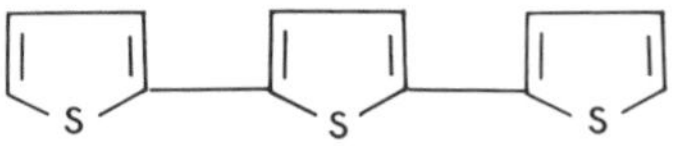

VII. Terthienyl

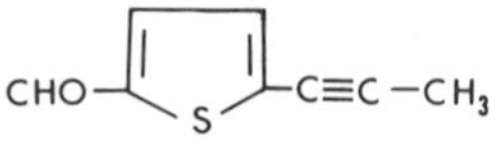

$CHO-C_4H_2S-C\equiv C-CH_3$

VIII. Junipal

Often, several general pathways are involved in the biosynthesis of a single natural product. It will suffice here to mention only a few cases. Flavonoids are formed by a combination of the shikimic acid pathway and the acetate (polyketide) pathway as indicated on the left side of Fig. 1.

Flavonoids are common in higher plants but for some reason they appear to be absent in bacteria, fungi, and algae. For a long time it was thought that only vascular plants were able to produce flavonoids, but it has recently been shown that certain mosses such as *Bryum* (Bendz et al., 1962; Bendz and Martensson, 1961, 1963) are able to synthesize anthocyanins, while another moss genus, *Mnium* (Melchert and Alston, 1965), produces flavones,

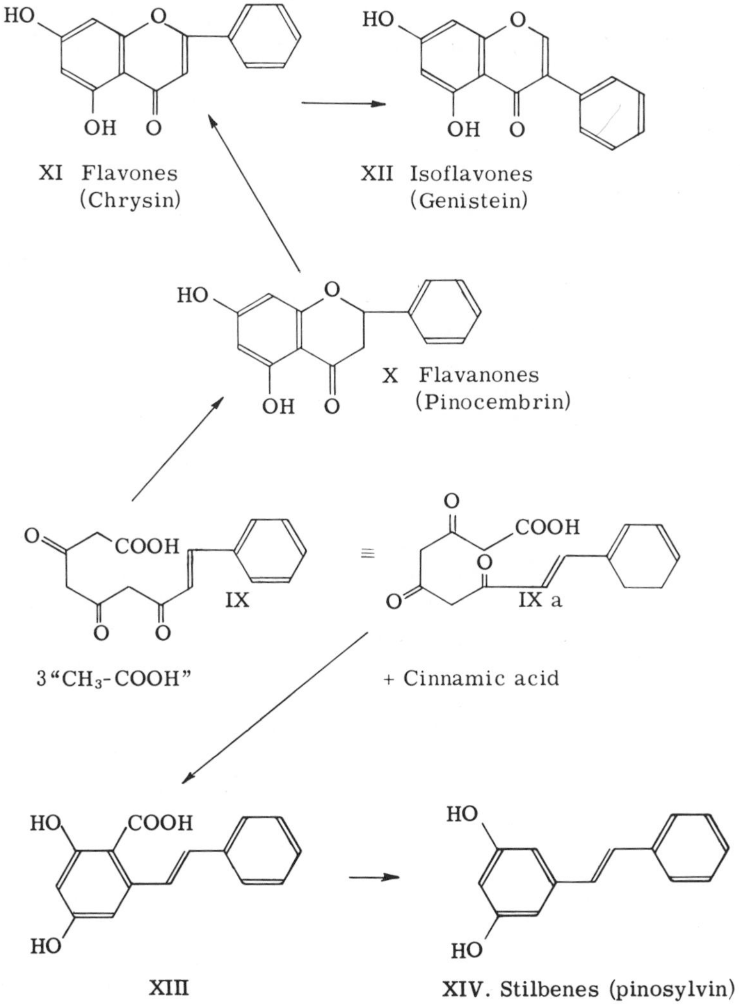

XI Flavones (Chrysin)

XII Isoflavones (Genistein)

X Flavanones (Pinocembrin)

IX

3 "CH_3-COOH"

IX a

+ Cinnamic acid

XIII

XIV. Stilbenes (pinosylvin)

Fig. 1. Connections between flavones and stilbenes.

even flavone *C*-glycosides. Microorganisms, although "experts" in polyketide synthesis, seem rarely to combine the shikimic acid and polyketide pathways.

In several families–for instance, Podocarpaceae, Iridaceae, Moraceae, Amaranthaceae, Rosaceae, and, particularly, Leguminosae–rearranged flavonoids, isoflavonoids, occur but not to the exclusion of compounds of the more generally encountered flavine type. The neoflavonoids such as dalbergin, the rotenoids, compounds of the peltygenol and brasilin types, and biflavonyls are examples of even more complex natural products of accepted or still disputed flavonoid origin, and their limited distributions add to their taxonomic value.

V. COEXISTING COMPOUNDS

It is of considerable biogenetic and taxonomic interest that, frequently, flavonoids and stilbenes coexist in plants. A relationship between them, particularly between isoflavones and stilbenes, has long been suspected. According to an ingenious idea of (Birch and Donovan, 1953) the connections between flavones and stilbenes are explained as shown in Fig. 1.

Flavone precursors, which can be formed from one molecule of cinnamic acid and three acetate units, can be folded in different ways; for example, IX and IXa. On cyclization, one route gives rise to flavonoids (X–XII); another, to stilbene-carboxylic acids (XIII), which undergo decarboxylation to stilbenes, for example, pinosylvin (XIV). This interesting branching of the biosynthetic route seems to be characteristic for the pines and has been studied by feeding experiments (Billek and Kindl, 1961; Rudloff and Jörgensen, 1963; Ibrahim and Towers, 1960,1962), the results of which were in general agreement with Birch's hypothesis.

The heartwood of *Laburnum alpinum* (Leguminosae) contains the stilbene piceatannol (XV) and the isoflavone genistein (XII) (Erdtman and Norin, 1963). They may be related in the same way as pinosylvin (XIV) and pinocembrin (X) are related.

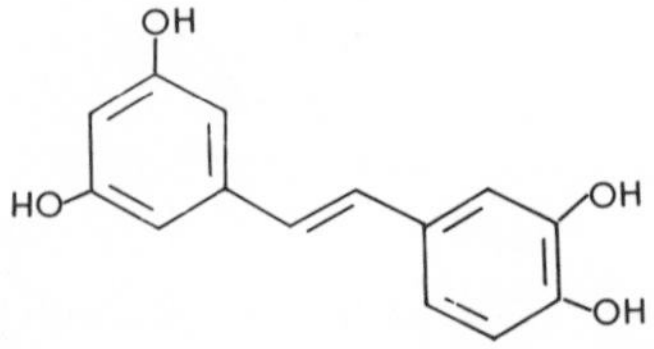

XV. Piceatannol

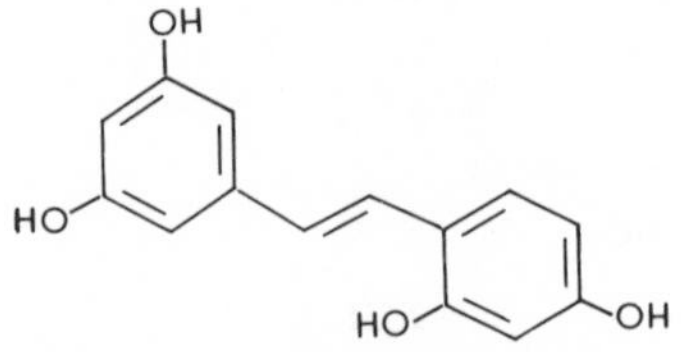

XVI. Oxyresveratrol

Other examples of the coexistence of flavones and stilbenes are provided by *Morus alba* and *Maclura pomifera* (both Moraceae). The wood of the former contains the stilbene oxyresveratrol (XVI) and the flavone morin (XVII), and that of the latter contains oxyresveratrol and the complex isoflavone pomiferin (XVIII).

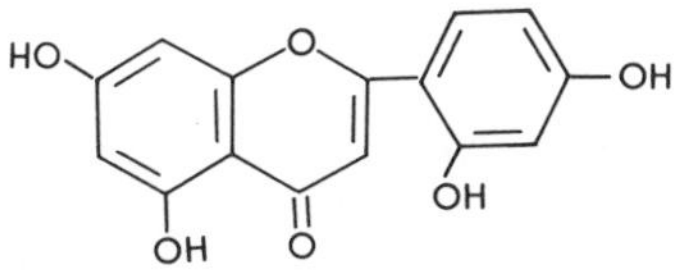

XVII. Morin

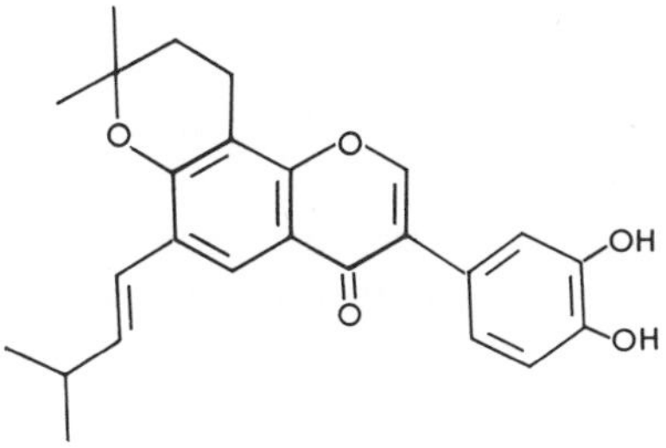

XVIII. Pomiferin

It is perhaps understandable that such branchings as those discussed here and those that require the participation of different enzyme systems may be profoundly influenced even by small genetical changes. For example, in *Laburnum anagyroides*, genistein was found, but no piceatannol was reported (Chopin et al., 1963). Closely related to *Laburnum* is *Sarothamnus*. The wood of one *Sarothamnus* species contains a pinocembrin glucoside, but again no stilbenes were found (Matas, 1960).

In pomiferin we have an example of the participation of three biosynthetic pathways (polyketide, shikimic, and mevalonic acid) and a similar case is the flavone derivative cycloartocarpin (XIX) from the wood of *Artocarpus integrifolia* (Moraceae).

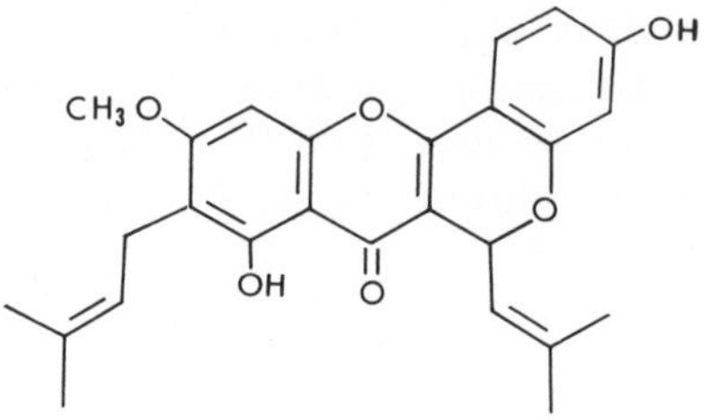

XIX. Cycloartocarpin

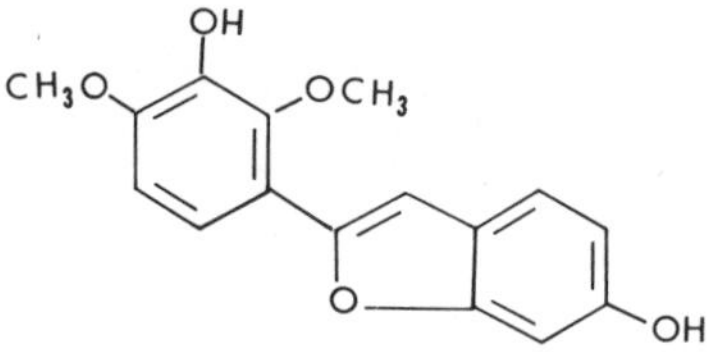

XX. Pterofurane

Pterofurane (XX), from the heartwood of *Pterocarpus indicus* (Leguminosae, Dalbergieae), is formally a stilbene derivative; but, owing to its congeners, angolensin, pterocarpin, isoliquiretigenin, formononetin, and *p*-hydroxyatropic acid, it is usually regarded as being biosynthetically related to the isoflavones.

VI. SEVERAL PATHWAYS TO IDENTICAL COMPOUNDS

A large number of anthraquinones are evidently of polyketide origin and the classical example is endocrocin (XXII), obtained from a lichen and some molds. Its biosynthesis has been interpreted as follows:

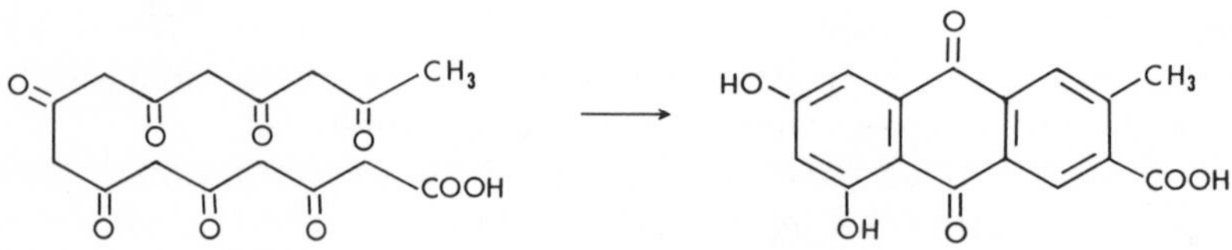

XXI.

XXII. Endocrocin

Decarboxylation yields emodine (XXI) with H instead of COOH), which is found in several fungi as well as higher plants.

The wood of the teak *Tectona grandis* (Verbenaceae) contains tectoquinone (XXIII) and several other anthraquinones (e.g., XXIII with CH_2OH or COOH instead of CH_3).

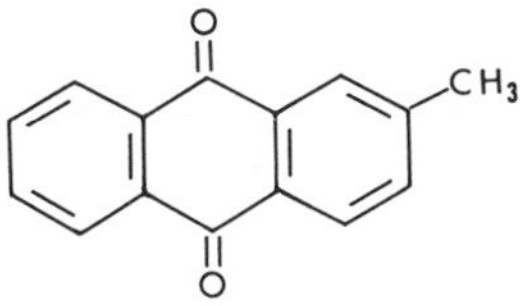

XXIII. Tectoquinone

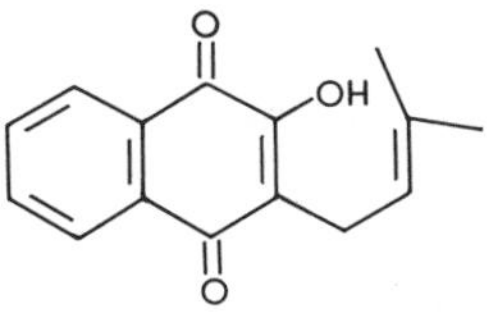

XXIV. Lapachol

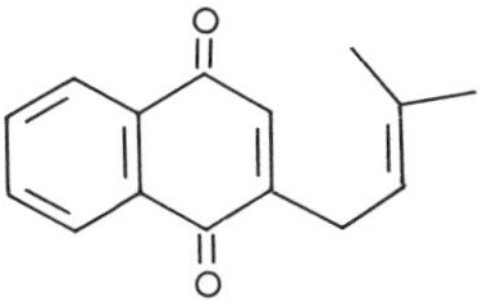

XXV. Deoxylapachol

Since tectoquinone does not contain any hydroxyl groups, the application of the acetate hypothesis is meaningless in this case. However, in the teakwood, tectoquinone occurs together with lapachol (XXIV) and deoxylapachol (XXV) and, as Sandermann and Simatupang (1963) have pointed out, tectoquinone may be regarded as derived from an isoprenoid naphthalene; for example, XXV. Of course, the actual stage at which the ring closure takes place is unknown. If Sandermann's hypothesis is correct, then the whole molecule of tectoquinone would probably be of acetate origin, the naphthalene moiety presumably being derived from acetate via the polyketide route and the isoprene unit from mevalonic acid.

Lapachol has been isolated from several woods of species belonging to the same order as *Tectona* (e.g., *Tecoma* and the mangrove, *Avicennia*) and the anthranol chrysarobin is known from *Tecoma*.

Anthraquinones are widely distributed and occur in many angiosperm orders such as Liliflorae, Polygonales, Rhamnales, Rubiales, Rosales, and Tubiflorae. It is possible that the anthraquinones arise in different ways in these taxa and that biosynthetic investigations of their genesis might be of taxonomic importance.[1] There are indications that xanthones may also arise in different ways in different plant species.

There are many other instances where one is left in doubt whether or not identical or very similar compounds have been formed by different biosynthetic routes. It is known that some simple compounds such as lysine and glutamic acid are formed in different ways in different plants, and, since such substances may be utilized in further reactions yielding secondary metabolites, it is obvious that the question of homology and analogy of natural products is difficult to answer. This problem does not make chemotaxonomy any easier, but the answers may enhance the value of chemotaxonomy.

VII. UNIQUE COMPOUNDS

It is clear that compounds found only in a single species do not possess any important taxonomic value. The alkaloid ricinine (XXVI) is a classical example of a natural product of unusual structure and such limited distribution. It was discovered a little over a hundred years ago in *Ricinus communis* and has never been found in any other plant. The genus *Ricinus* is monotypic. Ricinine possesses such a strange structure that *Ricinus* if it had been a large genus, would have been carefully scrutinized by chemists. A hundred years after the discovery of this unique substance, however, another compound bearing obvious structural similarities to ricinine, nudiflorine (XXVII), was isolated from *Trewia nudiflora* (Mukerjee and Chatterjee, 1964), again, unfortunately, the only species of its genus. The interesting fact is that *Ricinus* and *Trewia* belong to the same tribe, tribe Acalyphae, of the Euphorbiaceae, subfamily Crotonoideae.

[1]Compare E. Leistner and M. H. Zenk, 1967. Tetrahedron Letters, 5: 475.

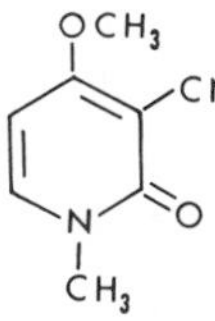

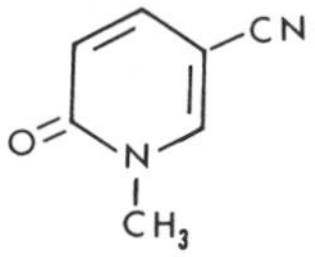

XXVI. Ricinine XXVII. Nudiflorine

It would indeed be astonishing if these compounds were not elaborated by similar enzyme systems.

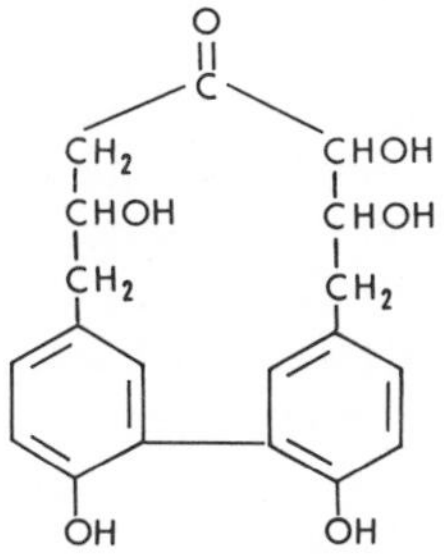

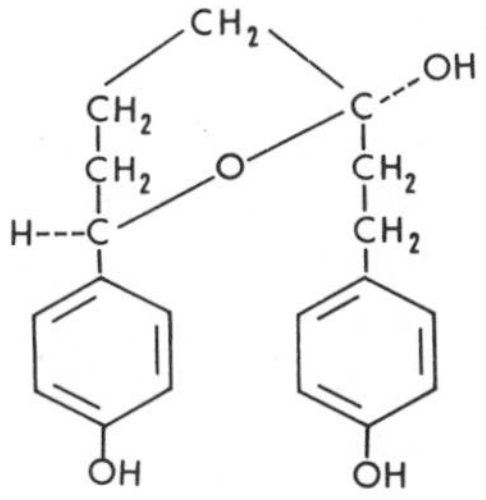

XXVIII. Asedanin XXIX. Centrolobin

Asedanin (XXVIII)(Yasue, 1965) from the wood of *Ostrya japonica* (Betulaceae), and centrolobin (XXIX)(Galeffi et al., 1965) from the wood of *Centrolobium robustum* (Leguminosae, Dalbergieae), are other examples of compounds of unique structures that, so far, are taxonomically of little utility. (The closest formal analogues are the $C_6C_7C_6$ phenols from *Curcuma.*) *Ostrya* and *Centrolobium* are not monotypic, and the other species of these genera should be investigated.

Many compounds are biochemically interconvertible and differ only in their state of oxidation or their methylation patterns. The lack of one such substance in one species where one would expect it to occur may, therefore, be compensated by the presence of a chemically closely analogous compound in a botanically related species.

VIII. COMPOUNDS OF WIDESPREAD OCCURRENCE

Many natural products are ubiquitous and are, for that reason, of little or no taxonomic interest. These substances generally take part in the basic metabolism of the plant and are essential for those biosynthetic reactions that are of particular interest to the biochemists. Often, these compounds do not accumulate as such in the cells but in the form of easily mobilized reserves, such as sucrose, starch, fat, and reserve proteins.

Proteins and nucleic acids are truly ubiquitous but are, nevertheless, of potentially great taxonomic value. Yet we are only beginning to get some information about their structures. It is obvious, however, that we will soon have something which might be called comparative protein, enzyme, and nucleic acid chemistry. We are at the beginning of a new era in biochemistry, and molecular biology will certainly contribute greatly to future taxonomy. Even so, as in the case of chemotaxonomy, it is improbable that molecular biology will become a substitute for classical taxonomy. Metzenberg's (1965) remarks in a recent review of the symposium volume, "Evolving Genes and Proteins" adequately account for the present situation. He wrote, "An uncommitted reader will almost certainly feel that amino acid and nucleotide sequences will add tremendously to our understanding of evolution; just as surely as he will hesitate to throw the classical taxonomy on to the pavement on the grounds that biochemists can now (laboriously) distinguish a fish from a bird."

In spite of their high molecular weight, compounds such as cellulose, proteins, and nucleic acids can be assigned definite structures. The lignins are less regularly constructed, probably being mixtures of polycondensation products. Nevertheless, the lignins appear to possess greater taxonomic value than cellulose. It must now be regarded as certain that Klason's old idea that lignins are built up from simple phenylpropane units was correct. The aromatic nuclei are of *p*-hydroxyphenyl, guaiacyl, and syringyl types. Oxidation of lignins with nitrobenzene and alkali, a method introduced by Freudenberg and Lautsch, gives, among other products, fairly high yields of the corresponding aldehydes, *p*-hydroxybenzaldehyde, vanillin, and syringic aldehyde. Although only a relatively small number of plants have been investigated with respect to the composition of their lignins, it appears that ferns and gymnosperms contain lignins of predominantly guaiacyl type. Monocotyledon lignins, particularly those of Gramineae, contain, in addition, relatively large amounts of *p*-hydroxyphenyl moieties, and the dicotyledons contain guaiacyl and syringyl elements. It is interesting taxonomically that *Ephedra* appears to produce lignins of dicotyledon type, while *Cycas* produces a lignin rich in guaiacyl groups. It has frequently been found that, during early lignification, *p*-hydroxyphenyl elements dominate.

Later, guaiacyl and syringyl units are added in increasing amounts. Some lower plants, such as certain algae, appear to contain lignins containing little or no methoxyl groups, but the nature of these products is unclear.

Some structural features of the lignins are now well known, but the structural schemes proposed for lignin merely illustrate possible combinations of units containing two or more C_6C_3 elements.

Lignin chemistry is a classical field of speculation. As the result of considerations regarding the possible biosynthesis of lignans, the present author suggested that lignans as well as lignins may be formed by oxidation of hydroxylated phenyl propanes (Erdtman, 1933a and b, 1950; Erdtman and Wachtmeister, 1957) as illustrated below for coniferyl alcohol (XXX).

Coniferyl alcohol was assumed to give an oxygen radical (XXXI) and the mesomeric radicals with the unpaired electrons located in positions indicated by dots in XXXII. A combination of two radicals of type XXXIII could yield compounds of the lignan type, and dimerization of other radicals (e.g., XXXI with XXXIII, or XXXIII with XXXIV) could yield various other coupled products, mostly quinone methides capable of addition and polymerization reactions.

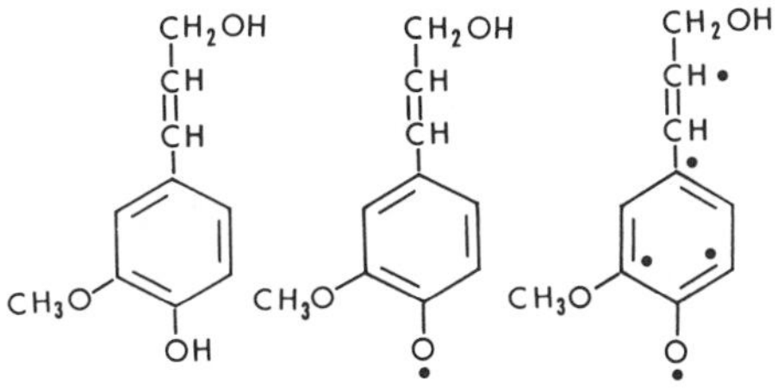

XXX XXXI XXXII

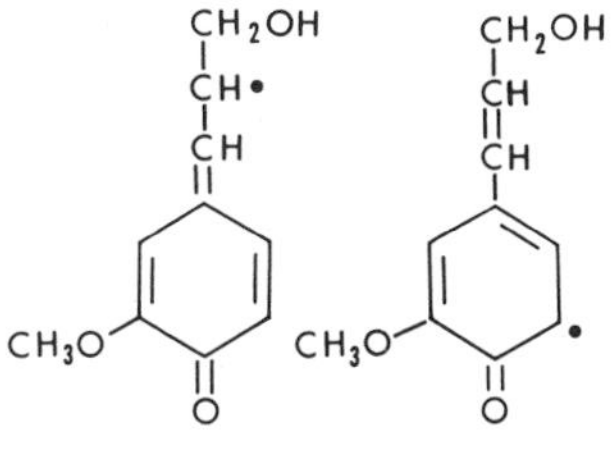

XXXIII XXXIV

A confirmation of this hypothesis was found in the structure of dehydrodi-isoeugenol (XXXVI, X = H), an oxidation product of isoeugenol (model for coniferyl alcohol). The most important step in lignin biosynthesis is perhaps the addition of the phenolic hydroxyl group to the quinone methide moiety in XXXV.

The quinone methide (XXXV, X = OH) or the alternative primary coupling product (XXXVII) could add water, coniferyl alcohol, or even another molecule of the same kind to yield products which, on further oxidation or polymerization, could give polymeric substances of ligninlike structure.

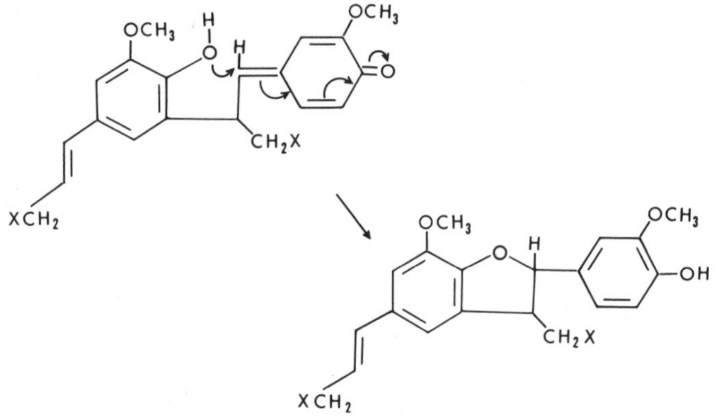

OCH_3 H O H OCH_3 O CH_2X XCH_2

OCH_3 O H OCH_3 OH CH_2X XCH_2

XXXV. Coupling of radicals XXXIV and XXXIII: (X = OH)

XXXVI. Dehydrodi-isoeugenol: (X = H)

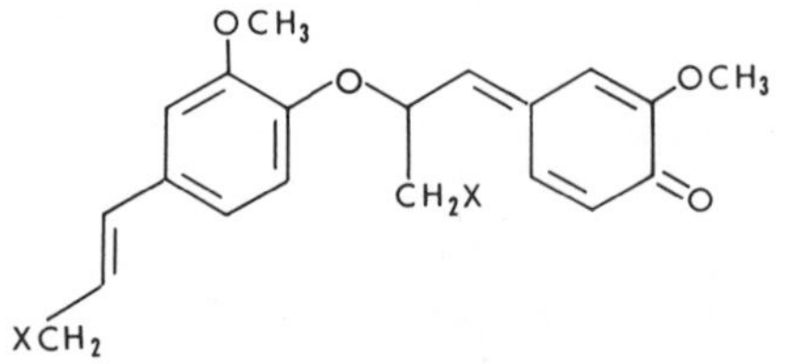

OCH_3 O OCH_3 CH_2X O XCH_2

XXXVII. Coupling of radicals XXXI and XXXIII to an ether.

When coniferyl alcohol became easily available, Freudenberg (1954, 1962) and his collaborators investigated its oxidation, and, in a series of brilliant experimental studies, succeeded in isolating a large number of oxidation products, all of expected structure. These investigators also obtained amorphous poly-condensation products which, in several respects, resembled the natural lignins.

Lignification takes place in living cells containing a multitude of enzymes including oxidases, and it may well be that oxidases take some part in the lignification process. However, there are other routes which might lead to similar results (Erdtman, 1959). Phenyl glycerols, perhaps, as phosphate esters, may react as indicated in the schematic formulas XXXVIII-XLIII.

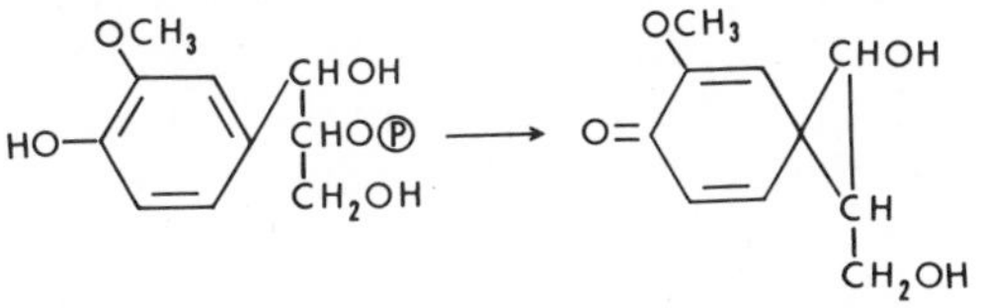

XXXVIII

XXXIX

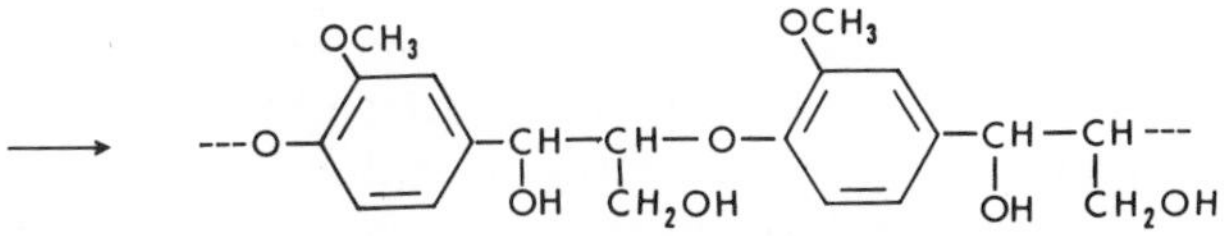

XL

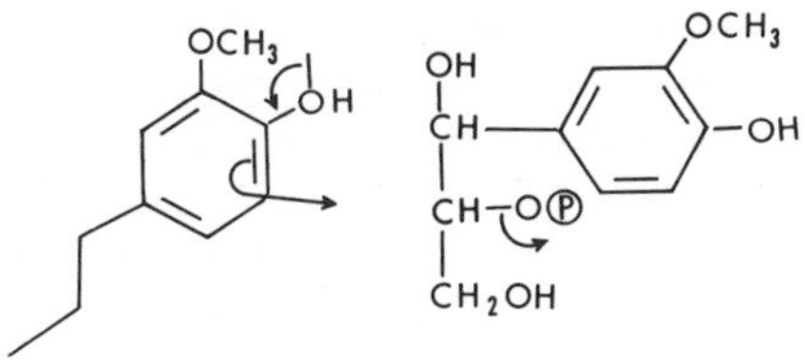

XLI

XLII

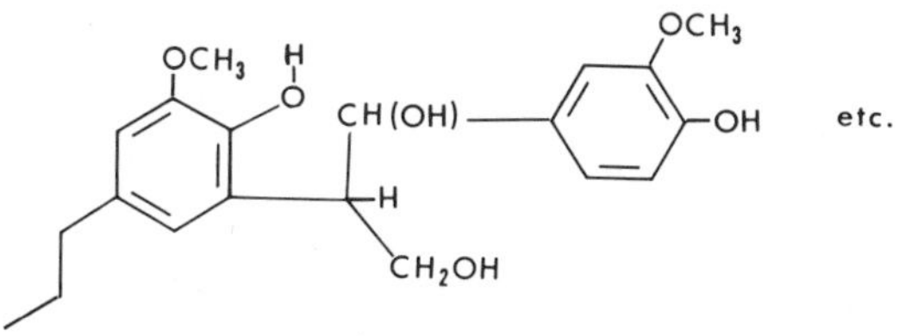

XLIII

Coniferyl alcohol, even coniferin, is apparently a rare natural product. Glucosides of guaiacyl glycerol have recently been isolated from pine needles. It is clear that lignins are formed via the shikimic acid pathway. Biosynthetic experiments have shown that several labeled aromatic substances may be incorporated into the lignins. Good candidates as natural precursors are phenylalanine, hydroxylated cinnamic acids, or alcohols, phenyl glycerols, and so on. In the past probably too much emphasis has been laid on coniferyl alcohol. It is interesting that, apparently, tyrosine can be used for lignin synthesis by some grasses, but not by several other plants (Brown, 1961; Brown and Neish, 1956; Higuchi and Barnoud, 1966). Much more biochemical work is needed to clarify the biosynthesis of lignins, but, owing to the complexity of the problem, their biosynthesis has unfortunately been unattractive to biochemists. It has not been possible to detect any optical activity in lignins or their degradation products, and this fact has been taken to indicate that the formation of lignins proceeds in a sterically unspecific manner as required by the dehydrogenation hypothesis.

Almost all naturally occurring lignans are optically active, however. According to the dehydrogenation hypothesis, the formation of lignans involves coupling of radicals such as XXXIII. Because of their optical activity, the coupling must be stereospecific and perhaps proceeds via phosphates, as discussed above. It is unfortunate that no work on the biosynthesis of lignans has been carried out. Some of the naturally occurring lignans possess hydroxyl groups attached to carbon atom 2 of the side chain of the presumed C_6C_3 precursor. The first example was gmelinol, but recently several such compounds have been found in the wood of *Thuja plicata;* examples are plicatic acid (XLIV) and XLV. This can be explained but only by piling one hypothesis on top of another. There are possibly several routes to lignans.

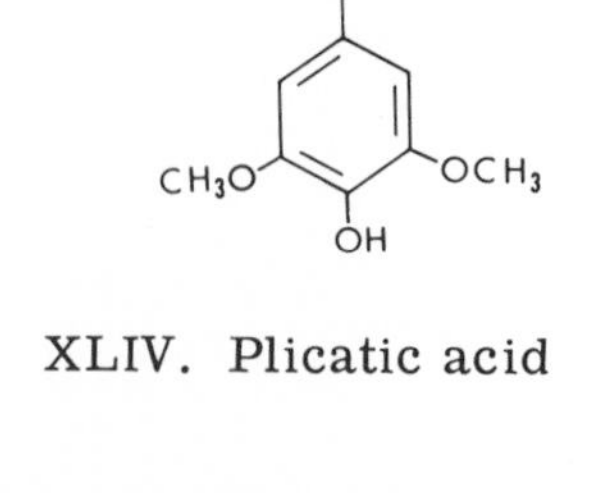

XLIV. Plicatic acid

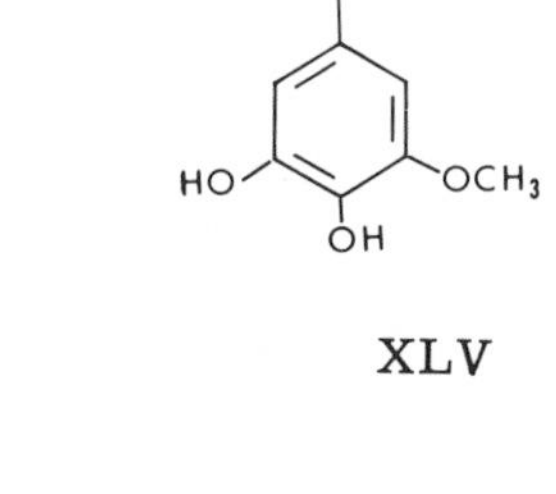

XLV

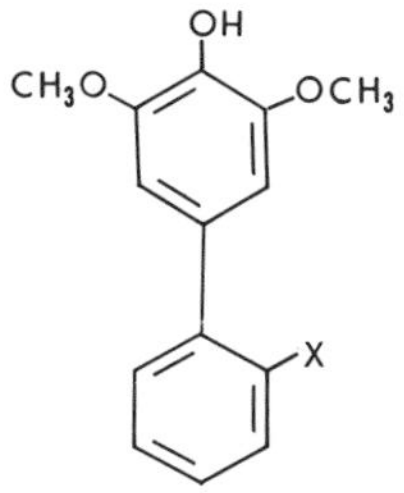

XLVI. X = H, aucuparin;
X = OCH_3, methoxyaucuparin

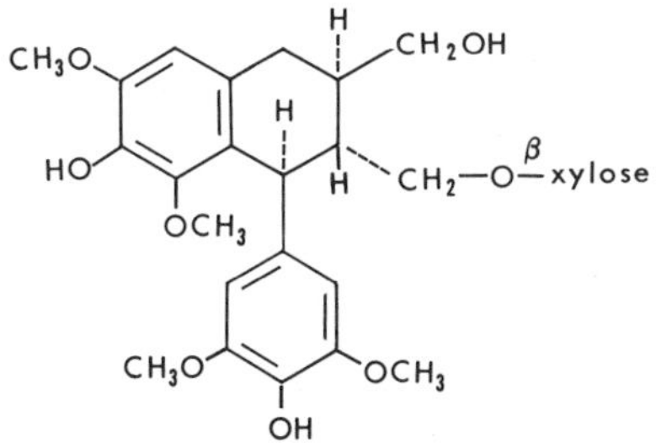

XLVII. Lyoniside

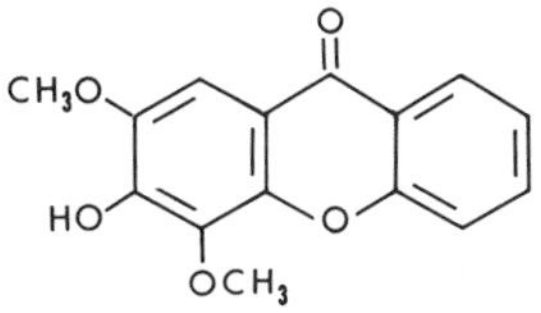

XLVIII

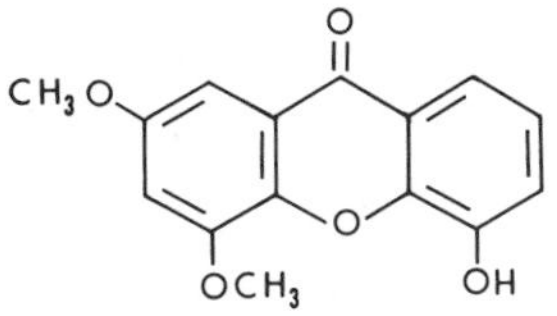

XLIX

IX. PATTERNS OF CONSTITUENTS

So far, we have only been discussing specific chemical characters of plants, compounds that are unique for a species or have a medium or widespread occurrence in the plant kingdom. No taxonomist would endeavor to classify a plant simply by means of a single botanical character. The critics of chemotaxonomy often emphasize the fact that very different plants frequently produce identical secondary metabolites. This criticism is not justified. It is true that, for example, nicotine has been isolated from several families, including Equisetaceae, Lycopodiaceae, Rosaceae, Leguminosae, Asclepiadaceae, Solanaceae, and Compositae. As long as we do not know that all these plants synthesize nicotine in the same manner, we cannot be sure whether or not these nicotines are, from a biological point of view, identical compounds; furthermore, independent evolutionary origins of secondary compounds by similar routes are possible. Moreover, nicotine is not the only compound produced by these plants, and there is no reason that nicotine should possess greater taxonomic weight than any one of the many other compounds which the plants are synthesizing. There are no *Lycopodium* alkaloids in *Nicotiana*, no macrocyclic diterpenes such as duvatrienediol in *Sedum*, no compounds of the sedidrine type in *Equisetum*, and no "*Equisetum* alkaloids" in *Lycopodium;* at least they have not been found.

It will be necessary for the chemotaxonomists to try to find patterns of compounds in various organs of a plant before such workers can claim that they have achieved anything like a systematically useful chemical characterization of species. Moreover, it is desirable that chemical constituents formed along very different biosynthetic pathways be included in such chemical plant descriptions whenever possible.

The fruits of *Sorbus aucuparia* (Rosaceae) contain sorbitol, iditol, parasorbic acid, cyanin, β-carotene, and cryptoxanthin; the bark contains several triterpenes (Lawrie et al., 1960); and the wood, like that of several other *Sorbus* species, contains the biphenyl derivatives aucuparin (XLVI, X = H)(Arya et al., 1962; Erdtman et al., 1961; Narasimachari and Rudloff, 1962) and methoxyaucuparin (XLVI, X = OCH_3) as well as the lignan xyloside, lyoniside (XLVII) (Yasue and Kato, 1960). The flowers are said to contain trimethylamine. Of course, there are also cellulose, lignins, and, obviously, a vast number of still unknown constituents. The known compounds may serve as a very incomplete, but reasonable chemical description of *Sorbus aucuparia. Sorbus intermedia* contains lyoniside but seems not to produce any aucuparins. Aucuparin has been found in the wood of *Kielmeyera coriacea* (Pimenta et al., 1964) where this compound cooccurs with a number of xanthones, such as XLVIII and XLIX. Lyoniside and three toxic principles of unknown nature occur in the leaves of *Lyonia ovalifolia* var. *elliptica* (Ericaceae), but no aucuparins occur. These plants are thus characterized not by their specific constituents, but by their different patterns of chemical characters.

It would be easy to enumerate many other similar patterns of compounds that are more or less characteristic for individual species — alkaloids coocurring with acidic or phenolic substances, cardiac glycosides coexisting with flavones, and so on — but this would serve no further purpose. Anyone can find out such patterns for himself by consulting Karrer's (1958) book or Hegnauer's magnum opus, "Chemotaxonomie der Pflanzen" (Hegnauer, 1962–1964). Usually, the most outstanding impressions that one gets from such literature searches are that we really know very little about the chemistry of most plants and that there are, in most families, much greater chemical similarities within the genera than between them. It is, however, sometimes possible to discern certain features that are more or less characteristic for some families.

X. BIOCHEMICAL PROFILES

In the large family Compositae, acetylenic compounds, as well as terpenoids of various kinds, have been found in most tribes. The sesquiterpenes are of the *cis*- as well as the *trans*-farnesyl types, although the latter type seems to dominate. *cis*-Farnesyls (e.g., cedrol) have been found in *Saussurea lappa* (Cynareae); cadinenes, in *Baccharis* and *Solidago* (Astereae). Groups of chemists are now screening genera producing sesquiterpenes of unusual, rearranged types, such as L in Heliantheae and Helenieae, and LI in Senecioneae (*Petasites*). Compounds related to germacrane (LII) and its cyclized analogues are common in Compositae. Diterpenes of labdane, pimarene, kaurane, and abietane types have been isolated. Apart from the *Senecio* alkaloids, which are fairly well studied (but over 1500 *Senecio* species exist), the Compositae alkaloids are little investigated. Some of these alkaloids are simple, the quinolone echinopsin, for instance, while others are complex. *Inula royleana* surprisingly produces esters of the *Aconitum-Delphinium* alkaloid, lycoctonine. Coumarins, flavones, chalcones, and aurones are well-known Compositae constituents. These and many other compounds give a partial chemical profile to the family Compositae.

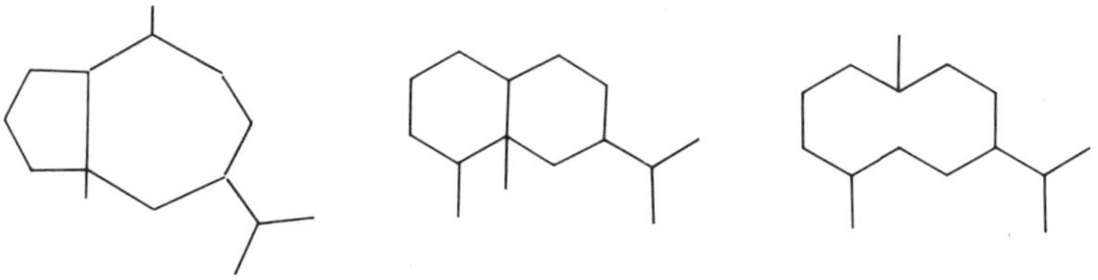

L LI LII

Concerning this family, like so many others, such as Leguminosae and Apocynaceae, there are many pieces of knowledge which, however, cannot yet be put together into anything like a clear picture. This situation exists partly because of the large sizes of these families, but also because of the fact that most of the various investigators have been very specialized. Those interested in flavonoids have been looking for flavonoids, and the alkaloid chemists have searched only for alkaloids. A common feature is that the chemists have been using the botanical classification in their search for new compounds. This is perhaps chemotaxonomy in reverse, and it is probably not too much of an overstatement to say that, in fact, few real chemotaxonomic investigations have yet been carried out. Perhaps it is more accurate to say that it has not been possible to undertake many real chemotaxonomic investigations because of our poor knowledge of the distribution of various plant constituents. Most chemotaxonomic investigations can be classifed as merely phytochemical studies within a genus or a family.

XI. CHEMOTAXONOMIC EXCURSIONS IN THE CONIFERAE

A. Pinales

It was known for a long time that the lignan pinoresinol is a constituent of the wound exudates of spruce and pine, and that another lignan, conidendrin, occurs in spruce wood. Later, it was found that the hemlocks (*Tsuga*) also contain conidendrin. This seemed to indicate that there may be some chemical similarities between the genera *Picea*, *Pinus*, and *Tsuga*. Moreover, the terpenoids pointed in the same direction, since abietic acid was known from *Picea* as well as *Pinus*, and many other conifers belonging to the order Pinales appeared to contain abietic acid or related diterpene acids. In 1939, it was found (Erdtman, 1939) that the heartwood of *Pinus silvestris* (Scot's pine) contained considerable amounts of a then rather unique stilbene phenol, pinosylvin (XV), a fungitoxic, insect-repellent substance that greatly contributed to the natural durability of the heartwood. Pinosylvin couples with bis-diazotized benzidine to give a red dye, and this reaction is used (Koch and Krieg, 1938) as a simple test to differentiate between pine heartwood and spruce wood. Later, it was observed that several other pines gave the same color reaction. A few pine woods were then more carefully investigated, and the results were so promising that it was decided to subject as many pine species as possible to a comparative phytochemical study. The heartwood appeared to be particularly suitable because it is dead and no seasonal variations occur.

The main results of the investigations (Lindstedt and Misiorny, 1951) were as follows. All pines belonging to the subgenus *Diploxylon*, characterized by having two vascular bundles in the leaves, contained pinosylvin (XIV), pinosylvin monomethyl ether, the flavonone pinocembrin (2,3-dihydrochrysin, X), and the flavanonol pinobanksin (3-hydroxypinocembrin). Thirty-one *Diploxylon* species were investigated. Pines belonging to the subgenus *Haploxylon*, characterized by the presence of a single vascular bundle in the needles, showed a richer pattern of heartwood constituents. Twenty-one species were investigated. The results were here more variable, and in two species, *P. lambertiana* and *P. peuce*,[2] no compounds of pinosylvin type were found, although they may occur in very small amounts. In other species, pinosylvin or its monomethyl ether or both could be isolated or detected by paper chromatography. In addition, these pines contained dihydropinosylvin or dihydropinosylvin monomethyl ether or both.

Flavonoids such as pinocembrin (X) and pinostrobin (pinocembrin 7-methyl ether) and the corresponding flavones, chrysin (XI, 2,3-didehydropinocembrin) and tectochrysin (2,3-didehydropinostrobin), as well as pinobanksin, were found to be characteristic *Haploxylon* constituents. The group *Strobi* of Shaw's subsection *Cembra* (Shaw, 1914) contained several carbon-methylated flavonoids, for example, strobopinin (6-methylpinocembrin) and cryptostrobin (8-methylpinocembrin). In addition, one of these pines, the white pine, *Pinus strobus*, was found to contain strobochrysin (6-methylchrysin) and a *C*-methylated pinobanksin, strobobanksin.

The group *Gerardianae* of Shaw's subsection *Paracembra*, comprising two species, *P. bungeana* and *P. gerardiana*, appeared to be aberrant, containing *C*-methylated flavonoids like the *Strobi* group of *Cembra* but not dihydropinosylvins, chrysin, or tectochrysin. Insufficient amounts of wood from these pines were available at that time, and the missing flavonoid constituents have now been found during a reinvestigation of the Gerardianae.

No chemical differences between Shaw's *Lariciones*, *Australes*, *Insignes*, and *Macrocarpae* of the subgenus *Diploxylon* could be discerned, however. In *Haploxylon*, there are striking chemical similarities between the *Strobi* and *Gerardianae*, which Shaw placed in different subsections. It is indeed very remark-

[2]A recent reinvestigation showed the presence of the stilbenes.

able that the botanical separation of the subgenera *Haploxylon* and *Diploxylon*, based essentially on an anatomical character, parallels so closely the chemical differences found in these subgenera. Botanists have put much taxonomical weight on the leaf anatomy of the pines, and they seem to have done this with good reason.

Independently, Mirov (1961) has long been involved in a study of the terpenes, particularly the monoterpenes, of the wood oleoresins of pines, and his investigations cover most of the nearly 100 species of *Pinus*. Mirov's results do not show the same distinct differences between the subgenera *Diploxylon* and *Haploxylon* that Lindstedt and Misiorny's investigations do, but they are of great interest for the characterization of the minor taxonomic groups. By studying the terpenes, Mirov has been able to demonstrate hybridizations between species growing in neighboring regions. These investigations have given rise to important studies on the higher pine terpenoids by Dauben and his collaborators resulting, for example, in the structural elucidation of the so-called cembrene (Dauben et al., 1965; Wienhaus, 1928) (thunbergene), a macrocyclic diterpene, and recently in the discovery of a new diterpene acid in *Pinus lambertiana*, lambertianic acid (Dauben and German, 1966).

The instrumental armament now available along with recent methodological advances makes it possible to investigate, in considerable detail, the terpenes and resin acids as well as other constituents of pine woods. As a counterpart to the broad investigation of the phenolic constituents of the pine heartwoods, Westfelt (1964) is now studying the sesqui- and diterpenoids of the wood of a single species, *Pinus silvestris*. So far, he has isolated the following known or new sesquiterpenes: longicyclene (LIII), α-longipinene (LIV), longifolene (LV), copaene (LVI), β-copaene (LVII), possibly β-ylangene, α-muurolene (LVIII), γ-muurolene (LIX), ϵ-muurolene (LX), γ-cadinene (LXI), δ-cadinene (LXII), calamenene, α-calacorene, and δ-cadinol. There are at least two or three other uncharacterized sesquiterpene alcohols present. Small amounts of eugenol methyl ether were also encountered.

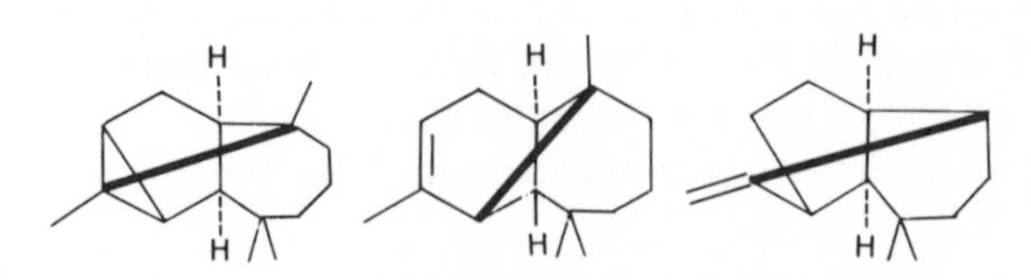

LIII. Longicyclene LIV. α-Longipinene LV. Longifolene

LVI. Copaene LVII. β-Copaene

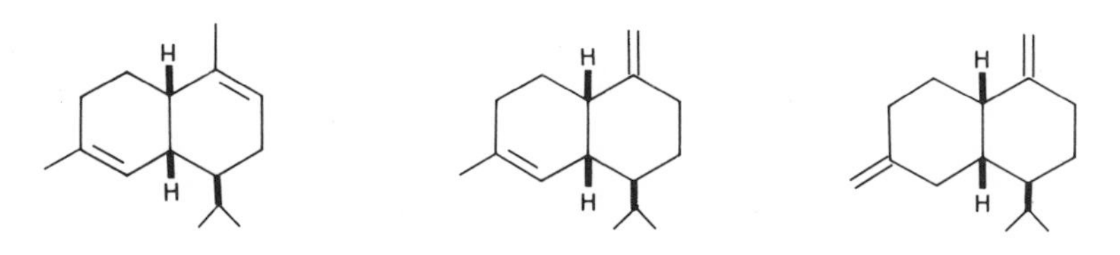

LVIII. α-Muurolene LIX. γ-Muurolene LX. ϵ-Muurolene

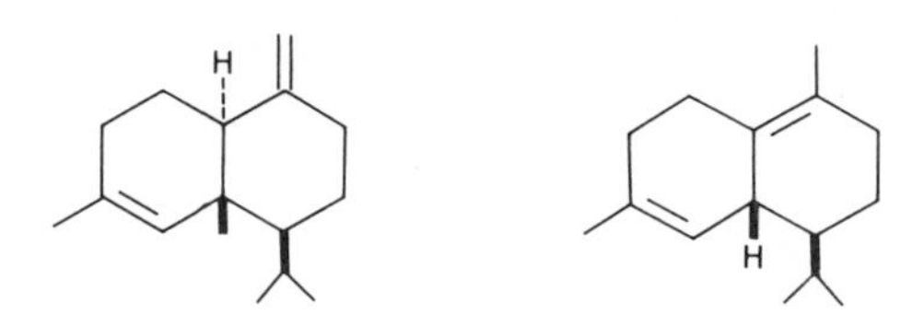

LXI. γ-Cadinene LXII. δ-Cadinene

The following diterpenes were isolated in pure crystalline form with the exception of isopimaradiene, which is an oil: pimaradiene, isopimaradiene, pimarinol, isopimarinol, pimarinal, isopimarinal abietinol, abietinal, and dehydroabietinal. There are also three unidentified diterpene acids in addition to those positively identified in wood of *Pinus silvestris*, which include pimaric, isopimaric, abietic, neoabietic, palustric, levopimaric, and dehydroabietic acid.

In pine needles, Theander (Enzell and Theander, 1962) has found a new diterpene dicarboxylic acid of the labdane type, pinifolic acid (LXIII), the structure of which was established in collaboration with Enzell. Theander also isolated two guaiacyl glycerol glucosides (LXIV, LXV)(Theander, 1965; Rudloff, 1965), and shikimic acid from the needles. These compounds appear to be subject to seasonal quantitative variations.

The bark of *Pinus silvestris* is now being investigated. Rowe (1964) has made the interesting discovery that several pine barks, from *Haploxylon* as well as from *Diploxylon* species, contain a series of triterpenes, one of which is identical with serratenediol (LXVI), the others being related to this diol. Some pine barks contain diterpenes of the labdane type.

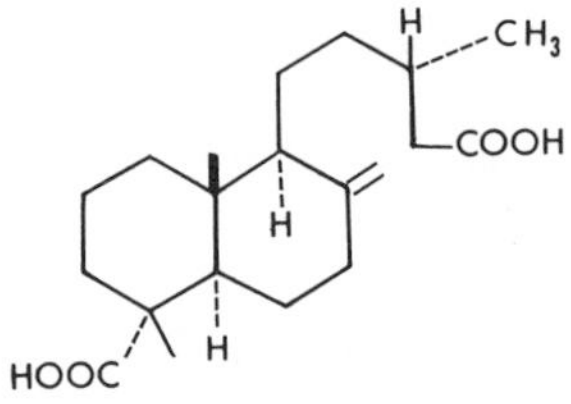

LXIII. Pinifolic acid

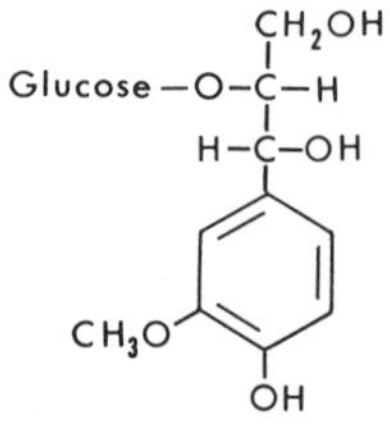

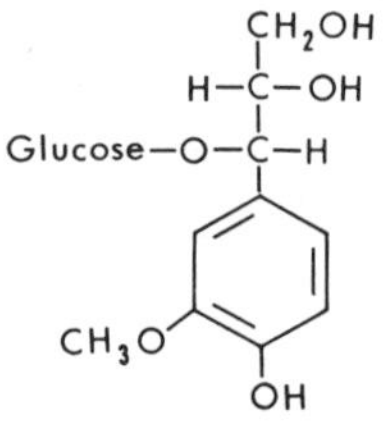

LXIV

LXV

Fatty acids of the more usual types have been found frequently in *Pinus silvestris*, but it is claimed that all-*cis*-5,9,12-octadecatrienoic acid is a major component of the fatty acids of wood and particularly the seeds of Scot's pine.

The alkaloids (+)-α-pipecoline and pinidine (LXVII) have been found in the needles of some pine species.

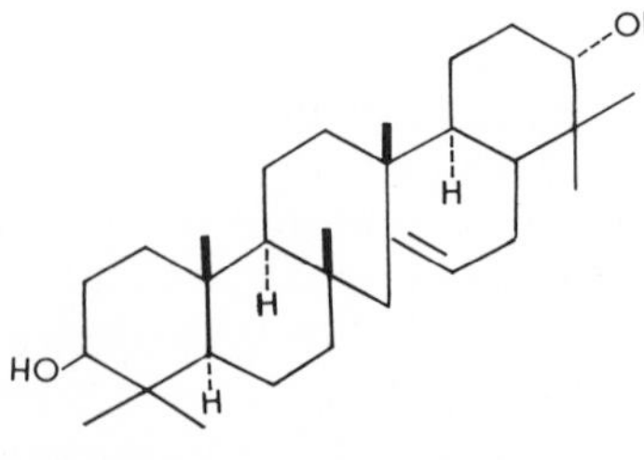

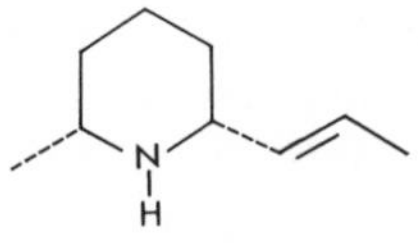

LXVI. Serratenediol

LXVII. Pinidine

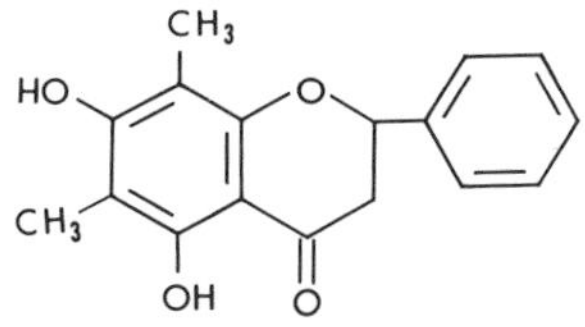

LXVIII. Demethoxymatteucinol

Many constituents of phenolic or terpenoid nature have been isolated from various other pine species. Many of these substances are identical with those from Scot's pine but there are numerous others. When a large number of pines have been stu-studied intensively, it will be possible to undertake a thorough comparison of the chemical and botanical characters of the pine genus. Such a comparison might give new insights into the evolution of this genus.

B. *Pinus* or *Ducampopinus*?

Pinus krempfii Lecomte, which grows in Viet Nam, differs to such an extent from other pines that some botanists regard it as a monotypic genus, *Ducampopinus krempfii* (Lecomte) Chevalier. This tree has smooth bark and fairly broad leaves (up to 5 mm) in pairs. There is only one vascular bundle in the needles, and consequently Florin (1931) considered this species to be simply an unusual *Haploxylon* pine. Naturally this pine, which had never been chemically investigated, aroused our interest. It appeared to be a suitable subject for a true chemotaxonomical investigation using our previous experience of the constituents of pine woods. Wood of this species has now been investigated (Erdtman et al., 1966a). The results are shown in Table 1 together with those from *Pinus strobus* (as a representative of the *Haploxylon* pines) and the general pattern of the *Diploxylon* pines. The results of a renewed investigation of the Gerardianae (*P. bungeana* and *P. gerardiana*) are also included.

It is obvious that, chemically, *Pinus krempfii* fits in very well indeed with the *Haploxylon* pines, and that it is definitely not a *Diploxylon*. The resin acids are all of the normal pine type. Whether or not it is justifiable to separate *Pinus krempfii* from the genus *Pinus* is a matter for the botanical taxonomists to consider. We are quite satisfied that, from a chemical point of view, this species cannot be distinguished from the *Haploxylon* pines,

TABLE 1

Constituents of *Pinus krempfii* and some *Haploxylon* pines

Compound	*Pinus krempfii* (*Ducampopinus krempfii*)	*Pinus strobus*	*Pinus bungeana*	*Pinus gerardiana*	*Diploxylon*
Pinosylvin	+	+	+	+	+
Pinosylvin monomethyl ether	+	+	+	+	+
Dihydropinosylvin	+				
Dihydropinosylvin monomethyl ether		+			
Pinocembrin	+	+	+	+	+
Chrysin	+	+	+	+	
Pinostrobin		+	+	+	
Tectochrysin	+	+	+	+	
Pinobanksin	+	+	+	+	+
Strobopinin	+	+	+	+	
Cryptostrobin	+	+			
Strobobanksin		+	+	+	
Strobochrysin		+			
Demethoxymatteucinol	+		+	+	

Diterpene acids of *Pinus krempfii*	Sapwood	Heartwood
Sandaracopimaric	2%	1%
Levopimaric + palustric	25%	7%
Dehydroabietic	63%	35%
Abietic	9%	57%
Neoabietic	1%	1%

at least not as regards the heartwood constituents. *Pinus krempfii* fits best with Shaw's subsections *Cembra* (*Strobi*) and *Paracembra* (*Gerardianae*). There was only a single compound in the wood of *P. krempfii* that was a new pine constituent, namely demethoxymatteucinol (LXVIII). This compound was not entirely unexpected, however, since it is 6,8-dimethylpinocembrin, and both 6- and 8-monomethylpinocembrin (cryptostrobin and strobopinin, respectively) are known *Haploxylon* flavanones.

During a reinvestigation of the *Gerardianae Pinus bungeana* and *P. gerardiana*, the presence of demethoxymatteucinol (LXVIII) in these species was also observed. Several other *Haploxylon* pines were then investigated, particularly species from Mexico. They appear to contain no trace of demethoxymatteucinol. Perhaps this dimethylated flavanone occurs only in a specific group of Asian pines (*P. bungeana*, China; *P. gerardiana*, Himalaya, northern Afghanistan; *P. krempfii*, Annam), and this invites a thorough search for demethoxymatteucinol in all the southern and East Asian *Haploxylon* pines. Quite recently we have also found that *Pinus peuce* (Bulgaria and so on) contains demethoxymatteucinol (Erdtman et al., 1966a).

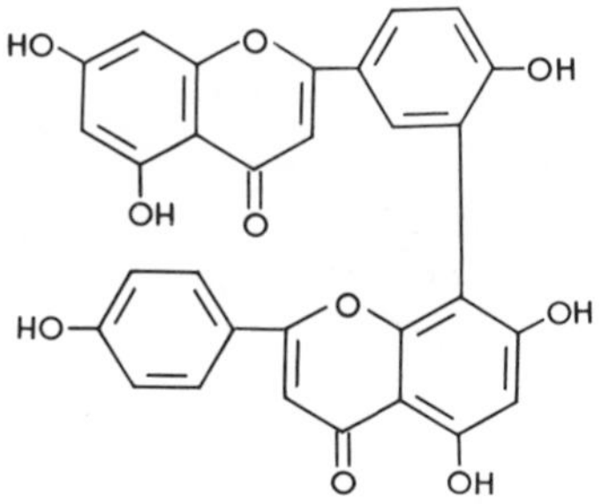

LXIX. Amentoflavone

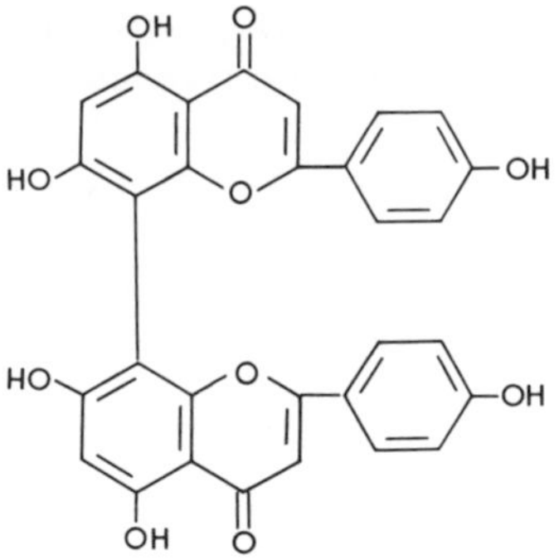

LXX. Cupressuflavone

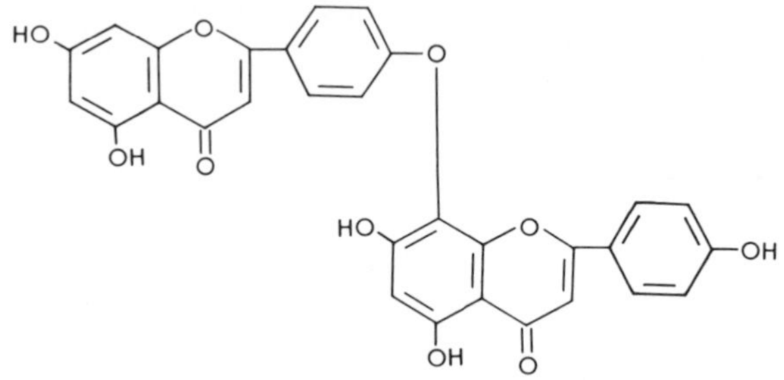

LXXI. Hinokiflavone

C. Other Genera of Pinales

Several other genera of the Pinaceae have been investigated but none as extensively as the pines. It appears that no biflavonyls occur in this family, although the compounds are common leaf constituents of other conifers. Amentoflavone (LXIX) and its various methyl ethers, cupressuflavone (LXX), and hinokiflavone (LXXI) are not entirely restricted to the conifers. The leaves of *Casuarina stricta* (Casuarinaceae) contain hinokiflavone, and amentoflavone has been discovered in the bark of one *Viburnum* species, *V. prunifolium* (Caprifoliaceae). All known natural biflavonyls are derived from the flavone apigenin.

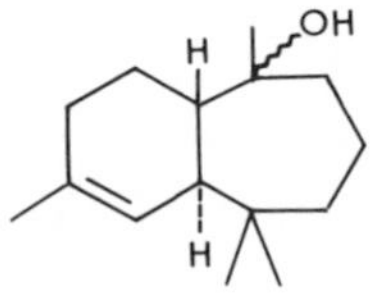

LXXII. Himachalol

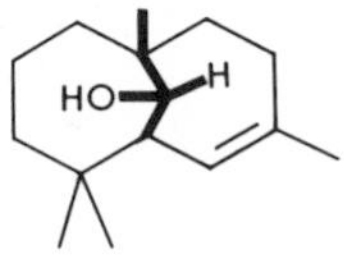

LXXIII. Allohimacholol

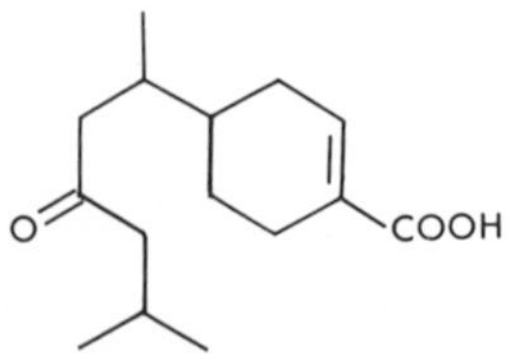

LXXIV. Todomatsuic acid

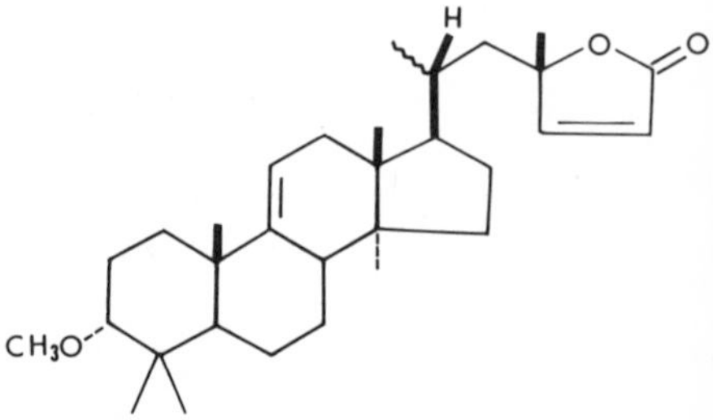

LXXV. Abieslactone

Other compounds isolated from Pinaceae (excluding *Pinus*) are lignans such as conidendrin and lariciresinol (*Picea, Larix*); flavanones such as aromadendrin and taxifolin (several genera); the stilbene piceatannol (XV)(*Picea*); and ferulic and phthalic acid esters of long-chain fatty alcohols. Terpenoids are, of course, quite common. Resin acids of the common pine types occur in all woods containing resin ducts. In *Cedrus* they occur in the cones (Kimland and Norin, unpublished). The pocket resin of Douglas fir (*Pseudotsuga mucronata*) contains large amounts of resin acid methyl esters as well as thunbergene (also in *Larix* and *Picea*) and a related alcohol, thunbergol (Kimland and Norin, unpublished). It was thought that resin acids of the rearranged abietane type, such as abietic acid, were restricted to Pinaceae, but this acid has recently been found to be a major constituent of Kauri copal (*Agathis australis*, Araucariales), and it may also occur in some *Dacrydium* species (Podocarpales), which produce resin acids of the pimarane type (Thomas, 1966). Diterpenes of the labdane type, manool, abienol, and larixol, have also been found in the Pinaceae. Sesquiterpenes are common. So far, they are all of *cis*-farnesyl type. The himachalols from *Cedrus*, LXXII and LXXIII, for example, are particularly interesting because they contain novel sesquiterpene skeletons. In the Coniferae, they have only been found in *Cedrus*. Botanists seem to be inclined to place *Larix* and *Cedrus* fairly close to each other (subfamily Laricoideae) but so far no obvious chemical similarities between these genera have been recorded. Todomatsuic acid (LXXIV) is an unusual sesquiterpene acid from an *Abies* species. In most cases only balsam or wood constituents of *Abies* species have been investigated. Abieslactone (LXXV), a triterpene lactone, has, however, been isolated from the bark of an *Abies* species.

Much more systematic chemical work has to be done before a detailed picture of the chemistry of Pinaceae can be presented. Some of the genera are large, *Abies* and *Picea*, for instance, and they are, therefore, suitable subjects for chemotaxonomic investigations.

D. Cupressales*

Much is known about the chemistry of the order Cupressales, but again there have been only a few systematic investigations. Cupressales is a heterogeneous group of conifers, and the order is subdivided into two families, Taxodiaceae and Cupressaceae.

*Compare Erdtman, H. and T. Norin, 1966.

TABLE 2

Some Types of Sesquiterpenes and Tropolones Found in the Wood of Cupressales Genera

Genus	Hemispheric Distribution	C_{15} Terpenoids of *trans*-Farnesyl Type				C_{15} Terpenoids of *cis*-Farnesyl Type				Tropolones		
		Cedrane	Thujopsane	Cuparane	Cadinane	Selinane	Occidane	Eremophilane	Guaiane	C_{10} Tropolones	C_{15} Tropolones	Carvacrol
Taxodiaceae												
Taxodium	N					+						
Cryptomeria	N				+	+						
Cunninghamia	N	+										
Sciadopitys	N	+										
Athrotaxis	S	+			+							
Taiwania	N				+							
Cupressaceae												
Cupressoideae												
Cupresseae												
Chamaecyparis	N	+	+	+	+	+		+[a]		+	+[a]	+[a]
Cupressus	N	+	+	+	+					+	+	+
Juniperae												
Juniperus	N	+	+	+	+					+	+	+

Thujopsideae											
Calocedrus	N									+	+
Fokienia	N					+					
Platycladus	N		+	+						+	
Thujopsis	N	+	+	+		+				+	+
Thuja	N					+	+			+	+
Callitroideae											
Tetraclinis	N	+								+	+
Austrocedrus	S									+	+
Pilgerodendron	S				+						
Papuacedrus	S									+	
Widdringtonia	S	+	+	+		+					
Callitris	S					+			+		
Neocallitropsis	S					+			+		

[a] Only in *Chameacyparis nootkatensis*.

Taxodiaceae embraces several small or monotypic genera. They are the remnants of old, only distantly related genera and are generally grouped together in Sequoieae, Metasequoieae, Taxodieae, Cryptomerieae, Cunninghamieae, Sciadopityeae, and Athrotaxeae.

The second family, Cupressaceae, is divided in different ways. According to one classification it is split in two subfamilies, Cupressoideae and Callitroideae. Cupressoideae is further divided into three tribes Cupresseae, Junipereae, and Thujopsideae. The genera of these tribes are northern hemispheric. The Callitroideae are southern hemispheric but include the northern, monotypic genus *Tetraclinis*.

A large number of constituents are known from Cupressales, including simple phenols, flavonoids, lignans, and mono-, sesqui-, di-, and triterpenes, as well as tropolones. Here we can only consider some of the better-known and, from a taxonomic point of view, particularly interesting wood constituents (see Table 2). In order to make the data more easily surveyed, only the carbon skeletons of the various terpenoids are mentioned. The various tropolones are given simply as C_{10} and C_{15} tropolones. It is immediately apparent that no tropolones have been found in the Taxodiaceae. With the exception of *Fokienia* (three species), all the northern genera of Cupresseae, Junipereae, and Thujopsideae have been found to contain tropolones. The C_{10} tropolones (e.g., thujaplicins, LXXVI) and hydroxylated thujaplicins are dominant. The C_{15} tropolones (e.g., nootkatin, LXXVII) are restricted to *Cupressus* and *Juniperus*. Only a single *Chamaecyparis* species, *Ch. nootkatensis*, has been found to contain C_{15} tropolones (e.g., nootkatin). In several other chemical respects, this species differs distinctly from the other *Chamaecyparis* species. *Callitris* and *Neocallitropsis* differ from other Callitroideae in that they produce guaiol. *Widdringtonia* differs distinctly from *Callitris-Neocallitropsis*, and so does *Tetraclinis*, though the latter was once considered to be a *Callitris* species. This distinction is particularly evident if the diterpenes are also included in the comparison.

Sesquiterpenes containing cedrane, thujopsane, and cuparane skeletons ("*cis*-farnesyls") often cooccur in Cupressales, and these terpenes are considered to be biosynthetically related. Not infrequently, however, *cis*- and *trans*-farnesyl sesquiterpenes cooccur (e.g., in *Chamaecyparis*, in *Thujopsis*, and in *Widdringtonia*). It is obvious that we know far too little about the chemistry of Cupressales to be able to draw any taxonomic conclusions. However, taking cognizance of other constituents, e.g., diterpenes, it is apparent that *Sciadopitys* differs greatly from the other Taxodiaceae. Many botanists consider the genus to constitute a monotypic family of its own.

Most Cupressales genera are small, often monotypic: *Juniperus* is the largest (about 70 species); *Cupressus* contains 20 species; *Chamaecyparis*, 7 species; *Callitris*, about 16 species; and *Widdringtonia*, 5 species. As far as is known, *Juniperus*, *Cupressus*, and *Callitris* have fairly characteristic patterns of wood constituents. Other genera, such as *Chamaecyparis*, vary considerably from species to species.

Of course, one should not speak about characteristic patterns if a genus comprises only one or two species. The heartwood constituents of the monotypic *Tetraclinis* and *Widdringtonia* appear to indicate relationships to the northern Cupressoideae and not to the southern Callitroideae. However, it is perhaps more correct to say that *Callitris* and *Neocallitropsis* occupy a unique position among the Callitroideae.

The tropolone character is certainly taxonomically important despite the fact that some species belonging to tropolone-producing genera lack these compounds. The wood of *Juniperus virginiana* has been investigated by several workers, but no traces of tropolones have ever been detected, not even in the commerical "cedar oil" from this species. The oil is obtained by steam distillation of wood from many individual trees, and the lack of tropolones cannot be explained away by using the convenient term "biological variation." Rather, it seems that several factors are needed for tropolone synthesis, and that one link in the chain is easily eliminated or blocked.

The needles of all Cupressales species seem to contain biflavonyls, but the distribution of simple flavonoids in this group has been little investigated. The needles of some species, particularly *Juniperus* species and also a few others, have been found to contain the lignan podophyllotoxin. The woods of some species have been found to contain phenols of novel structure, which may prove to be of taxonomic value when all of them have been structurally elucidated and their distributions investigated. From the wood of *Chamaecyparis obtusa*, the so-called hinokiresinol (LXXVIII) has been isolated (Hirose et al., 1965). A similar compound, sugiresinol (LXXX), was found in the wood of *Cryptomeria japonica*, sugi (Funaoka et al., 1963), together with hydroxysugiresinol of still uncertain structure. It contains a catechol group. Sugiresinol, hydroxysugiresinol, and "sequirin

C'' have been isolated from *Sequoia sempervirens* (Balogh and Anderson, 1965). *Athrotaxis selaginoides* contains athrotaxin (Erdtman and Vorbrüggen, 1960) and agatharesinol (LXXIX) (Erdtman et al., unpublished). The latter has also been isolated recent from *Agathis* of the order Araucariales (Enzell and Thomas, 196 The structure of athrotaxin, a dienone, is not yet known in all details, but it is undoubtedly related to agatharesinol. Fresh extrac of *Athrotaxis* wood do not seem to contain sugiresinol, although th old extracts do. Agatharesinol is transformed into sugiresinol by the action of acids. All of these compounds contain 17 carbon atoms. The discovery of this group of phenolic compounds is welcome, since it gives a new feature to the otherwise somewhat monotonous terpene themes of the Cupressales. Some of these new phenols are rather labile, and they are easily overlooked or lost when wood extracts are being worked up. They are perhaps not uncommon in woods turning pink on exposure to air and light. The biosynthesis of these $C_6C_3C_2C_6$ compounds poses an interesting problem.

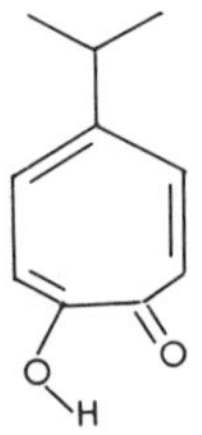

LXXVI. γ-Thujaplicin

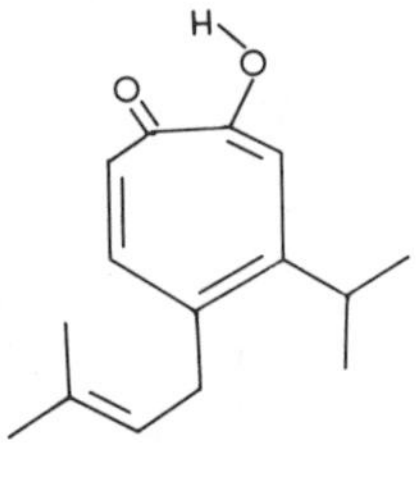

LXXVII. Nootkatin

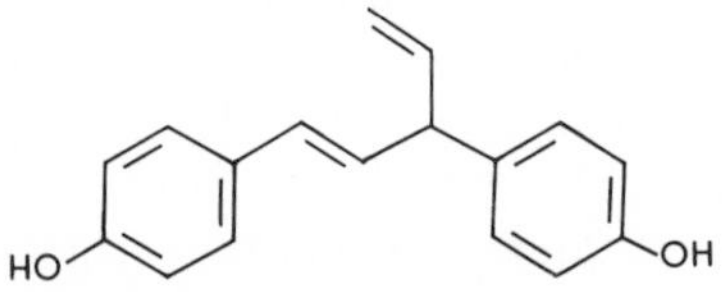

LXXVIII. Hinokiresinol

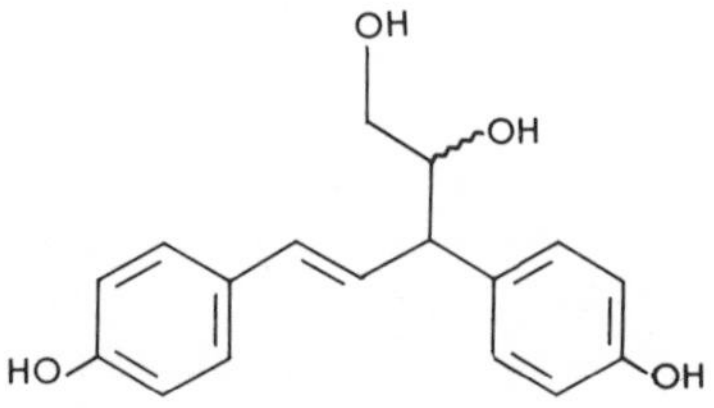

LXXIX. Agatharesinol

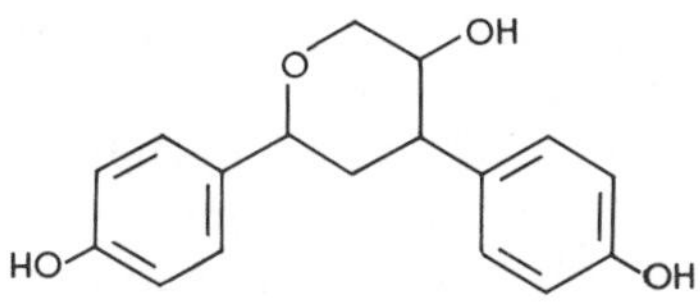

LXXX. Sugiresinol

The other conifer orders, Araucariales and Podocarpales, have so far been much less investigated than Pinales and Cupressales. It is mostly leaf diterpenes or diterpenes from the woods that have been studied, and it is, therefore, not possible to discern any specific patterns of constituents of different biosynthetic origin. Many of the diterpenes resemble those of the Pinales and Cupressales.

One could have chosen practically any of the plant classes or orders for a discussion of the chemistry of a taxon. The result would have been almost the same. One can only imagine certain more or less clear trends and patterns here and there which arouse one's curiosity.

XII. CHEMISTRY OF NATURAL HYBRIDS

Chemical studies of natural hybrids have been successful in demonstrating the effectiveness of such methods in appropriate situations. It was shown in hybrids of *Eucalyptus cinerea* and *E. macarthuri* (Pryor and Bryant, 1958) that oils that were characteristic of one species or the other occurred together in the hybrid. This result, if generally true, is a useful one in the study of hybridization in nature. Other investigators, with other material, have also found that the chemistry of a hybrid tends to be the sum of the components of the two parental species, and the idea of cumulative inheritance of secondary compounds seems to be established as a valid generalization. Therefore, it may be possible, in favorable circumstances to prove by chemical methods, the existence of a hybrid which might only be suspected on morphological or other grounds. This type of approach has been applied extensively in the study of the genus *Baptisia* (Alston and Turner, 1963a; Alston and Hempel, 1964), in which hybridization is occasionally so complex as to involve as many as four species and various types of hybrids present in a single field. Fortunately, in some pairs of *Baptisia* species involved in hybridization their flavonoid chemistry may be quite different and thus allow an analysis by chemical methods of the make-up of the population (Alston et al., 1965; Turner and Alston et al., 1959)

Another aspect of the study of hybrids which is of theoretical interest concerns the ability of the hybrid to produce "new" compounds which neither parent could make alone. The most completely investigated example of such a phenomenon is from the hybrid *Baptisia leucantha* × *B. sphaerocarpa* (Alston et al., 1965). *Baptisia leucantha* produces rutin (quercetin 3-β-D-rhamnoglucoside) but does not form 7-glycosides of flavonols. *Baptisia sphaerocarpa* makes, among other flavonoids, quercetin 3-β-D-glucoside 7-β-D-rhamnoglucoside. Neither parental species synthesizes quercetin 3-β-D-rhamnoglucoside 7-β-D-glucoside, but this compound is present in the hybrid. This then is truly a hybrid compound although it is not an especially rare type of flavonoid. These and other studies of a similar kind may prove to be a means of demonstrating a unique contribution of chemical methods to systematic investigations.

XIII. THE FUTURE OF CHEMOTAXONOMY

Chemotaxonomy is surely in its infancy, and it is probably correct to say that, insofar as taxonomy is concerned, plant chemistry now stands where botanists were 100–150 years ago. This is not at all surprising. In their fascinating book "Biochemical Systematics," Alston and Turner (1963b), discussing the major historical or developmental periods of systematic biology, reckoned that the fifth period, "the biochemical period," commenced about 1950. According to them, the period is "characterized in its early stages by the establishment of biochemical profiles for various plant taxa and their comparative use in solving taxonomic problems; in later stages by a comparative biochemical approach that takes into consideration metabolic pathways, protein evolution, and comparative enzymology."

It is obvious that there will be chemists keenly exploring various taxa more or less at random, but, let us hope, more and more systematically. They will remain chemists, albeit with taxonomic ambitions. Those botanists who are primarily interested in taxonomic problems will increasingly take cognizance of the chemical characters and eventually, perhaps, carry out most of the chemical screening of taxa. Both parties concerned will doubtless create intriguing problems for each other to solve.

An intimate collaboration between botanists and chemists will be necessary in this field. Chemists need the advice of botanists to choose the most rewarding problems for their investigations, and botanists need the aid of chemists in methodological questions. Chemists are, unfortunately, often careless with the identification of the plant material used, and, in turn, they are sometimes rather skeptical about the validity of the botanists' chromatographic identifications of compounds, so ever more effective interdisciplinary communication is badly needed.

New universities are now being established in many developing countries, and it is natural that botanists and chemists in such countries should collaborate in the exploration of their native flora. Chemotaxonomy could serve as very useful starting points for organic chemistry and botany in such schools.

As in many other branches of science, information retrieval is one of the worst headaches in chemotaxonomy. One badly needs a continuous registration of those compounds, new or known, which are being isolated. They should be arranged according to a botanical system. It is not a matter of prime importance which taxonomic system one is using, but major changes which seem to be well founded on biological principles should be included as needed. It is admittedly a difficult task to realize such a project. The editorial board of such an organization must consist of botanists as well as chemists, and it seems natural that this is a matter for the relevant international unions to consider jointly. Nothing would serve the future development of chemotaxonomy better than the publication of annual indexes on plant chemistry. Kariyone (1957–1961) has made a brave attemp in this direction with his Annual Indexes.

A great responsibility will always rest on the authors of chemical papers. They must endeavor to define accurately the plant material that they have been investigating. A pine is sometimes simply called "pitch pine," and no one, except perhaps the author himself, knows which "pitch pine" he was actually studying.

I have met chemists who have told me that they had isolated a compound from a certain plant, but, since the substance had already been isolated from another plant and the structure completely elucidated, they did not care to publish the results, thus showing themselves to be better chemists than scientists.

REFERENCES

Alston, R. E., and K. Hempel. 1964. J. Hered., 55: 267.

_____, and B. L. Turner. 1963a. Amer. J. Bot., 50: 159.

_____, and B. L. Turner. 1963b. Biochemical Systematics, N.J. Prentice-Hall, p. 37.

_____, H. Rösler, K. Naifeh, and T. J. Mabry. 1965. Proc. Nat. Acad. Sci. U. S. A., 54: 1458.

Arya, V. P., H. Erdtman, M. Krolikowska, and T. Norin. 1962. Acta Chem. Scand., 16: 518.

Balogh, B., and A. B. Anderson. 1965. Phytochemistry, 4: 569.

Bendz, G., and O. Mårtensson. 1961. Acta Chem. Scand., 15: 1185.

_____, and O. Mårtensson. 1963. Acta Chem. Scand., 17: 266.

_____, O. Mårtensson, and L. Terenius. 1962. Acta Chem. Scand., 16: 1183.

Billek, G., and H. Kindl. 1961. Mh. Chem., 92: 483.

_____, and H. Kindl. 1962. Österr. Chem. Z., 63: 273.

Birch, A. J. 1957. Fortschr. Chem. Org. Naturst., 14: 186.

_____, and F. W. Donovan. 1953. Aust. J. Chem., 6: 360.

Brown, S. A. 1961. Canad. J. Bot., 39: 253.

_____, and A. C. Neish. 1956. Canad. J. Biochem. Physiol., 34: 769.

Chopin, J., M. L. Bouillant, and P. Lebreton. 1963. C. R. Acad. Sci. (Paris), 256: 5653.

Dauben, W. G., and V. F. German. 1966. Tetrahedron, 22: 679.

_____, W. E. Thiessen, and P. R. Resnick. 1965. J. Org. Chem., 30: 1696.

Enzell, C., and O. Theander. 1962. Acta Chem. Scand., 16: 607.

_____, and B. R. Thomas. 1965. Acta Chem. Scand., 19: 913.

_____, and B. R. Thomas. 1966. Tetrahedron Lett., 2395.

Erdtman, H. 1933a. Biochem. Z., 258: 177.

_____. 1933b. Liebig Ann. Chem., 503: 283.

_____. 1939. Liebig Ann. Chem., 539: 116.

_____. 1950. Research, 3: 63.

_____. 1959. *In* Kratzl, K., and Billek, G., eds., Biochemistry of Wood, vol. II, 1, New York, Pergamon.

_____, and T. Norin. 1963. Acta Chem. Scand., 17: 1781.

_____, and T. Norin. 1966. *In* Zechmeister, L. ed. Forschitte Chemie Organisher Naturstoffe, XXIV: 206. Springer Verlag Wien.

_____, and H. Vorbrüggen. 1960. Acta. Chem. Scand., 14: 2161.

_____, and C. A. Wachtmeister. 1957. Festschr. Arthur Stoll:

_____, and L. Westfelt. 1963a. Acta Chem. Scand., 17: 1826.
_____, and L. Westfelt. 1963b. Acta Chem. Scand., 17: 2351.
_____, G. Eriksson, and T. Norin. 1961. Acta Chem. Scand., 15: 1796.
_____, B. Kimland, and T. Norin. 1966a.
_____, B. Kimland, and T. Norin. 1966. Phytochemistry, in press.
_____, P. J. L. Daniels, and K. Nishimura. Unpublished.
Florin, R. 1931. Kungl. Vetenskapsadameniens Handlingar [3], 10, no. 1: 336, 341.
Freudenberg, K. 1954. Fortschr. Chem. Org. Naturst., 11: 43.
_____. 1962. Fortschr. Chem. Org. Naturst., 20: 41.
Funaoka, K., Y. Kuroda, Y. Kai, and T. Kondo. 1963. J. Japan Wood Res. Soc., 9: 139.
Galeffi, C., C. G. Casinovi, and G. B. Marini-Bettolo. 1965. Gazz. Chim. Ital., 95: 95.
Geissman, T. A. 1963. *In* Bernfeld, P., ed., Biogenesis of Natural Compounds, Oxford, p. 589.
Hegnauer, R. 1962-1964. Chemotaxonomie der Pflanzen, Basel, Birkhäuser.
Higuchi, T., and F. Barnoud. 1966. J. Japan Wood Res. Soc., 12: 36.
Hirose, Y., N. Oishi, H. Nagaki, and T. Nakatsuka. 1965. Tetrahedron Lett.: 3665.
Ibrahim, R. K., and G. H. N. Towers. 1960. Canad. J. Biochem. Physiol., 38: 627.
_____, and G. H. N. Towers. 1962. Canad. J. Biochem. Physiol., 40: 449.
Kariyone, T. 1957-1961. Annual Index of the Reports on Plant Chemistry, Tokyo, Hirokawa Publ. Co.
Karrer, W. 1958. Konstitution und Vorkommen der Organischen Naturstoffe, Basel, Birkhäuser.
Koch, J. E., and W. Krieg. 1938. Chem. Z., 15: 140.
Lawrie, W., J. McLean, and G. R. Taylor. 1960. J. Chem. Soc.: 4303.
Liebman, A. A., F. Morsing, and H. Rapoport. 1965. J. Amer. Chem. Soc., 87: 4399.
Lindstedt, G., and A. Misiorny. 1951. Acta Chem. Scand., 5: 121.
Lynen, F. 1961. Fed. Proc., 20: 941.
Matas, L. C. 1960. Chem. Abstr., 54: 2502.
Melchert, T. E., and R. E. Alston. 1965. Science, 150: 1171.
Metzenberg, R. L. 1965. Science, 150: 471.
Mirov, N. T. 1961. Composition of Gum Turpentine of Pines.

U. S. Dept. Agr. Forest Serv. Tech. Bull. 1239.
Mukerjee, R., and A. Chatterjee. 1964. Chem. Industr.: 1524.
Narasimachari, N., and E. von Rudloff. 1962. Canad. J. Chem., 40: 1118.
Pimenta, A., A. A. L. Mesquita, M. Carney, E. Gottlieb, and M. T. Magalhães. 1964. An. Acad. Brasil. Ciências., 36: 283
Pryor, L. D., and L. H. Bryant. 1958. Proc. Linnean Soc. N. S. Wales, 83: 55.
Rowe, J. W. 1964. Tetrahedron Lett.: 2347.
Rudloff, E. von. 1965. Chem. Industr.: 180.
_____, and E. Jörgensen. 1963. Phytochemistry, 2: 297.
Sandermann, W., and M. H. Simatupang. 1963. Chem. Ber., 96: 2182.
Shaw, G. R. 1914. The Genus *Pinus*, Cambridge, Mass., Arnold Arboretum.
Spenser, I. D., and H. P. Tiwari. 1966. Chem. Commun., 55.
Theander, O. 1965. Acta Chem. Scand., 19: 1792.
Thomas, B. R. 1966. Acta Chem. Scand., 20: 1074.
Turner, B. L., and R. E. Alston. 1959. Amer. J. Bot., 46: 678.
Westfelt, L. 1964. Acta Chem. Scand., 18: 572.
Wienhaus, H. 1928. III. Nordiska Kemistmötets Förhandlingar 1926. Helsingfors. p. 211.
Yasue, M. 1965. J. Japan Wood Res. Soc., 11: 202.
_____, and Y. Kato. 1960. J. Pharm. Soc. Japan, 80: 1013.

part II

NITROGEN AND SULFUR COMPOUNDS

'We are surrounded with things which we have not made and which have a life and structure different from our own: trees, flowers, grasses, rivers, hills, clouds. For centuries they have inspired us with curiosity and awe. They have been objects of delight. We have recreated them in our imagination to reflect our moods. And we have come to think of them as contributing to an idea which we have called nature. Landscape painting marks the stages in our conception of nature. Its rise and development since the middle ages is part of a cycle in which the human spirit attempted once more to create a harmony with its environment..... All art is to some degree symbolic, and the readiness with which we accept symbols as real depends, to a certain extent, on familiarity...'

'Throughout this book, when I have used the word 'nature', I have meant that part of the world not created by man which can see with our unaided senses. Up to fifty years ago this anthropocentric definition did well enough. But since then the microscope and telescope have so greatly enlarged the range of our vision that the snug, sensible nature which we can see with our own eyes has ceased to satisfy our imaginations.... We know that every form we perceive is made up of smaller and yet smaller forms, each with a character foreign to our experience.... Leonardo da Vinci, who used to sign himself 'disciple of experience', left among his latest writings the sentence, 'Nature is full of an infinity of operations which have never been part of experience'.'

(Kenneth Clark, 'Landscape into Art',
Chapter 1 and Epilogue, John Murray Publishers, Ltd.)

3

SULFUR COMPOUNDS IN PLANTS

M. G. Ettlinger and A. Kjær
Organic Chemistry Laboratory
Royal Veterinary and Agricultural College
Copenhagen, Denmark

I. INTRODUCTION: CHEMISTRY AND TAXONOMY

Sulfur is indispensable for life. Some organic sulfur compounds, like coenzyme A, thiamine, thioctic acid, and biotin and cysteine, cystine, and methionine (free or combined in peptides and proteins), occur, we may suppose, universally in living organisms. Were one organic compound only to be picked as central to the biochemistry of sulfur, cysteine would probably be thought the most reasonable choice. Plants from algae to angiosperms also contain, however, sulfur compounds of restricted distribution, whose absence is compatible with the maintenance of all vital processes. Many of these variable (so-called secondary) constituents draw attention because they give rise in disrupted tissues to secondarily derived, volatile products with conspicuous smells. Thus, by enzymatic action, sulfonium salts in seaweeds lose dimethyl sulfide, *S*-substituted cysteines and their sulfoxides in garlic and other flowering plants become thiols and disulfides, glucosinolates (mustard oil glucosides) in mustard give isothiocyanates. The odors of divalent sulfur compounds are not always pleasant, but they confer on this area of natural products a homely appeal, and they are characteristic enough to show the widespread occurrence of special sulfur-bearing constituents in angiosperm families, like Bignoniaceae, Meliaceae, and Phytolaccaceae, that are yet to be studied chemically.

The term "natural sulfur compound" is commonly understood to exclude sulfate esters and imply the presence of a carbon-sulfur bond. The sulfur atom or atoms may be part of oxide, sulfone or sulfonate groups. In a useful arrangement of the compounds by type within a general biochemical framework, however, their placement frequently is governed not by the functionality of sulfur but by the character of the rest of the molecule. Challenger's narrative essays (1959) provide a vivid idea of the development of the field. Recent summaries (Kjaer, 1963a, 1966) broadly cover present-day knowledge of the variable sulfur compounds in higher plants and, through discussion of their distribution, indicate potential applications in botanical systematics.

Of the sulfur compounds that might concern us, one major sort is exemplified in Virtanen's account (1965) of amino acids and peptides in *Allium*. These nonprotein, *S*-substituted derivatives of cysteine are of great interest, but we will not discuss them here. We shall review sulfur-bearing alkaloids and polyacetylene derivatives, each subroup is closely associated in

Nature with a host of sulfur-free relatives. The other plant products to be included, the glucosinolates, are glucosides derived biogenetically from amino acids, but because sulfur forms part of a unique functional group that dominates their chemical behavior, they stand out sharply as a separate class.

Application of the study of sulfur compounds to botanical systematics is just beginning, and in this matter we turn first to general considerations. The current systematic framework includes much thoughtful reasoning from a large amount of observational information about plants, to which chemical analysis can powerfully contribute. We suppose that a scheme of classification whose aim is to reflect evolutionary relationships should take into account and correlate as much of the extant information as is humanly possible. The changes in biochemical processes during evolutionary descent are unknown a priori and need to be inferred from the collected evidence. A chemist who develops an interest in the plants that supply the natural products and enzymes with which he works is apt to have his attention fixed at first by the classification, which is usually based on what is hearsay to him. If he practices "comparative phytochemistry," he may well choose among existing classifications of the plants, without claiming to make an authoritative evaluation, in order to adopt the scheme that most conveniently accommodates the chemical data. He comes to more questionable footing if the currency of terms like "chemotaxonomy" and "biochemical systematics" leads him to suppose that the chemical data are meaningful for systematic purposes in isolation or inherently transcend other proceeds of observational analysis. The division of genera into sections of species grouped according to chemical composition, as done with the aid of studies partly on sulfur compounds for *Anthemis* (Bohlmann et al., 1965f) and *Chrysanthemum* (Bohlmann et al., 1964a) in Compositae and *Iberis* (Gmelin, 1963) in Cruciferae, is no more than a particular fashion of stating the chemical results. These arrangements can hardly be considered even sketches of classifications according to relationship, although they present valuable raw materials on which such classifications can be partly based.

The necessity of the mechanical features of systematic work, the collection and preservation of herbarium specimens and the assignment of correct names, tends to be disregarded by most chemists. They instinctively rebel, in the first place, against what is dimly imagined to be the arbitrary and prescriptive character of systematics. More particularly, they are used to going quickly beyond seemingly passive and simple observation, and find it difficult to believe that the information gained by looking carefully at a plant is not just as superficial as when they describe a compound as colorless needles. Chemists rely on structures as primary symbols, seldom pondering names; they object to botanical nomenclature both because it shifts with increasing data, which in hope lead to more enlightened judgment, and because the Latin language and rules of priority make it seem antiquarian. Few chemists are at peace with a sense of history. We remark only that the binomial of genus and species often cannot be traced without the author citation that completes the name (omitted, we might add, in some botanic garden seed lists) and that the problem of identification is separable to some extent from that of nomenclature, in that a complete name, even if invalid or out of fashion, can frequently be recognized—indeed, the system of intergeneric transfers is designed for this end. The name chosen, however, should convey meaning with a minimum of effort and possible confusion, and ordinarily a person competent to identify a plant will use valid, consistent nomenclature. The pitfalls, gross and subtle, in use of material from botanic gardens and the necessity for scrupulous identification of all specimens have been emphasized in other discussions (Kjær, 1960; Heywood, 1966a,b). We may point out that search in keys is only a first step in accurate identification, that comparison with authentic herbarium specimens is generally indispensable, and that many sheets are wrongly designated even in first-class herbaria. Plants, therefore, should be identified by someone experienced and critically interested in their particular group. The herbarium specimens are the permanent, versatile and reusable, experimental record.

The organic chemist has little trouble in deciding whether two structural formulas represent the same molecule or not or in grouping natural products according to composition, functionality, and biogenesis and is liable to instructive difficulty in realizing that the taxonomic issues and processes in biology are fundamentally different. Organisms are more complex than familiar molecules and are participants in the historical sequence of evolution, not a microcosm where time is repeatable with certainty; they enter our minds as direct visual images rather than abstractions. In the demarcation of a species, we look, as expressed in the formalism of particulate nomenclature, for correlated discontinuities in the pattern of variation, asking also that these discontinuities be maintainable in Nature by descent.

The question whether two distinguishable individuals should be considered to belong to the same species can therefore be answered only by reference to other specimens, and the answer may change when more samples become available. Put another way, a question that a chemist collecting and analyzing plants is likely to ask, "What morphological and chemical variations have systematic significance?," can be resolved in each case only by the outcome of experience. Species are empirically determined entities, not like a set of congruent boxes but like a set all of whose members differ in size, contour, and definiteness of outline. The individuals of one species may all be obviously the same, whereas a closely related species may encompass a wide range of variation, held together simply by the fact that it cannot be satisfactorily subdivided.

Occasional encounters with dichotomous keys for preliminary identification of plants might suggest that the taxonomic method is primarily a mechanical sieving process, but in fact a chemist whose idea of taxonomic cognition is derived from the keys is liable to be more misled than a botanist whose view of inorganic chemistry comes from acquaintance with schemes for qualitative elemental analysis. Attentive examination of a succession of plants yields a large number of correlated visual impressions, and the recognition or assessment of the complex pattern of perception that develops in a challenging taxonomic situation probably is less akin to mechanical sorting than to the comprehension of similarly intricate patterns during informed, practiced appraisal of art or, with another of the senses, music. Likewise, the conceptual images of taxa probably lie deeper in the preconscious than now is usual for corresponding images of chemical compounds. The applicability of rational discussion and interpretation, in harmony with the rest of biological knowledge, to taxonomic conceptions certainly is real and is extremely important, but a part may remain difficult to articulate. The division of certain families into natural tribes on the basis of minor technical characters, which sometimes puzzle chemists, can mean that these characters are markers, chosen because they can be singled out, of much more ramified, subtle and correlated, resemblances and differences.

Chemical data may appear relatively unambiguous, but their systematic usefulness depends on active effort toward the purpose. One reason why they have auxiliary status in classification is that without analyses for all kinds of constituents—major chemical differences, in contrast to morphological ones, and excepting color changes—they can readily pass unknown. The detailed study of variation among individuals requires a good deal of work to achieve proportionate coverage by chemical as compared to morphological means, especially if several classes of compounds are to be examined, and puts an extra premium on rapid analytical methods like vapor-phase and thin-layer chromatography. The dependence of composition on environmental factors like the degree of shade calls for attention. It is well known that the representation of a class of compounds can change strikingly among diverse parts of the same plant and at different growth stages. Furthermore, the distribution can be abruptly altered by hybridization (Alston and Simmons, 1962; Alston et al., 1965). For glucosinolates, several examples are known in which the side chain of a dominant glucoside in a seed contains a sulfoxide or sulfone group, whereas a corresponding reduced form (sulfide or sulfoxide) occurs in green parts or root (Kjær, 1958; Benkert, 1966; Friis and Kjær, 1966). The chemical relationships are not, however, always direct. Allylglucosinolate, whose presence in *Capsella bursa-pastoris* (L.) Med. (Cruciferae) has been controversial (Kjaer, 1958), can be found in the flowers, but the rest of the plant and the seeds contain 9-methylsulfinylnonyl and 10-methylsulfinyldecyl glucosinolates. *Brassica* seeds contain glucosinolates that are ω-methylthioalkyl (propyl to pentyl) derivatives, with only traces of aromatic side chains (Ettlinger and Thompson, 1962), but 3-indolylmethylglucosinolate and its *N*-methoxy derivative are prominent in the fresh parts (Gmelin and Virtanen, 1961, 1962), and 2-phenylethylglucosinolate is plentiful in roots (Lichtenstein et al., 1964). In a full evaluation, furthermore, knowledge of the chemical composition of a plant, including groups of related compounds, becomes subsumed into knowledge of the enzyme-catalyzed reaction sequences operating, with their specificity and their genetic origin and control. Statements about the absence of a potential natural product from a plant generally mean no more than the experimental fact that it was not detected; they should hence be taken lightly in making comparisons.

Any of the variable sulfur-containing plant constituents is a candidate for taxonomic use. One simple character, the relative proportions of *S*-methyl, propyl, and allyl cysteine sulfoxides in specimens of Western American *Allium* (Liliaceae), measured by gas chromatography of the alkyl disulfides, is reported to be independent of habitat, growth stage, and organ, but to show significant differences between species (Saghir et al., 1965). A preliminary investigation on glucosinolates in systematically well-characterized *Arabis* (Cruciferae) indicated the possibility of helpful correlations, but only one of the compounds was identified (Kjær and Hansen, 1958). Vaughan et al. (1963) have studied volatile mustard oils, or in effect their parent glucosinolates, from seed of *Brassica juncea* Czern. and Coss. (Cruciferae), which is known mainly as a cultivated mustard, presumably influenced by artifical selection. The species, Asiatic in origins, is believed to be an amphiploid hybrid of *B. nigra* (L.) Koch, which furnishes allylglucosinolate, and *B. campestris* L. or a closely related Far Eastern species (*B. chinensis* L.), known to contain 3-butenylglucosinolate and frequently another major component, the 2-hydroxy derivative, which here went unconsidered because it yields no volatile product. The hybrids of *B. nigra* and *B. campestris* that have been prepared synthetically and tested contain largely or apparently exclusively the allyl compound. Results on an extensive set of *B. juncea* samples showed an interesting close correlation between composition and home territory. Samples from Japan, China, Nepal, Russia, and Eastern Europe, including all the forms with yellow seed coats or dissected leaves, contained allylglucosinolate and at most traces of its 3-butenyl homologue. Indian and Pakistani samples consistently yielded the 3-butenyl glucosinolate, almost always in amounts comparable to or greater than those of the allyl compound. Even within the mustard family, of course, sulfur compounds are not the only chemical markers; other substances for which some indication of systematic utility has been given include cardiac glycosides (in *Cheiranthus* and *Erysimum*; Kowalewski, 1960), triterpenoid bitter principles (in *Iberis*; Gmelin, 1963), proteins (in *Brassica*; Vaughan and Waite, 1967), and fatty acids (in *Lesquerella*; Barclay et al., 1962). Systematics, which leads to classifications invested with abstract qualities, has its spring as a comprehensive, intensified mode of perception of the world around us (cf. Heywood, 1966a).

II. SULFUR-CONTAINING ALKALOIDS

Among the numerous sulfur compounds encountered in higher plants a small but rapidly growing group of alkaloids can now be distinguished.

As is apparent from the following discussion the various types of sulfur-containing alkaloids thus far reported possess no obvious common structural features. In most cases, they rather seem to be sulfur-containing modifications of accompanying non-sulfurous alkaloids, suggesting their origin by more or less accidental incorporation of sulfur-containing moieties into intermediates on the general biosynthetic pathway to the major alkaloids characteristic for the individual taxon.

Currently, sulfoacetic esters (Folkers, et al., 1944) do not have the interest of alkaloids with less oxidized sulfur, first exemplified by 4-methylthiocantin-6-one (I) from *Pentaceras australis* Hook. f. (Rutaceae) (Nelson and Price, 1952). Its biosynthetic origin is unknown but it seems conceivable that the thioenol ether derives from a ketonic precursor through reaction with an unknown HS- donor followed by *S*-methylation.

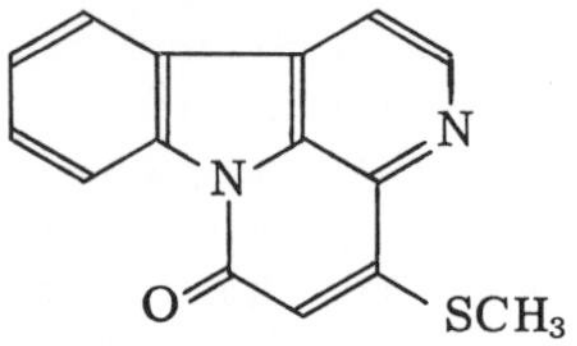

N O SCH_3
I

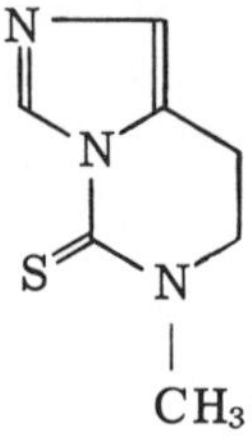

N N S N CH_3
II

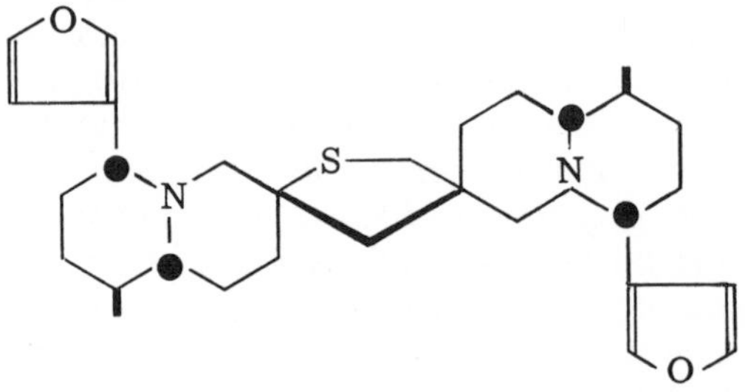

O N S N O
III

Of a structurally different type is zapotidine (II), a cyclic thiourea isolated as a minor constituent from seeds of another member of the family Rutaceae, *Casimiroa edulis* Llave & Lex. (Mechoulam et al., 1961; Mechoulam and Hirshfeld, 1967). Again, the in vivo synthesis of the unique natural thiourea remains obscure. Its formation by cyclization, and subsequent *N*-methylation, of 2-(5-imidazolyl)-ethyl isothiocyanate, a homologue of 5-imidazolylmethyl isothiocyanate, which could derive from the unknown and long sought glucosinolate corresponding to histidine, seemed to be an attractive possibility. It was not supported, however, by unpublished attempts (Kjaer) to establish the presence of glucosinolates in seeds of *C. edulis*.

A remarkable series of sulfur-containing alkaloids has been discovered in species of the genus *Nuphar* of the water lily family (Nymphaeaceae). Thus, Achmatowicz and Bellen (1962) isolated four crystalline S-alkaloids, from *N. luteum* (L.) Sm.: thiobinupharidine ($C_{30}H_{42}O_2N_2S$), allothiobinupharidine ($C_{30}H_{42}O_2N_2S$), pseudothiobinupharidine ($C_{30}H_{40}O_2N_2S$), and thiobidesoxynupharidine ($C_{30}H_{40}ON_2S$). Structure elucidation of these was undertaken but has not yet been reported. However, the recent isolation from the same source of neothiobinupharidine, a stereoisomer of thiobinupharidine (Achmatowicz and Wróbel, 1964), provided the key to the structural type. Detailed studies, mainly based on NMR and mass spectra, led to a structural proposal for neothiobinupharidine (Achmatowicz et al., 1964) which was slightly revised to structure III by Birnbaum (1965) on the basis of X-ray analysis. It would be of considerable interest to know the pathway by which sulfur is incorporated into S-containing *Nuphar* alkaloids which, upon inspection, seem to be of isoprenoid origin. In the absence of experimental evidence it seems reasonable to assume that coupling of a sulfur-containing quinolizidine intermediate with another, oxidized quinolizidine fragment is implied, particularly since the major, nonsulfurous *Nuphar* alkaloids contain the quinolizidine skeleton, in one case even as an *N*-oxide (nupharidine).

A recent note by Döpke (1965) reports the isolation of two unidentified S-alkaloids with the compositions $C_{12}H_{14}O_4N_2S$ and $C_{30}H_{42}O_2N_2S$ (the latter isomeric with thiobinupharidine and other *Nuphar* alkaloids) from "einer botanisch nicht einwandfrei identifizierten kongolesischen Baumrinde."

Recently, the mangrove family, Rhizophoraceae, also has been reported as a source of sulfur-containing alkaloids. Thus, Wright and Warren (1967a) isolated the bases cassipourine from *Cassipourea gummiflua* Tul. var. *verticellata* Lewis and gerrardine, gerrardamine, and gerrardoline from *C. gerrardii* Alston. The same authors (1967b) also established the structure IV for gerrardine, the stereochemistry of which was substantiated by an X-ray analysis performed by Gaffner. Similarly, cassipourine was studied in detail and its structure established to be as depicted in V (Cooks et al., 1967), again with recourse to X-ray analysis (Gaffner) for the stereochemical assignment.

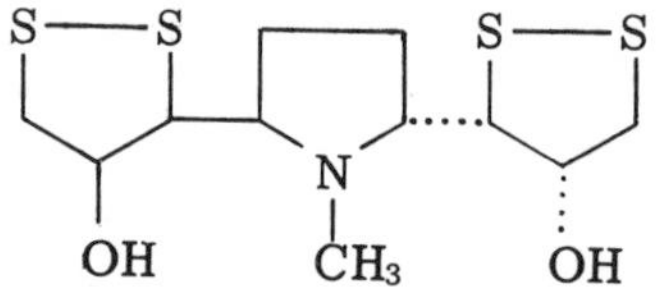

IV

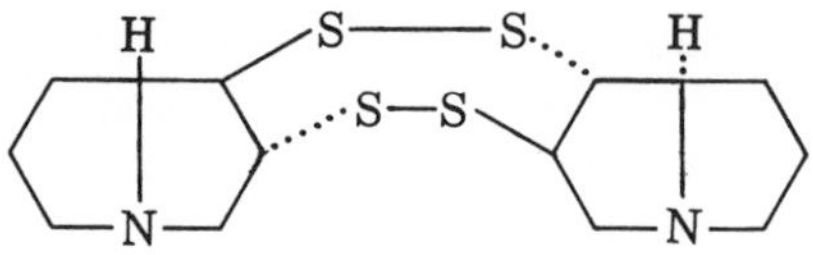

V

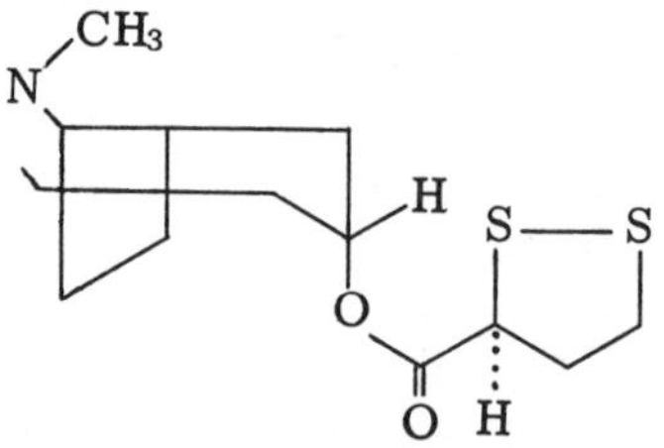

VI

A structurally related alkaloid, brugine, occurs in *Bruguiera sexangula* (Lour.) Poir., another member of the mangrove family. Its identity as VI, an ester of tropine and 1,2-dithiolane-3-carboxylic acid, was recently established by Loder and Russell (1966). Much interest is, of course, associated with an understanding of the pathways by which sulfur is incorporated into these novel types of natural compounds.

Less surprising with regard to structure but of considerable interest is the recent discovery of a sulfur-containing pyrrolizidine ester alkaloid in species of the genus *Planchonella* (fam. Sapotaceae). Hart and Lamberton (1966) isolated a new base, planchonelline, as the major alkaloid from leaves of *P. thyrsoidea* C. T. White and *P. anteridifera* (White & Francis) H. J. Lam, and elucidated its structure (VII) as the *trans*-β-methylthioacrylate ester of laburnine [(+)-trachelanthamidine]. The finding of methylthioacrylic acid as part of VII recalls the previous demonstration of *cis*- and *trans*-β-methylthioacrylic acids

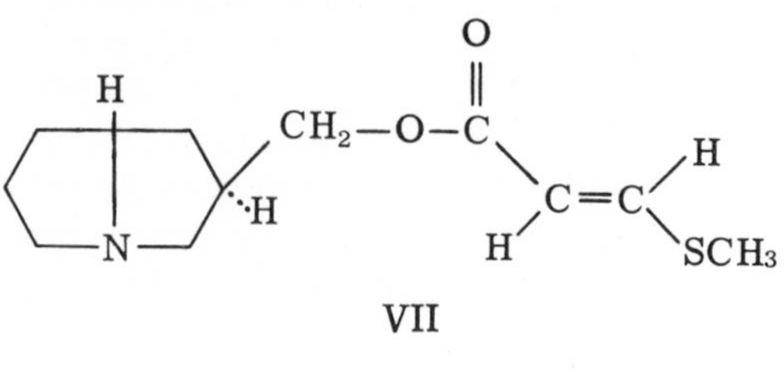

VII

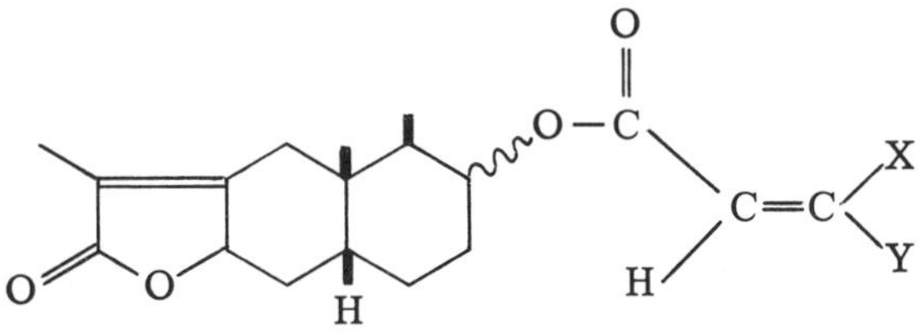

O
||
O−C
X
C=C
Y
H
O O
H

VIII. X = CH_3S, Y = H: S-petasitolide A
X = H, Y = CH_3S: S-petasitolide B

as the esterifying acids in the *S*-petasitolides A and B (VIII), isolated from *Petasites officinalis* Moench (Compositae)(Novotný et al., 1964). Whereas in the latter case stereomutation of the *cis* to the *trans* acid during hydrolysis was considered likely, evidence was presented in favor of planchonelline being the trans ester, although stereomutation of the *cis* isomer during isolation remains conceivable. Both in *Petasites* and in species of *Planchonella*, esters of the methylthioacrylic acids occur along with the corresponding esters of angelic and tiglic acids, suggesting a related biosynthetic pathway of the two pairs of acids, and hence also a wider distribution of the sulfur-containing acids in the plant kingdom. Within the series of acetylenic methyl thioenol ethers in Compositae, Bohlmann et al. (1965e) have demonstrated the feasibility of in vivo addition of methanethiol to acetylenic precursors. Here an alternative but still unknown biosynthesis possibly by catabolism of methionine in parallel with isoleucine (cf. Leete and Murrill, 1967), seems to be implied. Its clarification remains a matter of considerable interest.

The rapidly increasing number of reported sulfur-containing alkaloids will undoubtedly stimulate routine assays for S-containing derivatives in future alkaloid studies. It is therefore to be expected that our knowledge of the structural types and the botanical distribution of S-containing alkaloids will rapidly increase to the point where profitable conclusions can be drawn with regard both to the biosynthetic pathways and to the possible significance of these compounds for the purpose of classification.

III. SULFUR COMPOUNDS, DERIVED FROM POLYACETYLENES, IN COMPOSITAE

The sulfur-containing derivatives of polyacetylenes form a well-defined group compact in origin and behavior, which includes a remarkably large number of individual compounds. Their natural occurrence is highly concentrated. (One satellite of the acetylenes in fungi is a thiophene, junipal, which we omit from consideration.) Acetylenic compounds have been found in a dozen or more families of higher plants, but are most common and varied in the largest family, Compositae, and it is exclusively from Compositae that the substances which we are going to dis cuss, formed by addition of bivalent sulfur to a preexisting acety- lenic skeleton, so far have been isolated. The tribe Anthemideae of Compositae is a conspicuous source of sulfur compounds, and members of the tribes Cynareae, Inuleae, Heliantheae, and Helenieae have also furnished this kind of constituent. Knowledge of these natural products is wholly recent and is rapidly developing. The isolation of α-terthienyl (82)[1] from marigold (*Tagetes*) petals by Zechmeister and Sease (1947) now seems like the first snow to fall before an avalanche. Historicall Challenger and Holmes (1953) showed good judgment in suggestin as the aftermath of studies on the formation of thiophenes from acetylene and boiling sulfur, that

> It may be more than a coincidence that the only instance so far recorded of the occurrence of a true thiophen derivative in plants should be found in a family so many members of which contain polyacetylenes. The α-terthienyl may arise by interaction of hydrogen sulphide with a straight-chain compound containing an acetylenic-olefinic system.... Other reactions such as oxidation, decarboxylation, or dehydrogenation might be involved. It could be argued that a long-chain paraffin or fatty acid might serve equally well as the starting point. Ring closure would, however, undoubtedly be facilitated by the presence of olefinic and acetylenic linkages.

When Sørensen and Sørensen (1958) discovered the second thiophene from higher plants, (55), in *Coreopsis*, the compound turned out to occur with methylphenyltriacetylene (IX), from which (55) could formally be derived by addition of hydrogen sul-

[1]Throughout this section, Arabic numerals in parentheses refer to structural formulas in or from Table 1, Roman numerals to other structures in the text.

fide. [The in vitro addition of hydrosulfide to IX happens, however, to give not (55) but its isomer X (Schulte et al., 1962), which has not been found in nature.] In 1960, four specialized

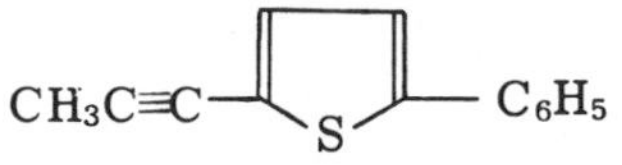

$CH_3C{\equiv}C$–(thiophene-2,5-diyl, S)–C_6H_5

(55)

$CH_3(C{\equiv}C)_3C_6H_5 \xrightarrow{HS^-}$

IX

CH_3–(thiophene-2,5-diyl, S)–$C{\equiv}CC_6H_5$

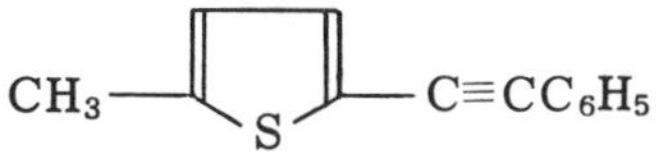

X

sulfur compounds, all thiophenes, had been reported from Compositae; the number had risen to approximately 27 by the end of 1963 and as of December 1966 stood, by our reckoning, omitting 24a-25 at 92. The work of Bohlmann and his collaborators in particular has been a steady source of new results. Obviously, Table 1 could not long be complete.

Our purpose in this survey is to sharpen our concept of the sulfur compounds in Compositae by the artifice of setting them apart from associated polyacetylenes, but also to explain what is known about the biogenetic relationship. We first invite reflective perusal of Table 1. The references for each compound or set of closely similar compounds are chosen to authenticate their structure and presence in nature, but do not attempt to give a picture of botanical distribution; they include only the synthetic work that we found most striking, with omission of the majority of it in the thiophene series.

The compounds in Table 1 are easy to categorize according to the functionality of sulfur. One conspicuous kind is made up of β-monosubstituted or α,β-disubstituted vinyl methyl sulfides, formal adducts of methanethiol and acetylenes. The butatrienes (5-6), like their prototropic isomer (4), could result from the thiol and a diacetylene. The assignment of *cis-trans* configurations to (5-6) is notable. Four of the methyl sulfides appear also in oxidized form as the sulfoxides (24-24a) and (28-29), the first pair of which is known to possess optical activity caused by the sulfinyl group. The chirality remains to be established. The sulfone (25), however, differs from its less oxygenated counterparts (22-24) by reversal in direction of the apparent addition of the sulfur to a triple bond. Compare, for another example of alterna-

TABLE 1

Sulfur Compounds from Compositae, Related to Polyacetylene

I. Methyl Thioethers (and Their Sulfoxides or Sulfones)

a. Derivatives of dehydromatricaria ester ($CH_3(C{\equiv}C)_3CH{=}CHCOOCH_3$) and relatives

(1-2)	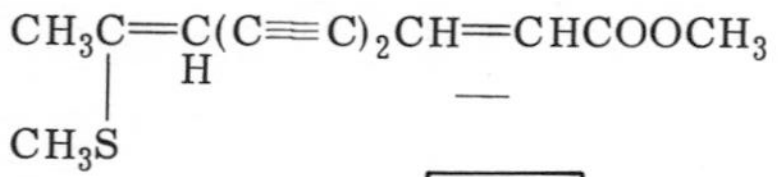 $CH_3C(SCH_3){=}CH(C{\equiv}C)_2CH{=}CHCOOCH_3$	Two stereoisomers[a]	Bohlmann et al. (1963a, 1965e and f)
(3)	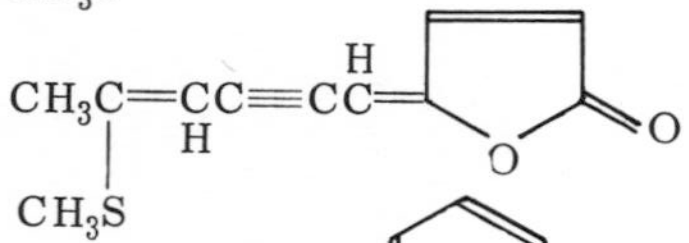 $CH_3C(SCH_3){=}CHC{\equiv}CCH{=}$(furanone ring, O, =O)		Bohlmann et al. (1965f)
(4)	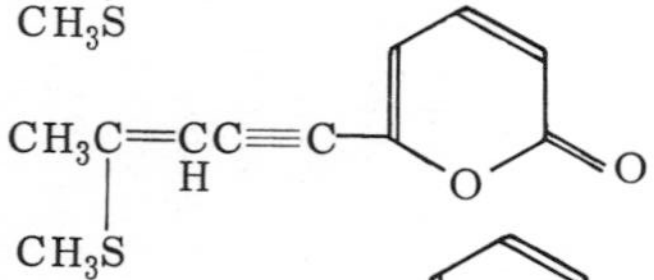 $CH_3C(SCH_3){=}CHC{\equiv}C$–(pyranone ring, O, =O)		Bohlmann et al. (1966c)
(5-6)	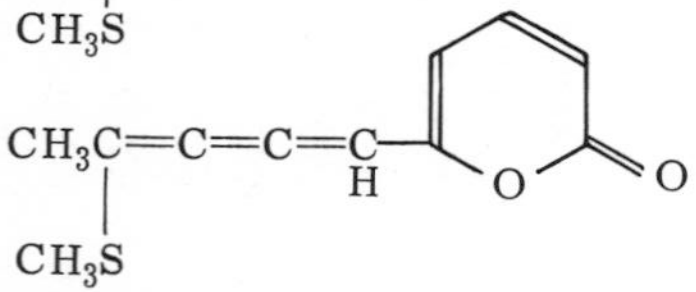 $CH_3C(SCH_3){=}C{=}C{=}CH$–(pyranone ring, O, =O)	And stereoisomer	Bohlmann et al. (1966c)
(7-10)	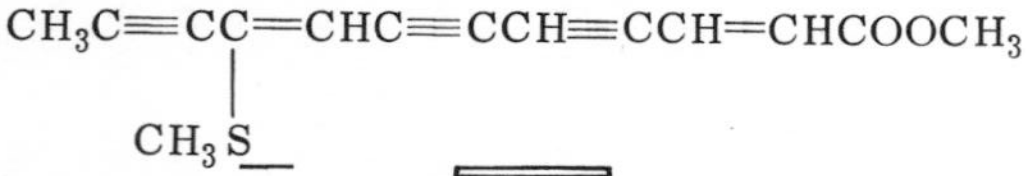 $CH_3C{\equiv}CC(SCH_3){=}CHC{\equiv}CCH{\equiv}CCH{=}CHCOOCH_3$	Four stereoisomers[a]	Bohlmann et al. (1963a, 1965f)
(11-12)	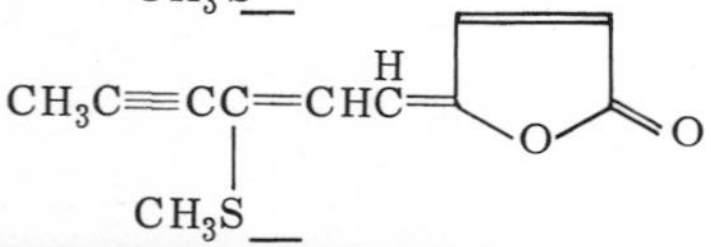 $CH_3C{\equiv}CC(SCH_3){=}CHCH{=}$(furanone ring, O, =O)	Two stereoisomers[a]	Bohlmann et al. (1965f)

(13)	CH_3S–(p-phenylene)–$C(CH_3)=CHCOOCH_3$		Bohlmann et al. (1965c)
(14–15)	$CH_3(C{\equiv}C)_2C(SCH_3){=}CHCH{=}CHCOOCH_3$	Two stereoisomers[a]	Bohlmann et al. (1963a); Bohlmann and Kleine (1966); Bohlmann and Seyberlich (1966)
(16–17)	$CH_3CH{=}CHC{\equiv}CC(SCH_3){=}CHCH{=}CHCOOCH_3$	Two stereoisomers[a]	Bohlmann et al. (1966c)
(18)	$CH_3C{\equiv}CCH{=}C(SCH_3)CH{=}CHCOOCH_3$		Bohlmann et al. (1963a)
(19–21)	$CH_3(C{\equiv}C)_2CH{=}C(SCH_3)CH{=}CHCOOCH_3$	Three stereoisomers,[a] lacking all-*trans*	Bohlmann and Kleine (1966)

b. Derivatives of tridecenepentayne ($CH_3(C{\equiv}C)_5CH{=}CH_2$)

(22–23)	$CH_3(C{\equiv}C)_2C(SCH_3){=}CH(C{\equiv}C)_2CH{=}CH_2$	Two stereoisomers[a]	Bohlmann and Kleine (1963); Bohlmann and Seyberlich (1965)

TABLE 1 (continued)

(24–24a)	$CH_3(C\equiv C)_2C(SOCH_3)=CH(C\equiv C)_2CH=CH_2$	(−); and *trans* (+) isomer	Bohlmann and Kleine (1963); Bohlmann et al. (1967)
(25)	$CH_3(C\equiv C)_2CH=C(SO_2CH_3)(C\equiv C)_2CH=CH_2$	*trans*	Bohlmann (1966b); Bohlmann et al. (1967)
	c. Derivatives of a demethylated, oxygenated C_{13} chain		
(26–27)	$CH_3SCH=CHC\equiv CCH=$ (spiroketal: dihydrofuran ring spiro-linked to tetrahydrofuran)	Two optically active *cis-trans*[a] isomers	Bohlmann et al. (1964a, 1965g)
(28–29)	$CH_3SOCH=CHC\equiv CCH=$ (spiroketal: dihydrofuran ring spiro-linked to tetrahydrofuran)	Two *cis-trans*[a] isomers	Bohlmann et al. (1964a)
(30)	$CH_3SCH=CHC\equiv CCH_2$–(furan-2,5-diyl)–$CH=CHCHO$		Bohlmann et al. (1964a)
	d. Derivatives of capillene ($CH_3(C\equiv C)_2CH_2C_6H_5$)		
(31)	$CH_3SCH=CHC\equiv CCH_2C_6H_5$		Bohlmann et al. (1963c)

(32–33)	$CH_3SCH{=}CHC{\equiv}CCOC_6H_5$	Two stereoisomers[a]	Bohlmann et al. (1963c)

II. Thiophenes (and Dithiadienes or Enedithiones)

a. Simple thiophenes

(34)	S, $-COCH_3$		Bohlmann et al. (1964b)

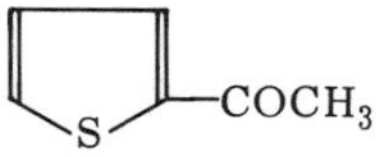

b. Unconjugated thiophenes

(35)	$(CH_3)_2CHCH_2NHCO(\overset{H}{C}{=}\underset{H}{C})_2CH_2-$ S		Winterfeldt (1963)

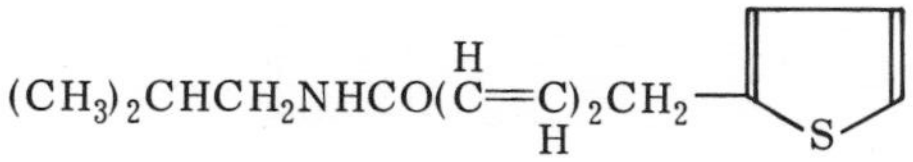

c. Derivatives of dehydromatricaria ester and relatives

(36–37)	CH_3- S $-C{\equiv}CCH{=}CHCOOCH_3$	Two stereoisomers[a]	Bohlmann et al. (1962b)
(38)	CH_3- S $-\overset{H}{C}{=}$ O, $=O$		Bohlmann and Zdero (1966)

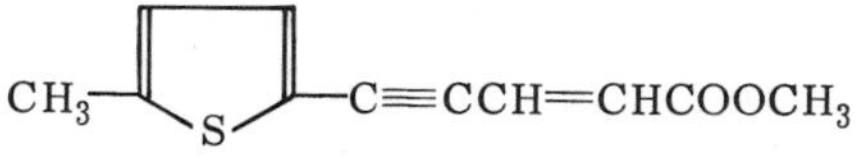

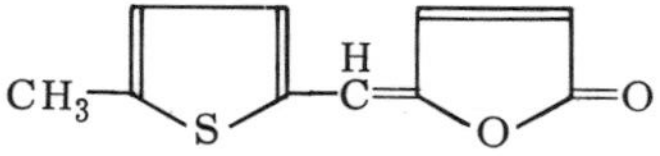

TABLE 1 (continued)

(39)	$S{-}C{\equiv}CCH{=}CHCOOCH_3$		Bohlmann et al. (1965d)
(40)	$CH_3C{\equiv}C{-}S{-}CH{=}CHCOOCH_3$		Guddal and Sørensen (1959)
(41)	$CH_3CH{=}CH{-}S{-}CH{=}CHCOOCH_3$		Bohlmann et al. (1966c)
(42-43)	$CH_3C{\equiv}C{-}S{-}COCH_3$, OH	And methyl ether	Bohlmann et al. (1962c)

d. Derivatives of C_{13} chains and analogues

(44-45)	$S{-}C{\equiv}C(CH{=}CH)_2CH{=}CH_2$	And stereoisomer	Sørensen (1961); Sørensen et al. (1954); Bohlmann and Kleine (1964); Bohlmann et al. (1964e)

(46–47)	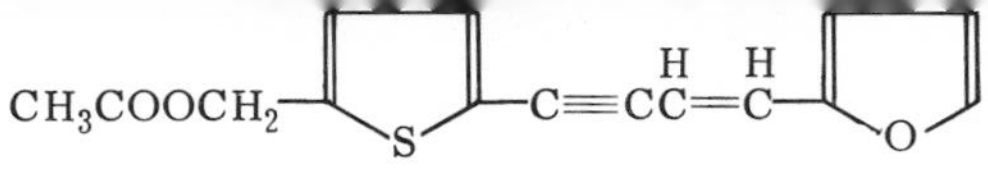 CH_3COOCH_2–(S ring)–$C{\equiv}CC(H){=}C(H)$–(O ring)	And alcohol, of unstated stereochemistry	Bohlmann and Arndt (1966); Atkinson and Curtis (1964)
(48–49)	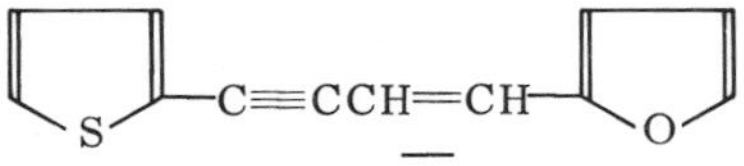 (S ring)–$C{\equiv}CCH{=}CH$–(O ring)	Two stereoisomers[a]	Bohlmann et al. (1962b, 1964a).
(50)	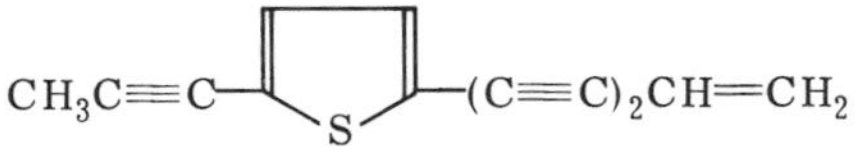 $CH_3C{\equiv}C$–(S ring)–$(C{\equiv}C)_2CH{=}CH_2$		Bohlmann et al. (1964e)
(51)	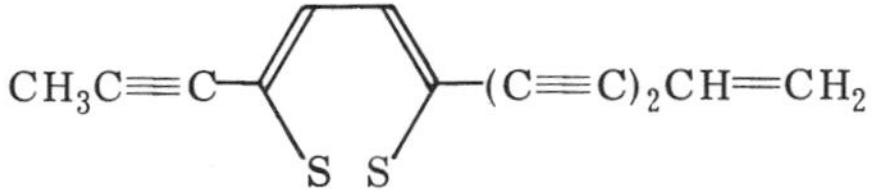 $CH_3C{\equiv}C$–(S S ring)–$(C{\equiv}C)_2CH{=}CH_2$		Bohlmann and Kleine (1965)
(52)	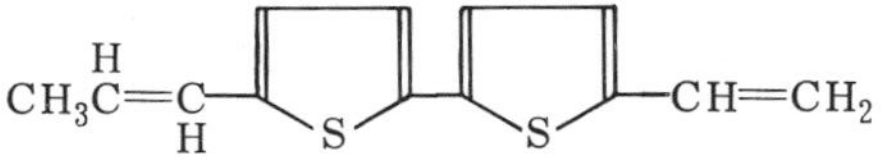 $CH_3C(H){=}C(H)$–(S ring)–(S ring)–$CH{=}CH_2$		Bohlmann et al. (1965b)
(53)	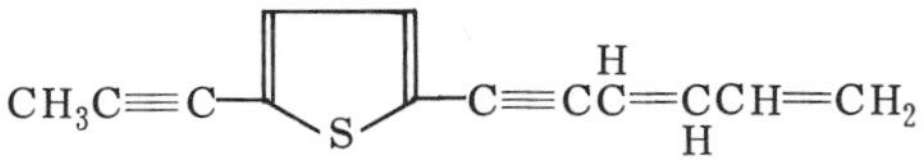 $CH_3C{\equiv}C$–(S ring)–$C{\equiv}CC(H){=}C(H)CH{=}CH_2$		Bohlmann et al. (1964e); Atkinson and Curtis (1965)
(54)	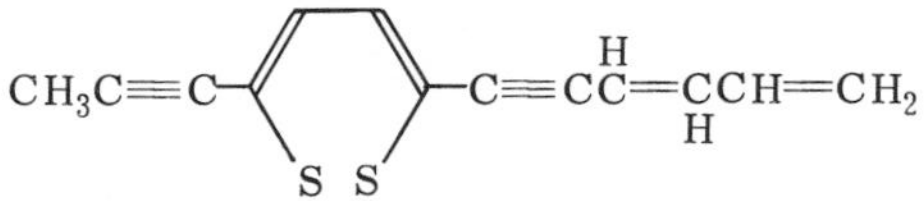 $CH_3C{\equiv}C$–(S S ring)–$C{\equiv}CC(H){=}C(H)CH{=}CH_2$		Bohlmann and Kleine (1965)

TABLE 1 (continued)

(55)	$CH_3C{\equiv}C$–[S ring]–C_6H_5		Sørensen and Sørensen (1958)
(56)	$CH_3CH{=}CHC{\equiv}C$–[S ring]–$C{\equiv}CCH{=}CH_2$		Bohlmann et al. (1964e); Atkinson and Curtis (1965)
(57-58)	$HOCH_2CH{=}CHC{\equiv}C$–[S ring]–$C{\equiv}CCH{=}CH_2$	And acetate	Bohlmann et al. (1964e)
(59)	$OCHCH{=}CHC{\equiv}C$–[S ring]–$C{\equiv}CCH{=}CH_2$		Bohlmann et al. (1964e)
(60)	$CH_3(C{\equiv}C)_2$–[S ring]–$C{\equiv}CCH{=}CH_2$		Sørensen et al. (1964); Bohlmann et al. (1965a)
(61)	$CH_3(C{\equiv}C)_2$–[S–S ring]–$C{\equiv}CCH{=}CH_2$		Mortensen et al. (1964); Bohlmann and Kleine (1965)

(62-63)	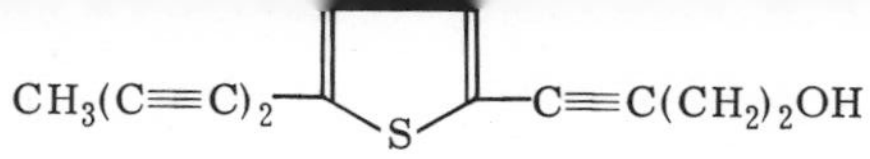 $CH_3(C\equiv C)_2$—(S)—$C\equiv C(CH_2)_2OH$	And acetate	Bohlmann et al. (1965a)
(64-65)	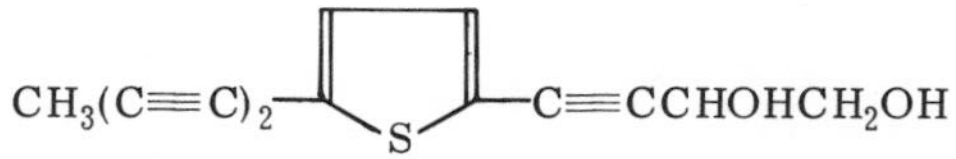 $CH_3(C\equiv C)_2$—(S)—$C\equiv CCHOHCH_2OH$	And diacetate	Bohlmann et al. (1965a)
(66-67)	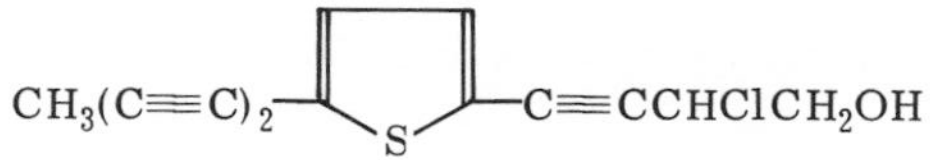 $CH_3(C\equiv C)_2$—(S)—$C\equiv CCHClCH_2OH$	And optically active acetate	Bohlmann et al. (1965a)
(68)	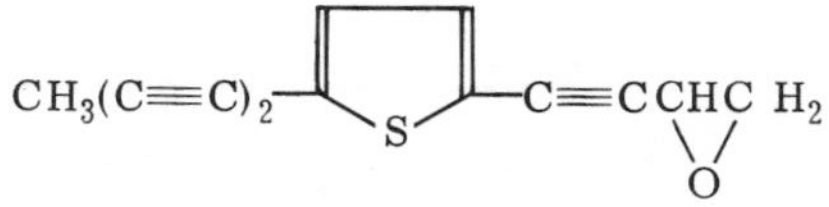 $CH_3(C\equiv C)_2$—(S)—$C\equiv CCHCH_2$ (epoxide, O)		Bohlmann et al. (1965a)
(69)	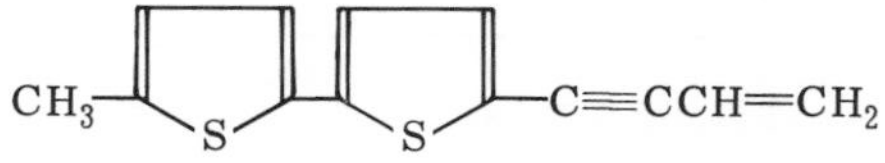 CH_3—(S)—(S)—$C\equiv CCH=CH_2$		Bohlmann and Berger (1965)
(70)	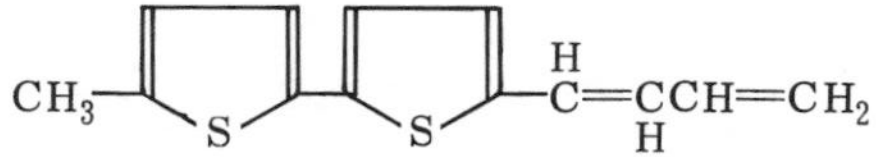 CH_3—(S)—(S)—$CH=CCH=CH_2$		Liaaen Jensen and Sörensen (1961)
(71-72)	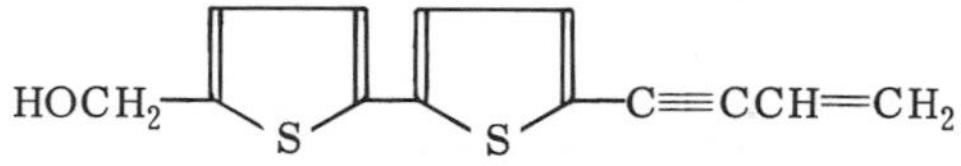 $HOCH_2$—(S)—(S)—$C\equiv CCH=CH_2$	And acetate	Bohlmann and Kleine (1963); Bohlmann et al. (1965a)
(73)	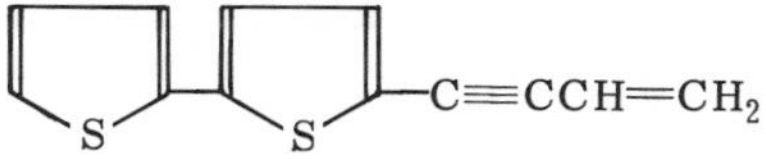 (S)—(S)—$C\equiv CCH=CH_2$		Uhlenbroek and Bijloo (1959); Bohlmann and Herbst (1962)

TABLE 1 (continued)

No.	Structure	Remarks	References
(74–75)	(C_4H_3S)–(C_4H_2S)–$C{\equiv}C(CH_2)_2OH$	And acetate	Bohlmann et al. (1964e, 1965a); Atkinson et al. (1964, 1965)
(76–79)	(C_4H_3S)–(C_4H_2S)–$C{\equiv}CCHOHCH_2OH$	(+), and primary and secondary monoacetates and (+) diacetate	Bohlmann et al. (1965a)
(80)	(C_4H_3S)–(C_4H_2S)–$C{\equiv}CCHOHCH_2Cl$		Atkinson et al. (1964, 1965, 1966)
(81)	$HOCH_2$–(C_4H_2S)–(C_4H_2S)–(C_4H_3S)		Krishnaswamy et al. (1966)
(82)	(C_4H_3S)–(C_4H_2S)–(C_4H_3S)		Zechmeister and Sease (1947)

e. Derivatives of artemisia ketone ($CH_3(C{\equiv}C)_3\overset{H}{C}{=}\underset{H}{C}(CH_2)_2COC_2H_5$) and other C_{14} chains and relatives

No.	Structure	Remarks	References
(83–84)	(C_4H_3S)–$C{\equiv}C\overset{H}{C}{=}\underset{H}{C}CH{=}CHCOC_2H_5$	Two stereoisomers[a]	Sørensen (1961); Sørensen et al. (1954); Bohlmann et al. (1966b)
(85)	(C_4H_3S)–$C{\equiv}C(\overset{H}{C}{=}\underset{H}{C})_2CHOHC_2H_5$		Bohlmann et al. (1966b)

No.	Structure	Remarks	Reference
(86)	2-thienyl $-C{\equiv}CCH{=}CH(CH_2)_2COC_2H_5$		Sørensen (1961); Bohlmann et al. (1964d)
(87)	2-thienyl $-C{\equiv}CCH{=}CH(CH_2)_2CHOHC_2H_5$		Bohlmann et al. (1966b)
(88)	2-thienyl $-C{\equiv}CCH{=}CHCH_2CH(OCOCH_2CH(CH_3)_2)COC_2H_5$	(−)	Bohlmann et al. (1966b)
(89)	2-thienyl $-C{\equiv}CCH{=}CHCH(OCOCH_2CH(CH_3)_2)CH_2COC_2H_5$	(+)	Bohlmann et al. (1966b)
(90)	2-thienyl $-C{\equiv}CCH{=}CHCHO$		Bohlmann et al. (1966b)
(91-92)	2-thienyl $-C{\equiv}C(CH{=}CH)_2(CH_2)_3OH$	And acetate	Bohlmann et al. (1964e)
(93)	2-thienyl $-CH{=}$ (spiroketal enol ether)	(+)	Bohlmann et al. (1962b, 1964d)

[a]The double bond or bonds that are sites of stereoisomerism are underlined.

tive orientations at the same distance from the *C*-methyl group, the esters (19-21) with (14-15). The most extensive class in Table 1 consists of the 2,5-disubstituted thiophenes, including those with two or three thiophene rings, which correspond to sums of hydrogen sulfide and 1,3-diyne systems. The most peculiar and least known compounds are the three (51, 54, 61) that may be formulated as disubstituted 1,2-dithiadienes (XI). These substances, which have also been written as *cis*-enedithiones (XII), are especially characterized by their red color and by the ease with which they lose sulfur, during thin-layer chromatography, for example, and pass into the corresponding thiophenes (XIII; 50, 53, 60)(see also Bohlmann and Bresinsky, 1967; Schroth et al., 1967.)

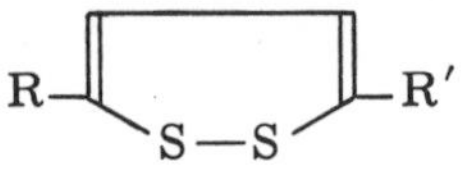

XI 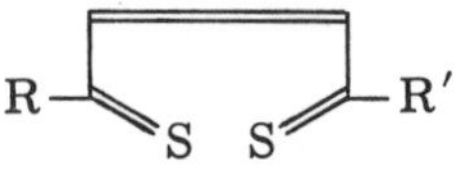 XII 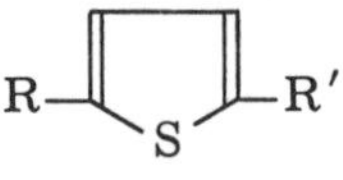 XIII

The natural polyacetylenes, including their sulfur-containing derivatives, were reviewed exhaustively four years ago by Bohlmann et al. (1962a; see also Bohlmann, 1966a). Sørensen (1963, Chap. 6, this volume) and Bu'Lock (1964, 1966) have given further accounts, and Bu'Lock in particular has discussed paths of biogenesis (cf. Bu'Lock and Smith, 1967). Since approximately three hundred structurally known polyacetylenes without sulfur have been obtained from flowering plants and fungi (Bohlmann, 1966b), we can give only a scanty outline of pertinent generalities here. The natural compounds cited in the subsection headings (Ia, b, and d and IIe) of Table 1 indicate the existence of sulfur-free prototypes with the same chain length and basic structural pattern as many of the compounds in the body of the table. Furthermore the specialized functional groups not involving sulfur that are seen in Table 1—the oxygen-containing rings, including the spiroketal systems in (26-29) and (93), the amide group of (35), the chlorohydrin groups of (66-67) and (80)—can all be paralleled in sulfur-free polyacetylenes. We shall therefore center our attention on the introduction of the sulfur substituents, but first adumbrate current opinion of the biogenetic features of polyacetylenes as a whole.

Typical natural polyacetylenes are formed by head-to-tail condensation of a single sequence of acetate and malonate units. One end of the system is marked by a *C*-methyl group coming from acetate; the chain is straight or cyclized near the other end to a phenyl group. The pathways closely resemble the biosynthesis of fatty acids, and many polyacetylenes may arise from oleic or linoleic acid. Crepenynic acid, *cis*-octadec-9-en-12-ynoic acid, which was isolated from the seed fat of *Crepis foetida* L. (Compositae)(Mikolajczak et al., 1964) and differs from linoleic acid only by the presence of the 12,13 triple bond instead of a double bond, is thought to be a prime example of the earliest stage of acetylene biogenesis. Proceeding onward from crepenynic acid or a similar compound, the usual transformations are chain shortening with loss of a segment remote from the *C*-methyl group, dehydrogenation, and oxygenation. Thus, a C_{16} or C_{18} fatty acid can yield a polyacetylene with the common chain length of 10 to 14 carbon atoms. When methyl-tritiated 10,12,14-hexadecatriynoic acid (XIV) was administered to a plant of *Artemisia vulgaris* L. (Compositae), *C*-methyl-tritiated *cis*-dehydromatricaria ester (XV) was obtained (Bohlmann et al., 1966a).

$$\underset{\text{XIV}}{TCH_2(C\equiv C)_3(CH_2)_8COOH} \longrightarrow \underset{\text{XV}}{TCH_2(C\equiv C)_3\overset{H}{C}=\overset{H}{C}COOCH_3}$$

$$\underset{\text{XVI}}{CH_3(C\equiv C)_3\overset{H}{C}=\underset{H}{C}(CH_2)_2COC_2H_5}$$

The artemisia ketone (XVI) isolated simultaneously was inactive. One type of oxygenation that is particularly important is oxidation of the terminal (acetate) methyl group when it is allylic or propargylic. Thus, 1,2-tritiated trideca-1,11-dien-3,5,7,9-tetrayne (XVII), in unlabeled form a natural constituent of *Bidens connatus* L. (Compositae), was efficiently hydroxylated by the plant to XVIII (Bohlmann et al., 1966d). The terminal methyl group can be lost, presumably by oxidation all the way to the acid stage and decarboxylation.[2] The oxidized methyl group

[2]In Table 1, all compounds are written with the site of the present or former *C*-methyl group to the left.

$$CH_3CH{=}CH(C{\equiv}C)_4CT{=}CHT \longrightarrow$$

XVII

$$HOCH_2CH{=}CH(C{\equiv}C)_4CH{=}CH_2]T$$

XVIII

appears in compounds (35) (maybe),—(46-47), (57-59), (71-72), and (81), and has departed from (26-33),(34) probably (39), (44-45), (48-49), (73-80), and (82-93). The methyl group that marks the acetate head in XVI is presumably the one attached to the triyne system, and this methyl is gone from compounds (83-89), though a methyl group formed after the initial condensation is present at the other end of the molecules.

The in vitro addition of a thiol to a carbon-carbon triple bond is exemplified by the base-catalyzed syntheses (Bohlmann et al., 1963b) of (31) and (32-33) from methanethiol and benzyldiacetylene or 1-phenyl-2,4-pentadiyn-1-one (XIX). Both sets of diacetylenes and corresponding sulfides occur together in *Chrysanthemum segetum* L., which indeed converted administered

$$HC{\equiv}CC{\equiv}CCOC_6H_5 \xrightarrow[\text{or in vivo}]{\text{in vitro}} CH_3SCH{=}CHC{\equiv}CCOC_6H_5 \xleftarrow{\text{in vivo}}$$

XIX (32-33)

$$CH_3C{\equiv}CC{\equiv}CCOC_6H_5$$

XX

XIX (carbonyl-^{14}C) to (32-33)(Bohlmann et al., 1964c). The plant transformed another constituent, the ketone XX, to (32-33) with similar efficiency, despite the need for removal of the terminal methyl group (Bohlmann et al., 1965e). *Flaveria repanda* Lag. converts tridec-1-en-3,5,7,9,11-pentayne (XXI; 1,2-tritiated), which notwithstanding its extreme unsaturation is widespread

$$CH_3(C{\equiv}C)_5CT{=}CHT \longrightarrow$$

XXI

$$CH_3(C{\equiv}C)_2\underset{\displaystyle SCH_3}{\underset{|}{C}}{=}CH(C{\equiv}C)_2CH{=}CH_2]T$$

(22-23) (*t*)

in Compositae, to the adducts (22-23)(Bohlmann and Hinz, 1965). *Anthemis tinctoria* L. produces (1-2)(*cis* isomer), with evidently easy stereomutation, from carbonyl-^{14}C *trans*-dehydromatricaria ester (XXII)(Bohlmann et al., 1965e). The same plant accom-

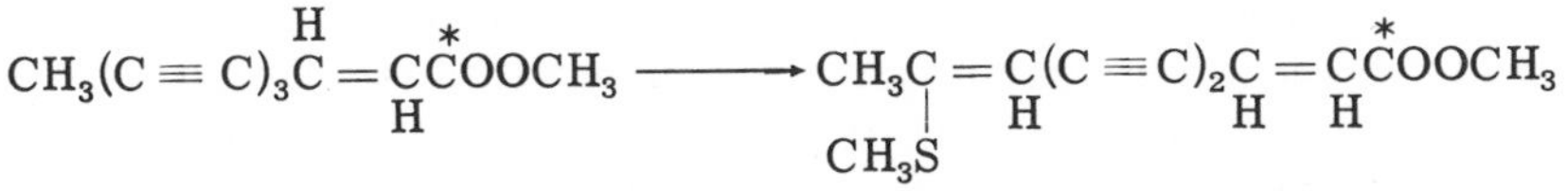

$$CH_3(C\equiv C)_3\overset{H}{C}=\underset{H}{C}\overset{*}{C}OOCH_3 \longrightarrow CH_3\underset{\underset{CH_3S}{|}}{C}=\underset{H}{C}(C\equiv C)_2\underset{H}{C}=\underset{H}{C}\overset{*}{C}OOCH_3$$

XXII (1-2)(*cis*, carbonyl-^{14}C)

plishes the remarkable transformation of XXII to (13)(carbonyl-^{14}C), in which the methylthio and ester groups are separated by the same number of carbon atoms as in the counterparts lacking two hydrogen atoms, the simple adducts (7-10)(Bohlmann et al., 1965c). An experiment with *C*-methyl-labeled *trans*-dehydromatricaria ester (Bohlmann and Laser, 1966) has shown that the

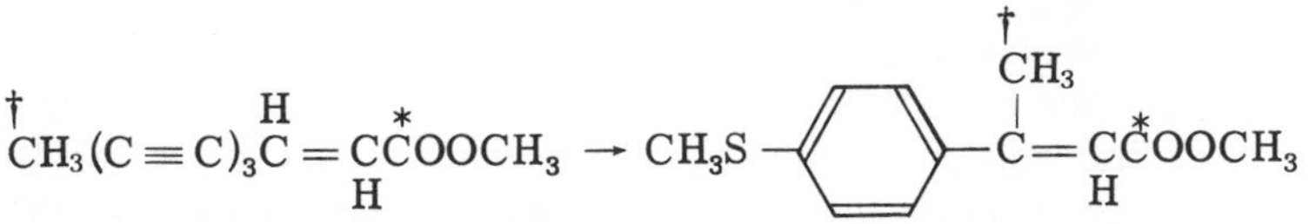

$$\overset{\dagger}{C}H_3(C\equiv C)_3\overset{H}{C}=\underset{H}{C}\overset{*}{C}OOCH_3 \rightarrow CH_3S-C_6H_4-\overset{\overset{\dagger}{C}H_3}{\overset{|}{C}}=\underset{H}{C}\overset{*}{C}OOCH_3$$

XXII (labeled alternatively at C-10) (13)(carbonyl- or *C*-methyl-^{14}C)

terminal methyl group of the straight-chain poly-yne indeed becomes the unique branching methyl of (13), with the equivalent, upon formation of the ring, of a 1,3 migration.

The in vitro addition of hydrogen sulfide to diynes in weakly basic medium, forming thiophenes, was studied by Schulte et al.

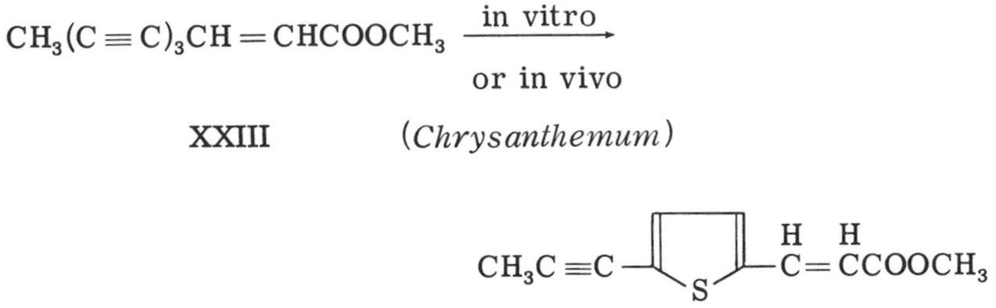

$$CH_3(C\equiv C)_3CH=CHCOOCH_3 \xrightarrow[\text{or in vivo}]{\text{in vitro}}$$

XXIII (*Chrysanthemum*)

$$CH_3C\equiv C-(C_4H_2S)-\overset{H}{C}=\overset{H}{C}COOCH_3$$

(40)

(1962), who found in particular that the dehydromatricaria esters (XXIII, *cis* and *trans*) react at the triple bonds closest to the ester group to give the natural product (40) and its *trans* isomer. Glutathione can act as a sulfur source in dilute alkali with *trans*-dehydromatricaria acid, furnishing the *cis* and *trans* acids corresponding to (40)(Schulte et al., 1963). Furthermore, *trans*-dehydromatricaria ester is converted to (40) by *Chrysanthemum vulgare* L. (Schulte et al., 1966). In *Anthemis nobilis* L., however, the addition to the ester (XXII) involves the terminal and central triple bonds and gives the thiophenes (36-37), isomeric with (40)(Rybak, 1965).

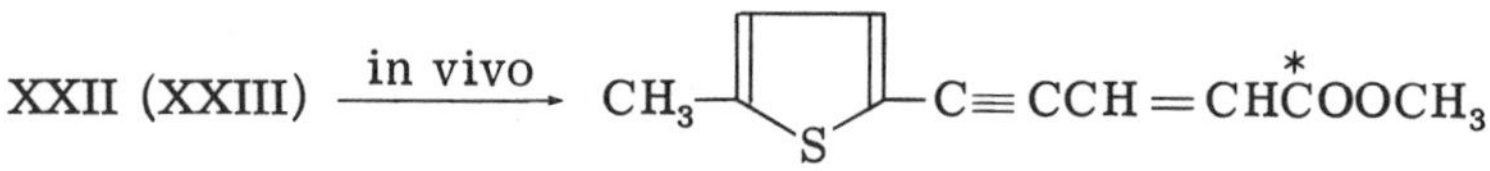

XXII (XXIII) $\xrightarrow{\text{in vivo}}$ CH_3–(thiophene, S)–$C\equiv CCH=CH\overset{*}{C}OOCH_3$

(*Anthemis*) (36-37)(carbonyl-^{14}C)

The tetrayne and pentayne hydrocarbons XVII and XXI have been established as precursors of a variety of thiophenes, including bithienyls and terthienyl. In *Bidens connatus* L., XVII furnishes the monothiophene (56) by addition to the central diyne system and the bithienyl (52) by reaction of the terminal triple bonds (Bohlmann et al., 1966d). *Buphthalmum salicifolium* L.

$CH_3CH=CH(C\equiv C)_4CT=CHT$ →

XVII

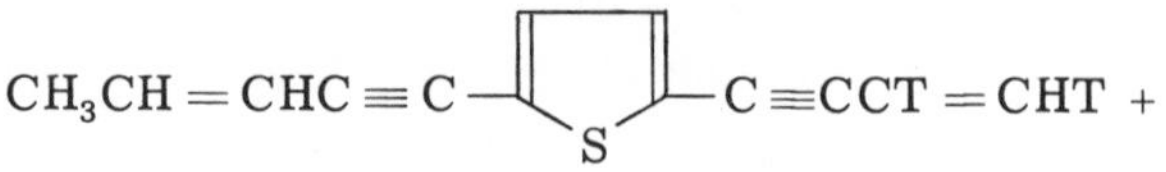

$CH_3CH=CHC\equiv C$–(thiophene, S)–$C\equiv CCT=CHT$ +

(56)(tritiated as shown)

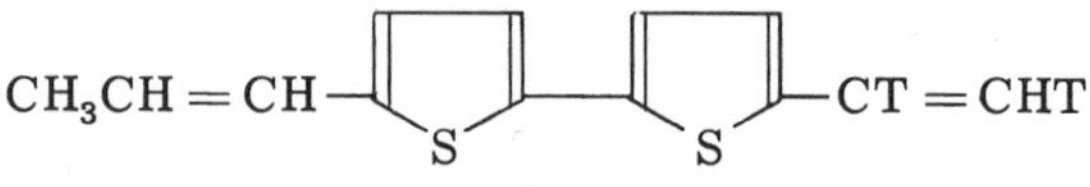

$CH_3CH=CH$–(thiophene, S)–(thiophene, S)–$CT=CHT$

(52)(tritiated as shown)

converts XXI to the double adduct (69) (no natural 1,2-dithienylacetylenes are known) and its oxidation product, the acetate (72) (Bohlmann et al., 1966d). In *Echinops sphaerocephalus* L., XXI

$CH_3(C\equiv C)_5CT=CHT \xrightarrow{\textit{Buphthalmum}}$

XXI

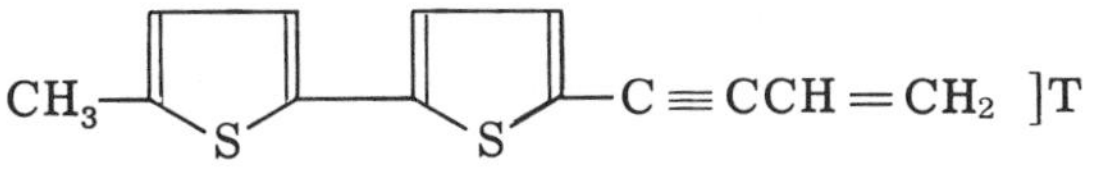

CH_3–[thiophene]–[thiophene]–$C\equiv CCH=CH_2$]T

(69)(*t*)

+ CH_3COOCH_2–[thiophene]–[thiophene]–$C\equiv CCH=CH_2$]T

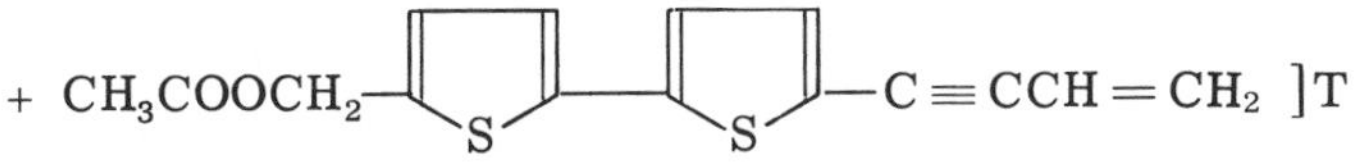

(72)(*t*)

is efficiently transformed to (73), with removal of the methyl group, and also to the glycol (76), the chlorohydrin acetate (67)

XXI $\xrightarrow{\textit{Echinops}}$ [thiophene]–[thiophene]–$C\equiv CCT=CHT$ +

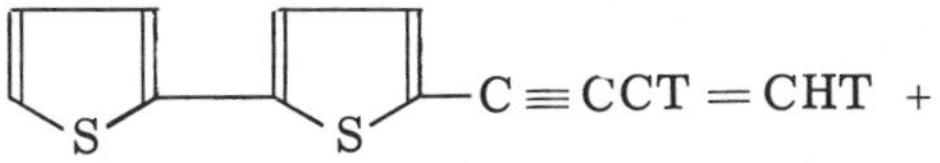

(73)(tritiated as shown)

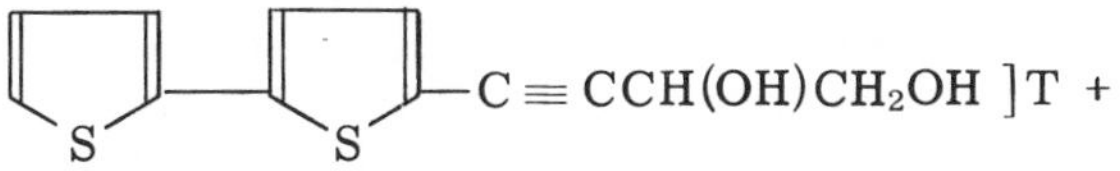

[thiophene]–[thiophene]–$C\equiv CCH(OH)CH_2OH$]T +

(76)(*t*)

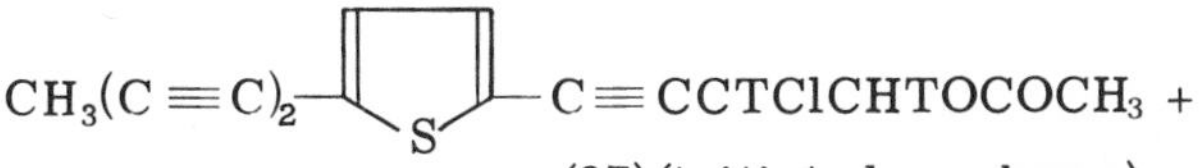

$CH_3(C\equiv C)_2$–[thiophene]–$C\equiv CCTClCHTOCOCH_3$ +

(67)(tritiated as shown)

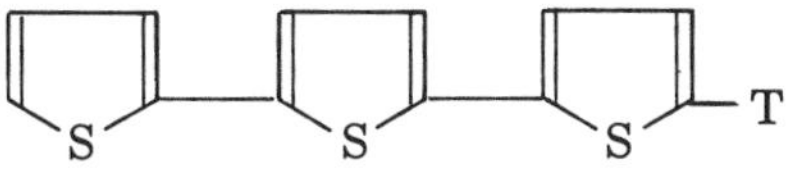

[thiophene]–[thiophene]–[thiophene]–T

(82)(tritiated as shown)

(a monothiophene retaining the terminal methyl group), and α-terthienyl (82)(Bohlmann and Hinz, 1965; Bohlmann et al., 1966e). The radioactivity in the terthienyl formed from XXI was entirely displaced by Friedel-Crafts acetylation of both free α positions, so all the 2-tritium in XXI has been lost during conversion to (82). The simplest explanation is that the double bond of XXI was dehydrogenated before attachment of sulfur to the 1 position. The bithienyl (73) is not on the route to (82) in *Tagetes* (Schulte and Foerster, 1966), but undergoes addition of acetic acid in that genus to furnish the acetate (75) (Bohlmann and Hinz, 1965).

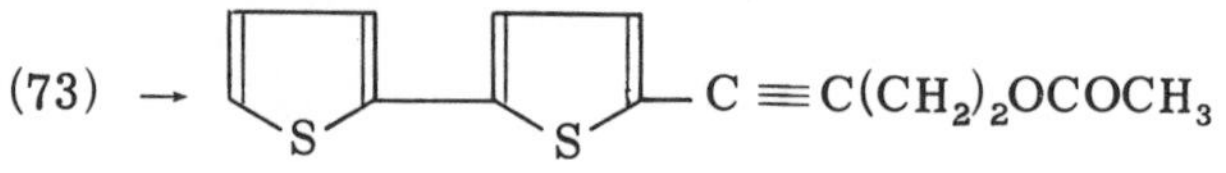

$$(73) \rightarrow \text{(bithienyl)}-C\equiv C(CH_2)_2OCOCH_3$$

(75)

The biosynthesis of the unconjugated thiophene amide (35), isolated from the roots of *Chrysanthemum frutescens* L. is puzzling to make out. When carboxyl-^{14}C acetate was administered to the plant and the distribution of radioactivity was determined in (35) and in benzyldiacetylenic constituents derivatives including capillol acetate (XXIV), the patterns were similar (Bohlmann and Jente, 1966). The N-isobutyl group of (35), presumably derived from valine, did not become active. The carbon atoms next to the rings of XXIV and (35) were each furnished by acetate carboxyl and may have been identically located in intermediates with respect to the terminal methyl group seen in XXIV. What is unclear is the original direction of the ten-carbon chain built of acetate units in (35). The methylbutadiyne system of XXIV might correspond in the amide to either the pentadienoyl or with loss of the methyl carbon, the thiophene section.

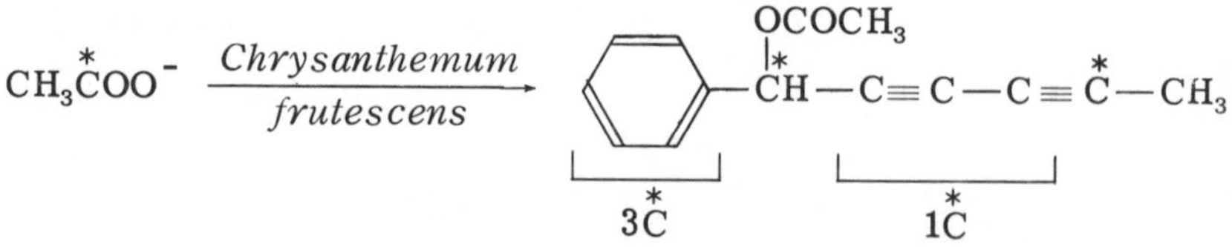

$$CH_3\overset{*}{C}OO^- \xrightarrow[\textit{frutescens}]{\textit{Chrysanthemum}} C_6H_5-\overset{*}{C}H(OCOCH_3)-C\equiv C-C\equiv \overset{*}{C}-CH_3$$

(phenyl ring: $3\overset{*}{C}$; diyne section: $1\overset{*}{C}$)

XXIV (acetoxyl activity omitted)

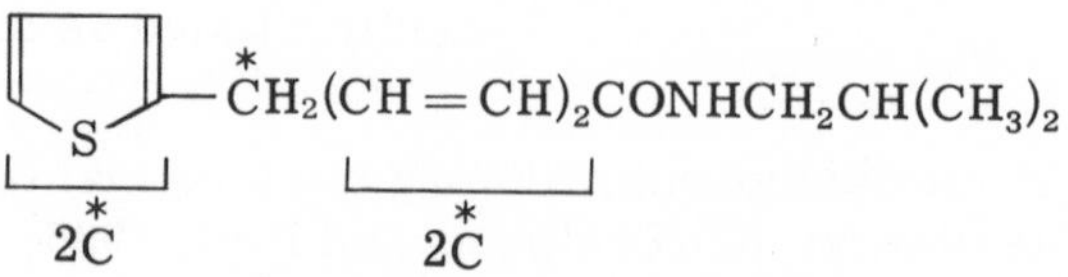

$$\text{(2-thienyl)}-\overset{*}{C}H_2(CH=CH)_2CONHCH_2CH(CH_3)_2$$

(thiophene ring: $2\overset{*}{C}$; diene section: $2\overset{*}{C}$)

(35) (labeled as indicated)

Very little is known about the stages through which sulfur passes before it appears in the finished metabolites. Probably sulfur enters the thiophenes from cysteine or an equivalent, but experiments in which cysteine-^{35}S was fed to *Chrysanthemum vulgare* (Schulte et al., 1966) have not yet led to a definite result. A preliminary study (Schuetz et al., 1965) with radioactive sulfur of the biosynthesis of α-terthienyl (82) in *Tagetes* root indicated only that hydrosulfide ion was taken up by the root but was not incorporated into (82), whereas the sulfur from sulfate or from D L-methionine was utilized for the designated product.

IV. GLUCOSINOLATES

A. Introduction and Nomenclature

The glucosinolates, as a structural class of anions, can be defined by the equivalent formulas XXVa or XXVb (R = any organic group). A characteristic general reaction of glucosinolates is their enzymatic hydrolysis to give, after separation of glucose

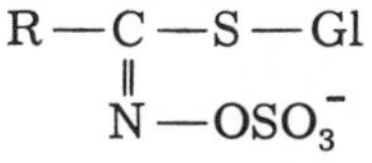

$$\begin{array}{l} R-C-S-Gl \\ \quad\;\; \| \\ \quad\;\; N-OSO_3^- \end{array} \quad (Gl-OH = \beta\text{-D-glucose})$$

XXV a

$$\begin{array}{ccccc} & HOCH_2 & & {}^{-}O_3SO-N & \\ & | & & \| & \\ & C & —O & S-C-R & \\ H & / \; | & \diagdown & | & \\ | & H & & C & \\ C & & & | & \\ | \diagdown & OH & H & / \; H & \\ HO & | & | & & \\ & C & —C & & \\ & | & | & & \\ & H & OH & & \end{array}$$

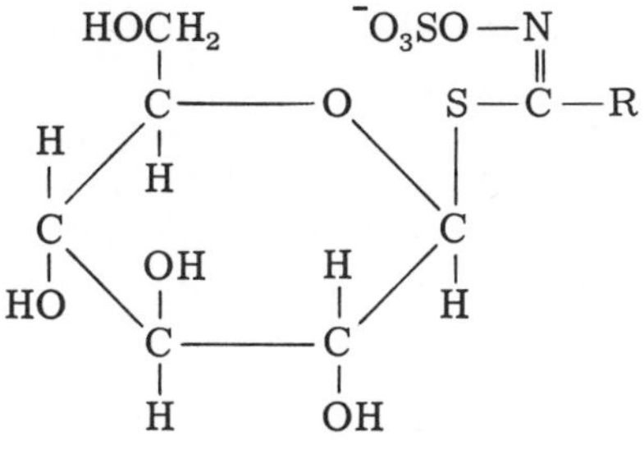

XXV b

and sulfate, isothiocyanates (equation 1).

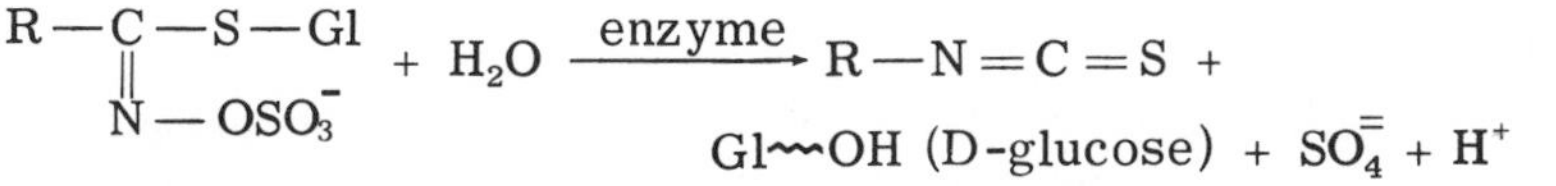

$$\begin{array}{l} R-C-S-Gl \\ \quad\;\; \| \\ \quad\;\; N-OSO_3^- \end{array} + H_2O \xrightarrow{\text{enzyme}} R-N=C=S + Gl\sim OH \text{ (D-glucose)} + SO_4^{=} + H^+ \quad (1)$$

XXVa

According to our present knowledge, the concentration of free isothiocyanates in any intact plant tissue able to yield them is negligible, and they are formed only after injury, by cleavage of precursors of the glucosinolate type. Yet although we shall treat the groups of natural mustard oil precursors and glucosinolates as if they are identical, it has not been proved that every known isothiocyanate or equivalent has the same precursor in all the plants from which it has been derived, or that one of its precursors is a glucosinolate. Strictly speaking, the myronate or allylglucosinolate ion is the only precursor for which the complete structure XXV has been proved, including the configuration about the carbon-nitrogen double bond. Besides the possibility of *syn-anti* stereoisomerism, which we should like to know more about, perhaps the most likely kind of natural variation on XXV would be alteration of the sugar residue, for example by introduction of an *O*-glucosyl group to give a thiogentiobioside comparable to amygdalin. Another conceivable change, of more importance, would be replacement of the sulfate by a phosphate ester group. No existing evidence, however, indicates deviation from the accepted structural pattern. The one known natural isothiocyanate that itself is a glycoside arises from a normal glucosinolate (XXV), in which the group R, hereafter referred to as the glucosinolate side chain, contains, to be sure, a L-α-rhamnopyranosyl group (compound 43, Table 2).[3] We have accordingly put forward a list of natural glucosinolates (Table 2) that embodies minor elements of conjecture, noting the instances in which the glucosinolate or an *O*-acetyl derivative has been adequately characterized. When a mustard oil precursor or an immediate derivative thereof has not been isolated and quantitative analyses are lacking, a basis for assuming the presence of the glucosinolate functional group can come from chromatographic behavior, qualitative identification of cleavage products, and considerations of enzymatic specificity.

[3]Throughout section IV, Arabic numerals in parentheses designate compounds in Table 2, or their structural formulas.

TABLE 2

Natural Glucosinolates (XXV)

I. Aliphatic Glucosinolates containing a *C*-Methyl Group, or Derived from Diacids

a. Alkyl, alkenyl, hydroxyalkyl, and acyloxyalkyl compounds

(1)[a,b] $CH_3—C(=NOSO_3^-)—SGl$

(2)[b] $C_2H_5—C(=NOSO_3^-)—SGl$

(3)[a,b] $(CH_3)_2CH—C(=NOSO_3^-)—SGl$

(4)[b] $HOCH_2—C(H)(CH_3)—C(=NOSO_3^-)—SGl$

(5) $C_6H_5COOCH_2—C(H)(CH_3)—C(=NOSO_3^-)—SGl$ Kjær and Christensen (1961)

(6)[a,b] $(CH_3)_2C(OH)—CH_2—C(=NOSO_3^-)—SGl$

(7)[a,b] $CH_3—C(H)(C_2H_5)—C(=NOSO_3^-)—SGl$

(8) $HOCH_2—C(H)(C_2H_5)—C(=NOSO_3^-)—SGl$ Kjær and Christensen (1962a)

TABLE 2 (continued)

(9)	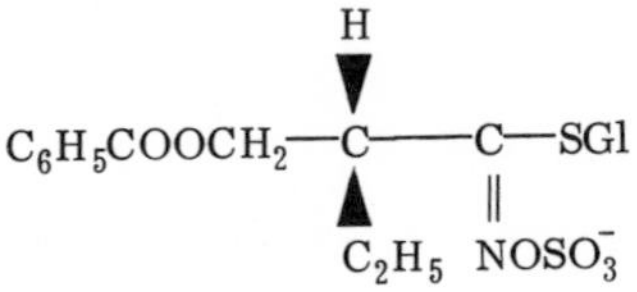 $C_6H_5COOCH_2-C(H)(C_2H_5)-C(=NOSO_3^-)-SGl$	Kjær and Christensen (1962b)
(10)	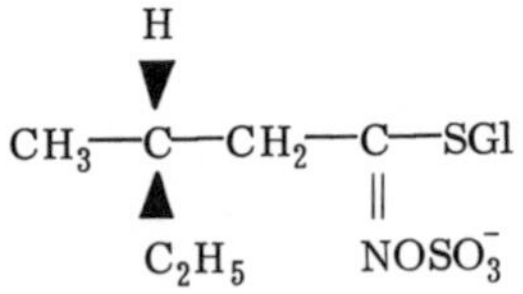 $CH_3-C(H)(C_2H_5)-CH_2-C(=NOSO_3^-)-SGl$	Kjær and Friis (1962); Delaveau and Kjær (1963)
(11)	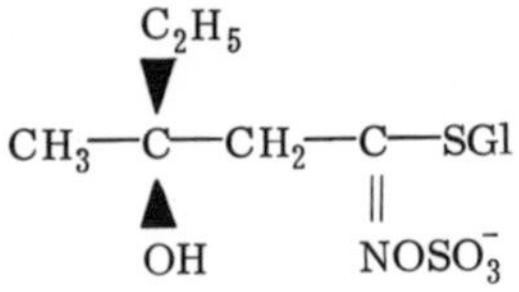 $CH_3-C(C_2H_5)(OH)-CH_2-C(=NOSO_3^-)-SGl$	Kjær and Thomsen (1962a); Christensen and Kjær (1963a)
(12)[a]	$CH_3-C(=CH_2)-(CH_2)_2-C(=NOSO_3^-)-SGl$	Kjær and Wagnières (1965)

b. Ketoalkyl compounds

(13)	$C_2H_5CO(CH_2)_4-C(=NOSO_3^-)-SGl$	Kjær and Thomsen (1963b)
(14)[a]	$CH_3(CH_2)_2CO(CH_2)_3-C(=NOSO_3^-)-SGl$	Kjær et al. (1960)
(15)[a]	$CH_3(CH_2)_2CO(CH_2)_4-C(=NOSO_3^-)-SGl$	Kjær and Thomsen (1962b)

c. Derivatives of diacids

(16)[a,b]	$CH_3OCO(CH_2)_3-C(=NOSO_3^-)-SGl$	

II. ω-Methylthioalkylglucosinolates and Derivatives

(17)[b]	$CH_3S(CH_2)_3-C(=NOSO_3^-)-SGl$	

TABLE 2 (continued)

(18)[a,b,c]	$CH_3{-}\overset{\blacktriangle O}{S}{-}(CH_2)_3{-}C({=}NOSO_3^-){-}SGl$	
(19)[a,b]	$CH_3SO_2(CH_2)_3{-}C({=}NOSO_3^-){-}SGl$	
(20)[a,b]	$H_2C{=}CHCH_2{-}C({=}NOSO_3^-){-}SGl$	
(21)[b]	$C_6H_5COO(CH_2)_3{-}C({=}NOSO_3^-){-}SGl$	
(22)[b]	$CH_3S(CH_2)_4{-}C({=}NOSO_3^-){-}SGl$	
(23)[b,c]	$CH_3{-}\overset{\blacktriangle O}{S}{-}(CH_2)_4{-}C({=}NOSO_3^-){-}SGl$	
(24)[b]	$CH_3SO_2(CH_2)_4{-}C({=}NOSO_3^-){-}SGl$	cf. Gmelin and Bredenberg (1966)
(25)	$CH_3SC(H){=}C(H)(CH_2)_2{-}C({=}NOSO_3^-){-}SGl$	Friis and Kjær (1966)
(26)[a,b,c]	$CH_3{-}\overset{\blacktriangle O}{S}{-}C(H){=}C(H)(CH_2)_2{-}C({=}NOSO_3^-){-}SGl$	cf. Balenović et al. (1966)
(27)[b]	$H_2C{=}CH(CH_2)_2{-}C({=}NOSO_3^-){-}SGl$	
(28)[a,b]	$H_2C{=}CH{-}C(\blacktriangledown OH)(\blacktriangle H){-}CH_2{-}C({=}NOSO_3^-){-}SGl$	cf. Greer (1962b)

TABLE 2 (continued)

(29)[a]	$H_2C{=}CH{-}C(H)(OH){-}CH_2{-}C({=}NOSO_3^-){-}SGl$	Daxenbichler et al. (1965)
(30)[b]	$CH_3S(CH_2)_5{-}C({=}NOSO_3^-){-}SGl$	
(31)[a,b,c]	$CH_3{-}S(O){-}(CH_2)_5{-}C({=}NOSO_3^-){-}SGl$	
(32)[b]	$H_2C{=}CH(CH_2)_3{-}C({=}NOSO_3^-){-}SGl$	
(33)	$H_2C{=}CHCH_2{-}C(OH)(H){-}CH_2{-}C({=}NOSO_3^-){-}SGl$	Tapper and MacGibbon (1967)
(34)	$CH_3S(CH_2)_6{-}C({=}NOSO_3^-){-}SGl$	Daxenbichler et al. (1961)
(35)[c]	$CH_3{-}S(O){-}(CH_2)_6{-}C({=}NOSO_3^-){-}SGl$	Christensen and Kjær (1963b)
(36)	$CH_3SO(CH_2)_7{-}C({=}NOSO_3^-){-}SGl$	See text
(37)[b,c]	$CH_3{-}S(O){-}(CH_2)_8{-}C({=}NOSO_3^-){-}SGl$	
(38)[b,c]	$CH_3{-}S(O){-}(CH_2)_9{-}C({=}NOSO_3^-){-}SGl$	
(39)[b,c]	$CH_3{-}S(O){-}(CH_2)_{10}{-}C({=}NOSO_3^-){-}SGl$	

TABLE 2 (continued)

III. Arylmethylglucosinolates

(40)[a,b] $C_6H_5CH_2—C(=NOSO_3^-)—SGl$

(41)[a,b] $HO—C_6H_4—CH_2—C(=NOSO_3^-)—SGl$ cf. Barothy and Neukom (1965)

(42)[b] $CH_3O—C_6H_4—CH_2—C(=NOSO_3^-)—SGl$

(43)[a] [rhamnopyranosyloxy: ring C(H)(CH_3)—O—C(H)(O—)—C(H)(OH)—C(H)(OH)—C(H)(HO)] $—O—C_6H_4—CH_2—C(=NOSO_3^-)—SGl$ Badgett (1964)

(44) $m\text{-}HO—C_6H_4—CH_2—C(=NOSO_3^-)—SGl$ Friis and Kjær (1963)

(45)[b] $m\text{-}CH_3O—C_6H_4—CH_2—C(=NOSO_3^-)—SGl$ cf. Friis and Kjær (1963)

(46) $(CH_3O)_2C_6H_3—CH_2—C(=NOSO_3^-)—SGl$ Ettlinger et al. (1966)

(47)[a] indol-3-yl$—CH_2—C(=NOSO_3^-)—SGl$ Gmelin and Virtanen (1961)

TABLE 2 (continued)

(48)[a]	1-methoxyindol-3-yl—CH_2—C(=$NOSO_3^-$)—SGl	Gmelin and Virtanen (1962)

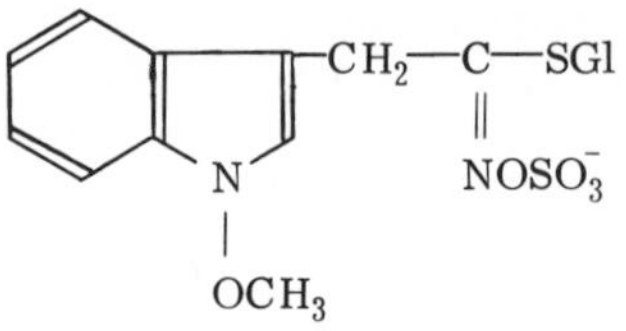

IV. 2-Arylethylglucosinolates and Derivatives

(49)[b]	$C_6H_5(CH_2)_2$—C(=$NOSO_3^-$)—SGl
(50)[b]	C_6H_5—C(H)(OH)—CH_2—C(=$NOSO_3^-$)—SGl

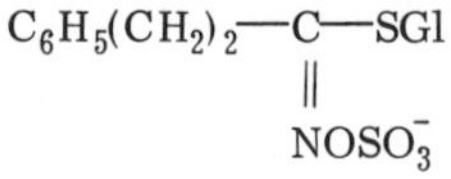

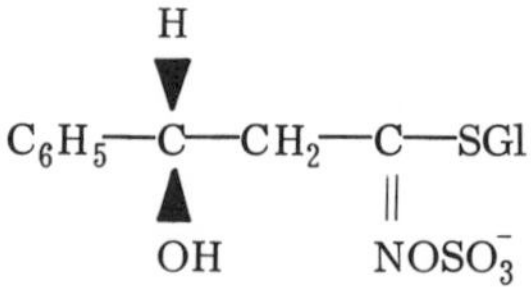

[a]The natural glucosinolate or an acetyl derivative has been isolated as a pure, crystalline salt.

[b]For references see Kjær (1960).

[c]For the absolute configuration of the sulfoxide group, see Cheung et al. (1965).

At the beginning of this decade, the natural glucosinolates and their varied aspects of scientific consequence were the subject of a painstaking review (Kjær, 1960), which we have used as the means of reference to prior work in Table 2 and do not intend to supersede now. This review was supplemented by further accounts (Kjær, 1961, 1963a,b). We now wish to approach these compounds, not exhaustively but in the light of the most recent advances, and with emphases that we hope will serve to convey a sense of unity and point out questions that hold our interest. We deal with a collection of 50 natural products, all possessing the same major functional group, whose chemical properties are therefore homogeneous to a considerable degree but have an intricacy and depth that are just coming to be realized. In 1839 and 1840, potassium myronate was isolated as the result of the first deliberate search backward from product to substrate of a reaction known to be enzymatic, and sinalbin (sinapine *p*-hydroxybenzylglucosinolate) was recognized to be an analogue. A tentative structure for the glucosinolates, containing a thioglucoside unit, was proposed by Gadamer in 1897, but thereafter for nearly 60 years no progress took place in understanding of the aglucone portion, although the work of Wilhelm Schneider, of Jena, culminating in 1928 to 1931, supplied our fundamental knowledge of 1-thioglucopyranose. Again, by 1912, eight different glucosinolates (7, 19, 20, 24, 27, 40, 41, 49) had been observed in plants, but from then until 1948 only one new side chain [from(6)] was discovered. In the three-year period 1954 to 1956, soon after O.-E. Schultz and others introduced chromatographic methods of analysis, the number of structurally known natural side chains more than doubled. The common structure XXV of the glucosinolates was elucidated, apart from *syn-anti* relationships, in 1956. X-ray crystallographic analysis of the myronate ion (20) later furnished its complete stereochemistry (Waser and Watson, 1963). Three additional major developments in the past 10 years, with implications that are yet to be explored fully, are the discovery that glucosinolates sometimes give rise to organic thiocyanates (Gmelin and Virtanen, 1959; Virtanen, 1962); the first characterization, in 1961, of a glucosinolate aglucone; and the study, beginning in 1962, of glucosinolate biosynthesis. We have therefore attempted a renewed understanding of the field.

The structures of glucosinolate side chains have usually been determined with the isothiocyanates or their derivatives as starting points, but the *p*-hydroxybenzyl glucosinolate was identified many years ago by means of the nitrile formed without rearrangement, and conversion to the nitriles, or analogous carboxylic acids or esters, has lately become an important avenue of structural analysis [so for (13-15, 44, 47-48)]. Examination of Table 2 will show that the natural side chains are diverse, but simple. An unusual feature for natural products is the large number of homologues differing by a single methylene group, frequently occurring together, with the notable exception of the benzyl and 2-phenylethyl compounds (40, 49). The prettiest example of this phenomenon is the consecutive sequence of ω-methylsulfinylalkyl glucosinolates (18, 23, 31, 35-39), the members of which show no evident alternation in abundance. Another regularity is the appearance of β hydroxylation [in (4, 6, 8, 11, 28-29, 33, 50)] in conjunction with the desoxy compounds [(3, 7, 10, 27, 32, and 49); the isobutyl glucosinolate corresponding to (6) has not yet been detected]. No α,β-unsaturated glucosinolates are known, and the origin of the hydroxyl groups is undetermined. Several glucosinolates have the skeletons of common amino acids, with replacement of carboxyl by the thioglucosyl group: thus the methyl (1), isopropyl (3), 2-butyl (7),, benzyl (40), *p*-hydroxybenzyl (41), and 3-indolylmethyl (47) derivatives correspond respectively to alanine, valine, isoleucine, phenylalanine, tyrosine, and tryptophan, whereas 3-methylthiopropylglucosinolate (17) is equivalent to methionine with an added methylene group.

The glucosinolates are, we know, versatile progenitors of organic isothiocyanates, thiocyanates, and cyanides, and of further transformation products of these, but we know little about the functions of glucosinolates in the plants that produce them. Schraudolf (1966b) has rightly emphasized the possible gap between enzymatic cleavages in disrupted tissues and catabolism in an intact organism; he has also pointed out that, on the other hand, the mechanical results of externally inflicted injury may be simulated by cell transformations taking place during normal growth (Schraudolf and Bergmann, 1965). Of great current interest is the role of the indole glucosinolate (47) which occurs in the growing, aerial parts of many plants of the Cruciferae and related families (Schraudolf, 1965) and is easily transformed in vitro to the powerful auxin 3-indoleacetonitrile (Gmelin and Virtanen, 1961). Plausible though the hypothesis is that 3-indolylmethylglucosinolate serves as a reserve pool of growth hormone, a conversion of (47) to simpler 3-indoleacetic acid derivatives in a cruciferous plant is remarkably hard to show unequivocally (Schraudolf and Bergmann, 1965; Schraudolf, 1966a). Likewise, the extent to which glucosinolates pass into isothiocyanates during plant metabolism with the fate of the latter compounds is unknown. A possible indication of a natural pathway from glucosinolates to isothiocyanates to amines has appeared in Olesen Larsen's findings (Larsen, 1965a,b) that *p*-hydroxybenzylamine occurs in seed of *Sinapis alba* L., a rich source of *p*-hydroxybenzylglucosinolate, and that seed of *Lunaria annua* L. (also Cruciferae) contains the γ-glutamyl derivative of isopropylamine, as well as the isopropyl glucosinolate.

The effects of glucosinolates and their cleavage products on microorganisms and animals continue to draw attention. The glucosides stimulate feeding of caterpillars that normally live on species of Cruciferae (David and Gardiner, 1966). Nitriles have toxic effects on mammals (Virtanen, 1963, 1965; Tookey et al., 1965). Isothiocyanates, which react generally with amine and thiol groups, have striking antibacterial, antifungal, and insecticidal properties (Dannenberg et al., 1956; Das et al., 1957; Virtanen, 1965; Lichtenstein et al., 1964). Much recent work, done primarily with synthetic compounds, on antibiotic and anthelmintic effects of isothiocyanates and precursors that generate them nonenzymatically, like dithiocarbamates, is not cited specifically here.

The major extrinsic source of the rising tide of interest in glucosinolates during the past 20 years has been the antithyroid, goiter-inducing effects that isothiocyanates and their transformation products have in mammals, including man. The active antithyroid agents are of three types: the isothiocyanates themselves; oxazolidine-2-thiones, formed by cyclization of β-hydroxy isothiocyanates; and thiocyanate ion, resulting from solvolysis of isothiocyanates with strongly electron-donating side chains, or perhaps directly from the corresponding glucosinolate aglucones. Organic thiocyanates give rise to thiocyanate ion in vivo (Virtanen, 1961). Cabbage, containing allylglucosinolate (20), 2(*R*)-hydroxy-3-butenylglucosinolate (28), and 3-indolylmethylglucosinolate (47), can furnish the three potential goitrogens allyl isothiocyanate (XXVI), (*S*) 5-vinyloxazolidine-2-thione or goitrin (XXVII), and thiocyanate ion (XXVIII). Several groups of workers have gathered information about the occurrence of these latent antithyroid agents in food and their liberation in the

(20) → $H_2C=CHCH_2NCS$

XXVI

(28) → $H_2C=CH-C$ (ring: H_2C–NH–CS–O–C, with H on C)

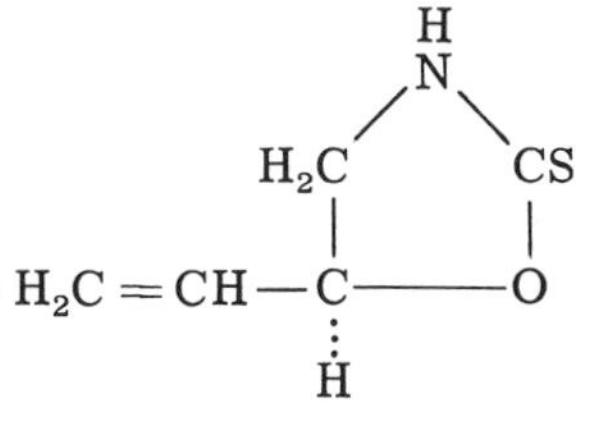

XXVII

(47) → SCN^-

XXVIII

mammalian body, transfer from cattle feed to milk, and physiological significance (Virtanen, 1961, 1963; Greer, 1962a; Bachelard et al., 1963; see also Podoba and Langer, 1964; Krusius and Peltola, 1966; Langer, 1966a,b).

The current nomenclature, according to which the parent ion (XXV, R = H) receives the trivial name "glucosinolate" and the chemical name of a side chain is used as a substituent prefix, was introduced by Dateo (1961; Ettlinger and Dateo, 1961, is identical in text) and has been accepted without demur. We should say a little about the older names, of which "glucotropaeolin" is a convenient prototype. The disadvantages that they followed but a ragged system and that 50 names of the sort, no matter how handy and endearing, require a lot of memory work to learn are obvious. The most serious trouble, however, is that the "-in" form, which is a matter of course for other glucosides, is not appropriate for an anion or its salts. Nobody knows exactly what "glucotropaeolin" now means: it may designate the benzylglucosinolate ion, or, in a generic sense, any of its salts, or a particular one, most likely the potassium salt, which in this instance has never been crystallized. As one evidence of the difficulty, we may cite the fact that competent workers in 1965 mentioned as significant the failure of a solution of potassium 2(*S*)-hydroxy-3-butenylglucosinolate to crystallize when seeded with what apparently was sodium 2(*R*)-hydroxy-3-butenylglucosinolate, paying no attention to the difference in cations—an oversight that is much easier if one calls the second salt merely "progoitrin." We believe, therefore, that "-in" names of glucosinolates have no place in formal discussion—not even sinigrin has a place, nor does sinalbin, other than in reference to the historic, crystalline salt—and should be coined no more.

B. Biosynthesis

The side chains of some natural glucosinolates are, we have seen, identical with the side chains of common α-amino acids. Even when only a few glucosinolates were known and these were thought of only in the guise of the corresponding isothiocyanates, the parallelism with amino acids was unmistakable and led to the postulate of a biogenetic relationship, used, it is worth recalling, to select a likely structure, the correct one, for methionine (Barger and Coyne, 1928; see also Challenger and Charlton, 1947). Not until the structure of the glucosinolates was elucidated, however, did the real nature of the possible relationship become clear. The glucosinolates (XXIX) could then occupy a rational place as potential derivatives of amino acids (XXX) in higher plants alongside the cyanohydrin glycosides. It is sufficient here to consider specifically the *O*-glucosyl cyanohydrins

(XXXI). Their structure has been known a long time, and experiments on their oxidative biosynthesis from amino acids go

$$\underset{\text{XXIX}}{RR'CH-\overset{\displaystyle SGl}{\overset{|}{C}}=NOSO_3^-} \longleftarrow \underset{\text{XXX}}{RR'CH-\underset{\underset{\displaystyle COOH}{|}}{\overset{\overset{\displaystyle H}{|}}{C}}-NH_2} \longrightarrow \underset{\text{XXXI}}{RR'\overset{\overset{\displaystyle OGl}{|}}{C}-C\equiv N}$$

back more than 40 years (Rosenthaler, 1923; cf. Gander, 1960). Biosynthetic studies with isotopic tracers, however, began only a few years earlier for the cyanogenetic glucosides than for the glucosinolates and, carried on independently for the two classes of compounds, have provided schemes for the formation of each that at the present moment are curiously parallel and simple. The general hypothesis that has resulted, although it has been tested for only a few compounds and plants, is that glucosinolates (XXIX) or cyanohydrin glucosides (XXXI) are indeed synthesized from the corresponding L-amino acids (XXX), or from amino acids differing from XXX solely in the functional groups of the side chain, and that these transformations of the amino acids occur with loss of the carboxyl group but preservation of all the linkages in the rest of the carbon-nitrogen skeleton. The conversions have been effected only with whole organs, and no intermediates have been identified, although some potential ones have been excluded. It will be exciting to learn how the pathways to XXIX and XXXI diverge. Work on the principal cyanogenetic glucosides can be summarized very briefly as follows. The aglucone carbon skeleton of linamarin (XXXI, R = R′ = CH_3) arises from valine (XXX, R = R′ = CH_3) and that of lotaustralin (XXXI, R = CH_3, R′ = C_2H_5; aglucone configuration unknown) from isoleucine [XXX, R′ = CH_3, R′ = C_2H_5, β-carbon (*S*)] (Butler and Butler, 1960; Butler and Conn, 1964). Prunasin [XXXI, R = C_6H_5, R′ = H, benzylic carbon (*R*)] comes from the phenylethylidene unit of phenylalanine (XXX, R = C_6H_5, R′ = H)(Mentzer et al., 1963), dhurrin [XXXI, R = *p*-HOC_6H_4, R′ = H, benzylic carbon (*S*) (Towers et al., 1964)] similarly from tyrosine (XXX, R = *p*-HOC_6H_4, R′ = H)(Koukol et al., 1962; Gander, 1962; Uribe and Conn, 1966). Furthermore, the nitrogen atom of L-valine is thought to be retained on passage to linamarin in *Linum usitatissimum* L. (Butler and Conn, 1964) and that of L-tyrosine upon conversion to dhurrin in *Sorghum vulgare* Pers. (Uribe and

Conn, 1966). As the account suggests, the known cyanohydrin glucosides are fewer than the glucosinolates, and a larger proportion of the former, including all the most familiar ones, correspond to protein amino acids. Amino acid equivalents of many glucosinolates are rare or unknown in plants, and with these glucosinolates a problem immediately raises itself, if amino acids with the same skeleton are progenitors, which is to trace the sequences to the amino acids from the ordinary constituents of cells. The matter was not always put so, to be sure, at the start of the biosynthetic investigations, when the overall derivation of the glucosinolates from ordinary cell constituents was the primary concern. Now, however, the results, even though incomplete, would lead us in each unknown instance to anticipate and search for an amino acid as a key intermediate from which the glucosinolate functional group has its origin.

The evidence showing that amino acids are direct precursors of cyanohydrin glucosides or glucosinolates has been gathered from biosynthetic experiments, with isotopically labeled starting materials, that in most respects are straightforward. Conversions of amino acids to cyanogenetic glucosides and to glucosinolates have been realized with incorporations as high as 25 to 40 percent and dilution factors as low as 10 to 15. The fate of individual carbon atoms has been studied by the usual methods of selective labeling and degradation. In one example, an amino acid (tyrosine) was administered as a mixture of two samples labeled one at the α, the other at the β position with ^{14}C, and the yields of activity were determined after separation of the corresponding carbon atoms of the cyanohydrin produced. The relative yield averaged unity, although it fluctuated by a factor of 2, and the conclusion was drawn that the carbon skeleton of the amino acid, aside from the carboxyl group, traversed the biosynthesis in one piece (Koukol et al., 1962). The same principle is involved in the use of an amino acid as a mixture labeled with ^{14}C and ^{15}N, but the conclusion is less a matter of course. Since an α-amino acid and the corresponding α-keto acid are in general readily interconverted, observed transformation of the carbon skeleton of an amino acid to a natural product with high efficiency leaves the possibility open that the keto acid may intervene and even that the main biosynthetic pathway runs through the keto acid but not the amino acid. Evidence, consistent with the structural requirements, that the nitrogen is retained therefore helps greatly to establish the amino acid as a natural biosynthetic intermediate. The simplest quantitative result of an experiment with isotopic carbon and nitrogen that fits the hypothesis of nitrogen retention is equality between the dilutions of the two elements in the product. The extent to which reversible transamination of the starting material, competing with the biosynthesis, can reasonably explain extra dilution of the nitrogen may be difficult to judge. Separation of carbon and nitrogen during biosynthesis by no means rules out the possibility of approximately equal incorporation for both. Three instructive experiments testing glucosinolate production from C,N-labeled amino acids are on record in which it is clear from structural restrictions or the actual result that the nitrogen was severed from the rest of the molecule. The dilution factor of nitrogen was of the order of a few hundred to a thousand at moderate dosages, and its incorporation ranged up to 3 percent. When 2-amino-4-pentenoic acid (XXX, R = $H_2C{=}CH$, R′ = H) was the starting material and allylglucosinolate (XXIX, R = $H_2C{=}CH$, R′ = H) the product in *Armoracia lapathifolia* Gilib. (Matsuo and Yamazaki, 1966), the carbon skeleton of the amino acid was probably degraded, and the nitrogen was more highly incorporated than the α-carbon by a factor of 15. On passage from phenylalanine (XXX, R = C_6H_5, R′= H) to 2-phenylethylglucosinolate (XXIX, R = $C_6H_5CH_2$, R′ = H) or 2-hydroxy-2-phenylethylglucosinolate (XXIX, R = C_6H_5CHOH, R′ = H)(Underhill, 1965a,b), the phenylethylidene group was transferred intact, and the nitrogen was diluted more than the carbon, but only by factors of 4 or 2.

Conversion of amino acids to glucosinolates with apparent preservation of nitrogen has been demonstrated for syntheses of benzylglucosinolate (XXIX, R = C_6H_5, R′ = H) from L-phenylalanine in *Tropaeolum majus* L. (Tropaeolaceae)(Underhill and Chisholm, 1964), 2-phenylethylglucosinolate from L-2-amino-4-phenylbutyric acid (XXX, R = $C_6H_5CH_2$, R′ = H) in *Rorippa nasturtium-aquaticum* (L.) Hayek (*Nasturtium officinale* R. Br.; Cruciferae)(Underhill, 1965a), and allylglucosinolate (XXIX, R = $H_2C{=}CH$, R′= H) from 2-amino-5-methylthiovaleric acid (XXX, R = $CH_3S(CH_2)_2$, R′= H) in *Armoracia lapathifolia* (Cruciferae)(Matsuo and Yamazaki, 1966). In these three examples, the dilution of ^{15}N in the glucosinolate equaled that of ^{14}C or exceeded it by no more than 20 percent. Comparison between experiments in which C, N – labeled 2-amino-4-phenylbutyric acid was administered to watercress plants as the pure L isomer or the racemate showed that carbon from the D isomer was utilized with the same efficiency (here, a dilution factor of 15 to 20) as that from the L compound but the nitrogen was diluted much more; hence, complete transformation of the amino acid from D to L through the keto acid was indicated. Only the racemate of 2-amino-5-methylthiovaleric acid was given to horseradish, and although the dilution of nitrogen equaled that of carbon after a 3-hour metabolic period, after 24 hours the dilution of nitrogen was the greater by a factor of nearly 2. The reasonable interpretation was made that biosynthesis of the glucosinolate from the L-amino acid with retention of nitrogen had proceeded faster than the transaminative conversion of D to L isomer.

One example where the reaction course still is open to doubt is the formation of 2-hydroxy-2-phenylethylglucosinolate from doubly labeled DL-2-amino-4-phenylbutyric acid in *Reseda luteola* L. (Resedaceae) (Underhill, 1965b). The incorporation of nitrogen, though more than 6 percent, was one quarter of that of carbon, and the dilution factor for nitrogen was nearly as great as when phenylalanine was converted to the same product. Underhill implied that extensive reversible transamination of the aminobutyric acid was less likely than hydroxylation by a mechanism requiring loss of nitrogen, presumably through keto acid intermediates. In fact, a judgment whether or not the L-amino acid is completely converted to the keto acid before transformation to the glucosinolate, and secondarily whether hydroxylation precedes or follows re-amination of the keto acid, can hardly be made from the existing data.

In biosynthesis of cyanohydrin glucosides, the dilution of ^{15}N was 25 percent greater than that of ^{14}C for conversion of L-tyrosine to dhurrin; it was 60 to 125 percent greater than that of ^{14}C for conversion of L-valine to linamarin in flax seedlings. The syntheses were assumed to proceed without detachment of nitrogen, and the losses of ^{15}N were consequently attributed to transamination, to the extent of, respectively, 20 percent and 40 to 55 percent. The results for flax (Butler and Conn, 1964) illustrate, however, a general kind of situation in which a test of mechanism could be applied that is more searching than the one numerical check for nitrogen retention used exclusively to date and is valuable even if the relative dilution of carbon and nitrogen comes out as unity. Flax contains roughtly equal quantities of linamarin and lotaustralin, and nitrogen transfer in the plant from ^{15}N-valine to isoleucine, a single member of the amino acid pool, was slow enough so that the final concentration of excess ^{15}N in isoleucine was an order of magnitude less than in linamarin. No evidence that the substantial amounts of added valine repressed lotaustralin synthesis was reported, and the extent of any inhibition could have been ascertained. It is curious that Butler and Conn seemingly made no attempt to monitor the appearance of nitrogen from valine in lotaustralin also. If nitrogen is detached and reincorporated during synthesis of linamarin from valine, the same would be true of the simultaneous formation of lotaustralin from isoleucine, and on a reasonable hypothesis the nitrogen from both amino acids would be set free in a common pool and enter both glucosides to a similar degree. If, therefore, the isotope content of lotaustralin formed in the presence of ^{15}N-valine is not enriched over that of isoleucine, the result would be a more stringent proof of nitrogen retention in synthesis of both linamarin and lotaustralin than any yet devised.

We turn to a more specific account of experimental studies on the origin of the carbon framework of the glucosinolate aglucones and consider initially glucosinolates corresponding to ordinary amino acids. The biosynthesis of a glucosinolate from an amino acid was first demonstrated by Kutáček et al., 1962b). When tryptophan (XXXII) labeled with ^{14}C was administered to cabbage (*Brassica oleracea* L.; Cruciferae), a chromatogram of the plant constituents two days later showed a peak of radioactivity at the place of 3-indolylmethylglucosinolate (47). The expressed plant juice was already known to contain radioactive, indolic products of enzymatic degradation of the glucosinolate

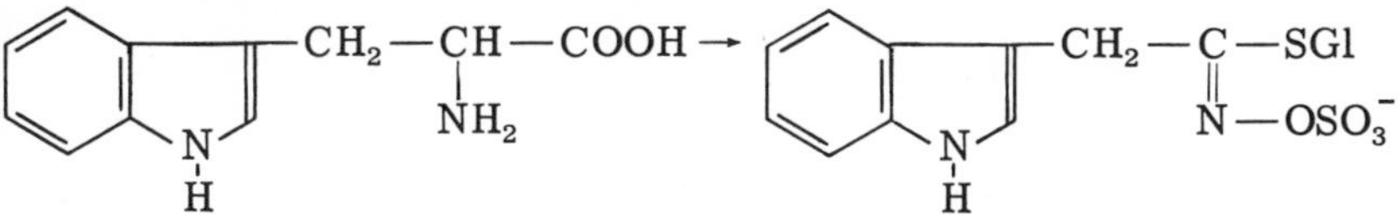

$CH_2—CH(NH_2)—COOH$ → $CH_2—C(=N—OSO_3^-)—SGl$

XXXII (47)

+ $CH_2—C(=N—OSO_3^-)—SGl$ (N—CH_3O)

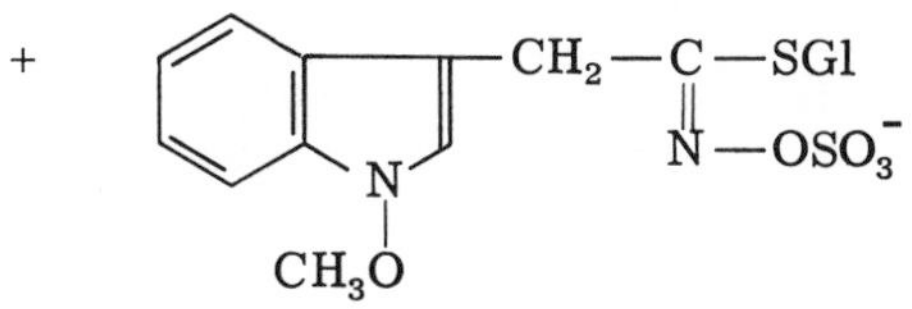

(48)

(Kutáček et al., 1962a). 3-Indoleacetonitrile-^{14}C did not give rise to labeled glucosinolate in the *Brassica*. Schraudolf and Bergmann (1965) studied chromatographically the metabolism of ring-labeled tryptophan in etiolated three-day old seedlings of *Sinapis alba* L. (Cruciferae). The D-amino acid was converted exclusively to its *N*-malonyl derivative, whereas L-tryptophan furnished the two indole glucosinolates (47) and (48). No other radioactive metabolites could be detected. The synthesis of the *N*-methoxyindole (48) was slower than that of (47), but whether (47) was a precursor of (48) was not decided. When indole-2-^{14}C was given to the seedling hypocotyls (Schraudolf, 1966a), the processes observed were a rapid synthesis of tryptophan (evidently L-), accelerated by L-serine, and the slower appearance of the same pair of glucosinolates.

Results of investigations of the biosynthesis of benzylglucosinolate (40) in *Tropaeolum majus* were published independently and almost simultaneously by Underhill et al. (1962) and by Benn (1962). After administration of phenylalanine (XXXIII) samples labeled at each carbon of the aliphatic chain, the aglucone portion of the glucosinolate was isolated as benzylthiourea and degraded to locate any significant radioactivity at the thiohydroximate carbon or in the benzyl group. The outcome showed that the carboxyl carbon of the starting material was lost and the other carbon atoms of the chain appeared to occupy the equivalent positions in the glucosinolate. Underhill et al. also reported

$$C_6H_5-\overset{\ddagger}{C}H_2-\overset{\dagger}{C}H(NH_2)-\overset{*}{C}OOH \longrightarrow C_6H_5-\overset{\ddagger}{C}H_2-\overset{\dagger}{C}(=N-OSO_3^-)-SGl$$

XXXIII (40)

that 2-hydroxy-3-phenylpropionic acid ($^{14}CHOH$) was converted to the glucosinolate ($^{14}C{=}N$), with fair efficiency, whereas cinnamic acid (labeled next the benzene ring) and phenylacetic acid ($^{14}COOH$) were very feebly utilized. Underhill and Chisholm (1964) added the amide, hydroxamic acid, and nitrile of phenylacetic acid, as well as 2-phenylethylamine, to the list of nonprecursors. For 2-oximino-3-phenylpropionic acid ($^{14}C{=}N$), the dilution factor and incorporation were of the order of geometric means between the values for phenylalanine and for the nonprecursors. The stability of the oxime during the experiment was uncertain, and only one stereoisomer, thought to have *anti* carboxyl and hydroxyl groups, was available. Whether or not an α-keto acid oxime can intervene on the way from amino acid to glucosinolate remains unknown.

One group of possible intermediates between amino acids and glucosinolates that has not been considered previously is the α-nitro carboxylic acids (XXXIV). The following remarks

$$\underset{\displaystyle\text{XXXIV}}{\mathrm{R{-}\underset{\displaystyle NO_2}{\underset{|}{C}H}{-}COOH}} \qquad\qquad \underset{\displaystyle\text{XXXV}}{\mathrm{C_6H_5CH_2CH_2NO_2}}$$

indicate why we mention the compounds in this sense. α-Nitro acids have never been isolated from natural sources and decarboxylate very easily in neutral solution, but the anions can be stabilized by chelation with metals (Pedersen, 1949; Finkbeiner and Stiles, 1963). *O*-Methyl-*aci*-nitroacetamides are among the *Streptomyces* antibiotics (Mizuno, 1961; Mitscher et al., 1965; Wiley et al., 1965). Two primary nitro compounds are known from higher plants, 3-nitropropionic acid, which has been found, often as glucose esters, in species of Corynocarpaceae, Leguminosae, Malpighiaceae, and Violaceae (Pailer, 1960; Finnegan and Mueller, 1965), and 1-nitro-2-phenylethane (XXXV) from Lauraceae (Gottlieb and Magalhães, 1959). A fungal synthesis of 3-nitropropionic acid from aspartic acid may involve oxidation of the amino group without its detachment (cf. Shaw, 1967, and Lancini et al., 1966). The speculation that these nitro compounds can arise in higher plants through oxidation of the amino acids, aspartic acid and phenylalanine (Gottlieb et al., 1961), to the α-nitro acids is attractive. Emery (1966) has considered the natural N-hydroxylation of amino acids. The base-catalyzed reaction of primary nitro compounds and thiols to give thiohydroximic acis (XXXVI) (Copenhaver, 1957) probably takes place by addition

$$\mathrm{RCH_2NO_2 \rightleftharpoons RCH{=}\underset{\displaystyle OH}{\underset{|}{\overset{+}{N}}}{-}O^- + R'{-}S^- \rightarrow \left[\underset{\displaystyle\ }{\overset{\displaystyle SR'}{\overset{|}{R C H}}}{-}\underset{\displaystyle OH}{\underset{|}{N}}{-}O^-\right]}$$

$$\mathrm{\xrightarrow{-OH^-} \left[\overset{\displaystyle SR'}{\overset{|}{RCH}}{-}N{=}O\right] \longrightarrow \underset{\displaystyle\text{XXXVI}}{\overset{\displaystyle SR'}{\overset{|}{RC}}{=}NOH}}$$

of the mercaptide to the *aci*-nitro derivative, analogously to the attack of nucleophiles on nitrones (Hamer and Macaluso, 1964). α-Nitro acids have α mobile hydrogen, and furthermore decarboxylation of their mono salts produces *aci*-nitro compounds (Pedersen, 1934). At any rate, examination of the reactivity of α-nitro acids toward thiols, including β-1-thioglucopyranose, seems worthwhile and is projected for the near future. (See also note at end of section IV.)

Preliminary evidence for the conversion of valine (XXXVII; methyl-^{14}C) to isopropylglucosinolate (3) in *Tropaeolum peregrinum* L. has been offered by Benn and Meakin (1965). After

$$\underset{\text{XXXVII}}{(CH_3)_2CH-\underset{\displaystyle NH_2}{\underset{|}{C}H}-COOH} \longrightarrow \underset{(3)}{(CH_3)_2CH-\underset{\displaystyle N-OSO_3^-}{\underset{\|}{C}}-SGl}$$

enzymatic hydrolysis of the glucosinolate produced in the exper-iment, followed by treatment with ammonia and chromatography a small amount of radioactive isopropylthiourea was detected. Kindl (1964, 1965) has investigated the biosynthesis of *p*-hydrox benzylglucosinolate in *Sinapis alba.* In two- to three-week old plants, administered tyrosine was evidently degraded very rapidly, and when the *p*-hydroxybenzylglucosinolate from DL-tyrosine (^{14}C-N) was isolated after a five-day metabolic period, 90 percent of the radioactivity in the glucosinolate was in the glucosyl group. In older plants, up to nine weeks of age, the amino acid was better utilized for the glucosinolate and scatteri of the activity, in the aglucone portion anyway, diminished, but n definite conclusion about the reactions taking place was reached The fact that, as we have seen, tryptophan can be efficiently con verted to glucosinolates in *Sinapis alba* makes it seem likely that a similar conversion of tyrosine, obscured by competing. specific cleavage, operates in the plant. *p*-Hydroxycinnamic aci served as a good precursor of the aglucone of *p*-hydroxybenzyl-glucosinolate, whereas cinnamic acid, in comparison with pheny alanine, is very little utilized for synthesis not only of benzyl-glucosinolate in Tropaeolaceae, but also of chain-extended glucosinolates by way of phenylpyruvic acid in Resedaceae and in *Rorippa* (Cruciferae). Phenylalanine was incorporated to a small extent in the *p*-hydroxybenzylglucosinolate aglucone, but labeled benzylglucosinolate given to the older *Sinapis* plants was not detectably hydroxylated.

Our knowledge of the biosynthesis of glucosinolates with

longer side chains than ordinary amino acids stems chiefly from the work of the able group at the Canadian Prairie Regional Laboratory, who also were the first to use ^{15}N labeling in the study of glucosinolate biosynthesis. Their clear-cut results have been achieved by a technique in which the test compounds are absorbed by the experimental plants within a few hours and it is practicable to limit the total metabolic period to 24 hours. Underhill (1965a) showed that *Rorippa nasturtium-aquaticum* converted 2-amino-4-phenylbutyric acid (XXXVIII) selectively labeled by ^{14}C to 2-phenylethylglucosinolate (49) with preservation

$$\underset{\text{XXXVIII}}{\mathrm{C_6H_5CH_2\overset{\dagger}{C}H_2\underset{\displaystyle NH_2}{\underset{|}{\overset{*}{C}}}HCOOH}} \longrightarrow \underset{(49)}{\mathrm{C_6H_5CH_2\overset{\dagger}{C}H_2\underset{\displaystyle N{-}OSO_3^-}{\underset{\|}{\overset{*}{C}}}{-}SGl}}$$

of the carbon skeleton as well as, according to the test previously discussed, the nitrogen atom. The amino acid XXXVIII has not, however, been isolated from nature, although a presumption now exists that the compound occurs, perhaps in very low concentration, in watercress. Experiments with other carbon sources (Underhill et al., 1962; Underhill, 1965a) have established that phenylalanine, 2-hydroxy-3-phenylpropionic acid ($^{14}CHOH$), and acetic acid are good precursors of the aglucone of (49), whereas cinnamic acid (labeled β to carboxyl), phenylacetic acid ($^{14}COOH$), hydroxyacetic acid ($^{14}CH_2$), and formic acid are very slightly utilized. Glycine ($^{14}CH_2$) and serine ($^{14}CH_2$) were fair precursors of the aglucone, but no better than glucose, and the activity from glycine was scattered through the phenylpropylidyne unit. With the use of selectively labeled phenylalanine (XXXIII) and acetic acid (XXXIX), it was found that the carboxyl

$$\underset{\text{XXXIII}}{\mathrm{C_6H_5\overset{\ddagger}{C}H_2\underset{\displaystyle NH_2}{\underset{|}{\overset{\dagger}{C}}}H\overset{*}{C}OOH}} + \underset{\text{XXXIX}}{\mathrm{\overset{\circ}{C}H_3\overset{x}{C}OOH}} \longrightarrow \underset{(49)}{\mathrm{C_6H_5\overset{\ddagger}{C}H_2\overset{\dagger}{C}H_2\underset{\displaystyle N{-}OSO_3^-}{\underset{\|}{\overset{\circ}{C}}}{-}SGl}}$$

groups of both components were lost, the phenylethylidene group of phenylalanine was incorporated intact, and the methyl group of acetic acid furnished the thiohydroximate carbon atom of the glucosinolate. The only qualifying remark that need be added to the statement of these beautiful and definitive results is that phenylalanine is presumably functioning here as the equivalent of phenylpyruvic acid. Underhill (1965b) also demonstrated the identical pattern of incorporation of carbon from 2-amino-4-phenylbutyric acid and from phenylalanine and acetic acid during formation of 2(*S*)-hydroxy-2-phenylethylglucosinolate (50) in *Reseda luteola*. Compounds that behaved insignificantly as precursors of the aglucone included glycine, serine, and formic, hydroxyacetic, and cinnamic acids, previously cited, and in addition, glyoxylic acid, 2-phenylethylamine (^{14}C-N), hydroxycinnamic and 3-hydroxy-3-phenylpropionic acids (benzyl- or benzylidene-labeled), and *threo*-β-hydroxyphenylalanine (^{14}C-N). 2-Amino-4-hydroxy-4-phenylbutyric acid was not tested. The evidence whether the nitrogen atom of 2-amino-4-phenylbutyric acid was incorporated into the hydroxy glucosinolate directly or not was ambiguous, as discussed previously, but in either event benzylpyruvic acid, formed from phenylalanine or phenylpyruvic acid and acetic acid, would appear to be a key biosynthetic intermediate.

The first experiments on the synthesis of allylglucosinolate (20) in *Armoracia lapathifolia* were performed by Underhill et al. (1962). The methyl group of acetic acid was an excellent, specific source of the thiohydroximate carbon atom, but the carboxyl group was not utilized for the glucosinolate. Glutamic acid (^{14}C-N) was not significantly incorporated; aspartic acid and glucose furnished, respectively and specifically, the aglucone and glucosyl portions of the glucosinolate; glycine ($^{14}CH_2$) was metabolized to give both moieties. Chisholm and Wetter (1964) then discovered that homocysteine and methionine, which can be formed from aspartic acid by way of initial reduction to homoserine, were by far the best precursors, apart from acetic acid, of the aglucone of allylglucosinolate. In comparison, formic and hydroxyacetic ($^{14}CH_2$) acids, serine ($^{14}CH_2$), alanine (^{14}C-N), threonine, and 2-aminobutyric acid ($^{14}CH_2$) were inefficient carbon sources. By use of selectively labeled methionine (XL), as well as acetic acid (XXXIX), they reached the elegant conclusion that allylglucosinolate (20) was formed from the two components with loss of the two carboxyl groups and the thiomethyl group, conversion of the ethylene unit and adjacent carbon of methionine respectively to vinyl and saturated methylene groups, and confirmed transformation of the acetate methyl group to the thiohydroximate carbon atom of allylglucosinolate. Matsuo and Ya-

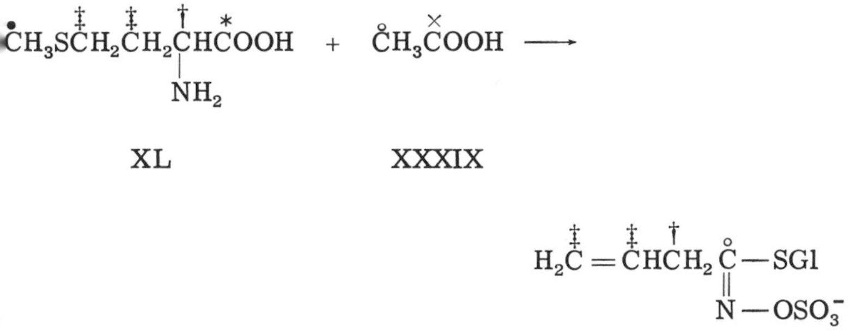

$$\overset{\bullet}{C}H_3S\overset{\ddagger}{C}H_2\overset{\ddagger}{C}H_2\overset{\dagger}{C}H(NH_2)\overset{*}{C}OOH + \overset{\circ}{C}H_3\overset{\times}{C}OOH \longrightarrow$$

XL XXXIX

$$H_2\overset{\ddagger}{C}=\overset{\ddagger}{C}H\overset{\dagger}{C}H_2\overset{\circ}{C}(=N-OSO_3^-)-SGl$$

(20)

mazaki (1966) reported that 3-methylthiopropionamide and 4-methylthiobutyramide (both 2-^{14}C) were not converted to allylglucosinolate aglucone and that acetic and malonic acids (^{14}CH) supplied the thiohydroximate carbon atom to the same extent. The competitive experiment by which they sought to prove that malonic acid was the immediate precursor does not appear decisive. They found also that 2-amino-4-pentenoic acid (^{14}C-N) was not appreciably incorporated, in contrast to methionine or 2-amino-5-methylthiovaleric acid (XLI), and that the carbon-nitrogen unit of the latter amino acid (^{14}C-N) became the thiohydroximate portion of the glucosinolate (20). Chisholm and Wetter (1966) independently, using preparations labeled with ^{14}C in the

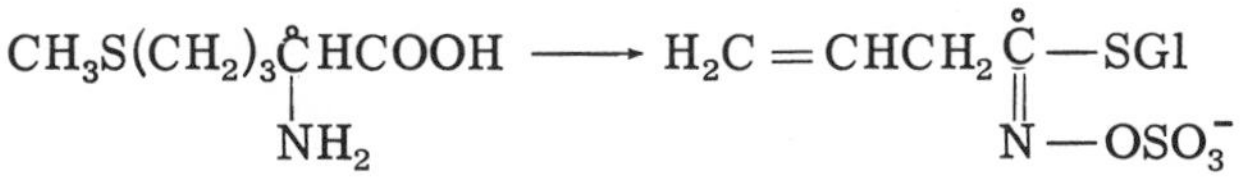

$$CH_3S(CH_2)_3\overset{\circ}{C}H(NH_2)COOH \longrightarrow H_2C=CHCH_2\overset{\circ}{C}(=N-OSO_3^-)-SGl$$

XLI (20)

2-position but not with ^{15}N, made essentially identical observations with the aminomethylthiovaleric and aminopentenoic acids and established that 2-amino-5-hydroxyvaleric acid was not converted to allylglucosinolate but homoserine was a good precursor, in accord with the expectation that it would give rise to methionine. They furthermore showed that after administration of methionine (^{14}C-N), the isotope in the amino acid fraction recovered dwelt almost equally in methionine and in 2-amino-5-methylthiovaleric acid. The latter amino acid was originally isolated (Sugii et al., 1964) from leaves of *Brassica oleracea*, which contain allylglucosinolate and a lesser amount of 3-methylthiopropylglucosinolate (17) (Clapp et al., 1959). The high specificity that the glucosinolate-forming system in horseradish possesses for the amino acid with a terminal methylthioethyl group as against the acids with terminal hydroxyethyl or vinyl groups is a striking result. There is as yet no experimental evidence, however, that 3-methylthiopropylglucosinolate is an intermediate between the amino acid XLI and allylglucosinolate. Matsuo and Yamazaki (1963, 1964) earlier studied the long-term (three or six weeks) incorporation of labeled substrates into allylglucosinolate deposited in seeds of *Brassica juncea*. Their observations, especially on the location of carbon atoms from acetic and malonic acids (^{14}CH) and from aspartic acid and other compounds connected with the tricarboxylic acid cycle, are compatible with the same biosynthetic sequence as in horseradish.

From the results that we have described, it is a fair deduction that the biosyntheses of 2-phenylethylglucosinolate from phenylalanine and of allylglucosinolate from methionine proceed in two stages: first, the synthesis of an α-amino acid with one more carbon atom, in which the original α carbon is β, the methyl carbon atom of acetic acid (or, it could be, the methylene carbon of malonic acid) appears as the new α carbon, and the carboxyl group presumably comes from acetic acid rather than the original amino acid; second, decarboxylative conversion of the new amino acid to a glucosinolate in a process identical, so far as can be told from the data, with the transformation of phenylalanine to benzylglucosinolate. The homologization, on which we now focus, can equivalently be regarded as a passage from a given α-keto acid to a higher. Systematic biochemical chain extension by one carbon atom is less familiar than the net addition

$$R\overset{\dagger}{C}H(NH_2)COOH + \overset{*}{C}H_3COOH \longrightarrow R\overset{\dagger}{C}H_2\overset{*}{C}H(NH_2)COOH$$

$$R\overset{\dagger}{C}OCOOH \quad \overset{*}{C}H_3COOH \longrightarrow R\overset{\dagger}{C}H_2\overset{*}{C}OCOOH$$

of both carbon atoms of acetic acid in fatty acid biosynthesis. The overall gain of one carbon atom in the fashion observed can be explained, however, by a straightforward reaction sequence (Chisholm and Wetter, 1964). The individual steps are identical with the steps from oxaloacetic acid (XLII, R = CH_2COOH) to α-ketoglutaric acid (XLIII, R = CH_2COOH) in the well-known tri-

$$\underset{\text{XLV}}{\underset{\displaystyle NH_2}{RCHCOOH}} \rightleftharpoons \underset{\text{XLII}}{\overset{\displaystyle COOH}{RCO}} + CH_3COSCoA \longrightarrow$$

$$\overset{\displaystyle COOH}{RC(OH)CH_2COOH} \longrightarrow \overset{\displaystyle COOH}{RC{=}\underset{H}{C}COOH} \longrightarrow$$

$$\overset{\displaystyle COOH}{RCHCHOHCOOH} \longrightarrow \underset{\text{XLIII}}{RCH_2COCOOH} \longrightarrow \underset{\text{XLIV}}{RCH_2\underset{\displaystyle NH_2}{C}HCOOH}$$

carboxylic acid cycle. Citric acid, however, the condensation product of oxaloacetic acid and acetyl coenzyme A, has two carboxymethyl groups, and in fact the carboxymethyl group originally from oxaloacetic acid is the one normally converted to the α-keto acid unit. A unique example has lately been discovered in which the steric course of the condensation is reversed and the full result is a true homologization of oxaloacetic to α-ketoglutaric acid, or of aspartic to glutamic acid (Gottschalk and Barker, 1966; contrast Stern et al., 1966). The conversion of α-ketoglutaric acid (XLII, R = $(CH_2)_2COOH$) to α-ketoadipic acid (XLIII, R = $(CH_2)_2COOH$) by the same sequence, which forms part of a microbiological synthesis of lysine through α-aminoadipic acid, has been resolved into the individual steps indicated (Strassman and Ceci, 1966, and references there cited). A particularly opposite example of this scheme of homologization is the synthesis of leucine (XLIV, R = $(CH_3)_2CH$) from valine (XLV, R = $(CH_3)_2CH$) through the corresponding keto acids, which has been studied in detail for microorganisms (Calvo et al., 1964, and references there cited). The same gross pathway from valine to leucine appears to operate in *Allium*, *Sorghum*, and *Linum* (Suzuki et al., 1962; Butler and Shen, 1963). One enzyme that condenses acetyl coenzyme A with α-ketoisovaleric acid can also condense the ester with pyruvic and α-ketobutyric acids (Webster and Gross, 1965; for other examples of the reaction with the last two keto acids, see Canovas et al., 1965). Yeast apparently is able to convert α-aminobutyric acid to α-ketovaleric acid and α-aminovaleric acid to α-ketocaproic acid, in both cases with incorporation of acetic acid (Ingraham et al., 1961); the experimental observations on the fusel oil produced neatly parallel some of those for glucosinolates. Therefore, although the proposed detailed pathway leading from an amino acid keto acid pair (XLII, XLV) plus acetate to the homologues (XLIII XLIV) during glucosinolate biosynthesis remains an unproved hypothesis, the overall transformation has ample precedent and the series of reactions assumed is a reasonable choice.

The two sulfur atoms in the glucosinolate functional group, one in sulfide, the other in sulfate linkage, are as far apart in oxidation level as possible. The incorporation of radioactive sulfur given in various simple forms has been followed by Wetter (1964) for allylglucosinolate in *Armoracia lapathifolia*, by Kindl (1964, 1965) for *p*-hydroxybenzylglucosinolate in *Sinapis alba*, and by Schraudolf and Bergmann (1965) for 3-indolylmethylglucosinolate in the same plant. All these workers employed sulfat ^{35}S, with the concordant results that after metabolic periods of several hours, 80 to 90 percent of the isotope in the glucosinolate was in the sulfate group, but after periods of a day to several days, the average content of radioactivity in the divalent sulfur had risen to a third of the total. In Wetter's experiments, the incorporation of sulfate quickly reached a level of 20 percen sulfide and thiosulfate (samples labeled in each position) were utilized for the glucosinolate more slowly than sulfate and with essentially the same ratio of preferential distribution into sulfate ester over sulfide linkage and so may well have been oxidize to sulfate before entrance into organic combination. On the othe hand, the sulfur from methionine-^{35}S (Wetter) and cysteine-^{35}S (Kindl, 1965), which also supplied the element efficiently to the glucosinolates, clearly was not oxidized, and at least 90 percent of what was incorporated appeared in the thioglucosyl group. Kindl observed that benzylglucosinolate that had been biosynthesized from cysteine-^{35}S, but in which the sulfur distribution had not been checked explicitly, did not release the isotope for synthesis of the *p*-hydroxy derivative. The result presumably does not exclude sulfate transfer between glucosinolates. Altogether, the data show that the sources and paths of incorporation of the divalent and hexavalent sulfur are separable, but dis close little more. The introduction of the sulfate group may be the very last step of glucosinolate biosynthesis (Kindl, 1964), but the matter has not been tested with thiohydroximic acids as substrates. The divalent sulfur may come proximately, as a guess, from cysteine (Wetter), but whether it first becomes attached to the thiohydroximate carbon atom (Kindl, 1964) or to glucose (Meakin, 1967) is scarcely decided. Much remains to be understood about basic sulfur metabolism in higher plants, and we hope that the challenge of understanding the incorporation of sulfur into glucosinolates will stimulate further work on this important problem.

The biosynthetic studies conducted thus far within the group of glucosinolates have provided a sound idea of some principal pathways by which their unique molecular framework is assembled. Much remains to be done, however, before we can experience the pleasure of a profound and detailed understanding of the entire complex of reactions that take part.

C. Types of Side Chains

In the light of the biosynthetic evidence, we return to Table 2 in order to comment more fully on the compounds in it and our arrangement of them. More glucosinolate side chains are sure to be discovered, some perhaps of types wholly unconsidered now, but the available knowledge of natural glucosinolates is rich already, and some features of the structural associations that can be perceived seem likely to endure. The glucosinolate side chains can of course be classified by their functional groups, and for certain purposes it would still be convenient to list together all β-hydroxylated compounds, which yield oxazolidine-2-thiones on enzymatic hydrolysis, as has been done in the past. The arrangement adopted now is based upon and intended to reflect similarities of total structure, biogenetic inferences, and consistent patterns of occurrence of the compounds in the same plant or related plants, as shown by intensive, partly unpublished experience. The coexistence of two glucosinolates in a species or genus is by itself no evidence of proximate common origin, but when structural logic can be discerned in accumulated data of this sort, the intuition of a close biochemical relationship acquires weight. An understanding of why glucosinolates are produced in a plant from certain amino acids and not from others and of shifts from one precursor to another during growth would require knowledge of enzymatic specificities and of competition among metabolic pathways that we have not begun to possess.

The empirical approach feasible now can be illustrated as follows. When 2(*S*)-hydroxy-3-butenylglucosinolate (29) was discovered a couple of years ago, the protein amino acid mentioned speculatively as a precursor was lysine. The suggestion may be correct and is indeed plausible on structural grounds in a sufficiently restricted context. Conversion of homoarginine to a γ-hydroxy derivative has been demonstrated in Leguminosae (Bell, 1964). The crucifer seed yielding (29) also contains the desoxy derivative (27) and its lower homologue (20) (Daxenbichler et al., 1964), seemingly paralleling lysine and ornithine. However, the epimeric hydroxybutenylglucosinolate (28) occurs in *Brassica* not only with (27), (20), and two 4-pentenyl derivatives (32, 33) but also with a group (17, 18, 22, 23, 31) of saturated glucosinolates having the same carbon chains and terminal methylthio or methylsulfinyl groups (see Ettlinger and Thompson, 1962). The association of ω-alkenyl and ω-methylthioalkyl or ω-methylsulfinylalkyl glucosinolates holds also in *Alyssum* (Kjær and Gmelin, 1956) and *Iberis* (Gmelin, 1963). The distributional data combined with the biosynthetic results for allylglucosinolate (20) justify a general rule that an unsubstituted terminal double bond in a straight-chain glucosinolate most likely arises from elimination of methanethiol or an oxidized or alkylated equivalent.

The concept that the precursors of glucosinolates are amino acids with the same carbon framework implies the presence of a variety of unusual or unknown amino acids in appropriate, glucosinolate-containing plants (Benn, 1962). So far, this expectation has been verified only for 2-amino-5-methylthiovaleric acid (cf. Challenger and Charlton, 1947). The prediction that more nonprotein amino acids will be found corresponding to glucosinolates appears reasonable, with two qualifications. First, no one can tell in advance the extent to which these amino acids will accumulate as steady state intermediates in biosynthesis, and their concentrations may be extremely small. Secondly, functional alteration of the side chain can take place after an amino acid is launched on a track leading to glucosinolates—besides elimination, changes might include oxidation, reduction, and *O*-methylation, glycosylation, or acylation—and there may be only one immediate amino acid precursor of several glucosinolates. In what follows, we have cited various instances in which unusual amino acids resembling glucosinolates have been isolated from plants and microorganisms, simply as a means of illustrating the synthetic capabilities of living cells.

In Table 2, we have divided the natural glucosinolates into

four principal groups. Group I, the most heterogeneous, can be loosely styled the group in which the aglucones originate from acetic and pyruvic acids. Group II consists of biogenetic derivatives of methionine, all with lengthened chains. Group III includes the glucosinolates corresponding directly to phenylalanine, hydroxylated phenylalanines, and tryptophan. The little group IV shares a common origin in phenylalanine with the first member of group III and the chain-extending biosynthetic history with group II. We proceed to closer scrutiny of individual representatives in the sequence listed.

Methylglucosinolate (1) corresponds to alanine, but seems to be almost entirely restricted in occurrence to Capparidaceae. The rarity of ethylglucosinolate (2) contrasts with the ubiquitous function of α- ketobutyric acid as an intermediate in isoleucine biosynthesis. The unusual occurrence of L-2-aminobutyric acid in proteins, from seed of *Salvia officinalis* L. (Labiatae), has recently been reported (Brieskorn and Glasz, 1964, 1965, 1966). The isopropyl (3) and *sec*-butyl (7) glucosinolates, with their β-hydroxy derivatives (4) and (8), and the β-hydroxylated isobutyl (6) glucosinolate are widely distributed. The benzoates (5) and (9) have only been found with the corresponding hydroxy compounds (4) and (8) and may be naturally derived from them (Kjær and Christensen, 1962b). The 2(*S*)-butylglucosinolate (7) has the same absolute configuration at the tertiary carbon atom as L-isoleucine, which is the obvious biogenetic parent of (7), as valine is of the isopropyl compound (3). Hydroxylated amino acids of this series, equivalent to (4), (6), and (8), are, however, rare. An isolated occurrence in Crassulaceae of a γ-hydroxyvaline of unknown configuration has been recorded (Pollard et al., 1958); the fungal *Amanita* toxins contain γ-hydroxyleucine and also a γ-hydroxyisoleucine, oxygenated not in the methyl group like (8) but in the methylene group (see Pfaender and Wieland, 1966). Isobutylglucosinolate, corresponding to leucine, is curiously unknown. The *sec*-alkyl glucosinolates (3) and (7) sometimes occur with the hydroxylated leucine derivative (6), and it may be that the introduction of side-chain oxygen during biosynthesis is particularly easy with leucine. The only known cyanogenetic glycoside with the leucine skeleton, acacipetalin, has a double bond α, β to the cyano group (Rimington, 1935). A simpler consideration is that the analytical methods employed to date may not have sufficed to recognize the isobutyl glucosinolate, especially in mixture with its isomer (7). Chromatographic discrimination between the isobutyl and *sec*-butyl glucosinolates or thioureas

(Kjær and Rubinstein, 1953) has probably not been achieved, and of the thioureas, the *sec*-butyl derivative is the higher melting by over 30° C (Kjær and Rubinstein, 1953; Kjær and Friis, 1962). Another conceivable natural product in this series is the β-methallyl glucosinolate, whose chromatographic distinction from the 3-butenyl compound may not be obvious (cf. Kjær and Rubinstein, 1953). 2(*S*)-Methylbutylglucosinolate (10), which has been found in company with the lower alkyl glucosinolates (3) and (7) and its hydroxy derivative (11), presumably comes from 2-amino-4-methylhexanoic acid, the amino acid related to isoleucine as leucine is to valine. The sequence of enzymes converting valine (or α-ketoisovaleric acid) plus acetate to leucine might reasonably be expected to have a little activity for the homologization of isoleucine (see Loftfield and Eigner, 1966), and in this sense L-2-amino-4(*S*)-methylhexanoic acid, though never isolated from natural sources, is a logical candidate to occur widely in small amounts. 3-Methyl-3-butenylglucosinolate (12) appears to arise by homologization of an intermediate with a leucine skeleton which perhaps is concerned also in biosynthesis of the hydroxy compound (6).

The little subgroup I b of keto glucosinolates, which is highly restricted in occurrence, corresponds to amino acids with chain lengths of nine to ten carbon atoms. In the absence of biosynthetic information about these compounds or close relatives, we can only conjecture their origin on the basis of structure. The alternative presence of terminal propionyl or butyryl groups in (13-15) suggests that the side-chain oxygen is introduced after assembly of the paraffin framework. Similarly, one plant contains the homologous pair of (13) and (15), which do not differ by the simple insertion of a methylene unit next the site of the glucosinolate function. In this species the separation of the carbonyl and the other group is fixed, and divergence of chain length appears to have preceded oxygenation during biosynthesis. Like 2-aminooctanoic acid, isolated from a mold (*Aspergillus;* Staron et al., 1965), 2-aminononanoic and 2-aminodecanoic acids, containing the skeletons of (13-15) could result from successive homologizations of 2-aminobutyric acid. These compounds can also be regarded, however, as nitrogenous α-oxidation products (Hitchcock and James, 1966; Morris and Hall, 1967) of fatty acids. On such a hypothesis, the glucosinolates might originate either from octanoic or from decanoic acid (ultimately, of course, from acetic acid). The aminononanoic acid could arise by chain shortening from decanoic acid.

3-Methoxycarbonylpropylglucosinolate (16)(subgroup Ic) corresponds to α-aminoadipic acid, which is widespread in higher plants (Hatanaka and Virtanen, 1962), including Cruciferae (Larsen, 1965a). No uniform biosynthesis of the diacid in higher plants has been established, however, and (16) stands in no manifest close relation to other compounds in Table 2. The glucosinolate is known only from a few species of *Erysimum* (Cruciferae), other species of which contain glucosinolates like the 4-methylsulfonylbutyl derivative (24). Possibly the ester side chain results from oxidation of an appropriate glucosinolate in group II or equivalent precursor, but we arbitrarily treat (16) as a derivative of acetic acid, since it may arise by chain elongation, without gain and loss of sulfur, from intermediates in the tricarboxylic acid cycle.

In group II, the largest known division of glucosinolates, the only compound whose placement is doubtful is the rare 3-benzoyl oxypropylglucosinolate (21). The substance, which for purposes of classification is equivalent to 3-hydroxypropylglucosinolate, might be, and perhaps will be found to be, generated in more than one way. The corresponding L-2-amino-5-hydroxyvaleric acid, isolated from Leguminosae (Thompson et al., 1964), quite likely arises there from glutamic acid, and in that relation the glucosinolate (21) could be set beside the longer-chained (16). The notion, however, that sulfur was attached to the terminal carbon atom of the side chain of (21) at some stage of biosynthesis is also reasonable and fits, without being required by, the known simultaneous occurrence of (21) and methylsulfonylpropylglucosinolate (19). The origin of the 22 members of group II other than (21) from methionine can be told with reasonable confidence, and they make an extraordinary ensemble. The simple counterparts of methionine, 2-methylthioethylglucosinolate, vinylglucosinolate, and so forth, have never been found. All the compounds of group II result from chain elongation, apparently by one carbon atom at a time, repeated in up to eight cycles. The relatively small difference in the efficiencies with which methionine and 2-amino-5-methylthiovaleric acid are utilized for allylglucosinolate indicates that diversions from the reaction sequence are small and cumulative extension truly is feasible. The acme known to date, 10-methylsulfinyldecylglucosinolate (39), does seem less common than its lower homologues, and the sequence may not run to much higher carbon numbers. The fact that the series of ω-methylthioalkyl, ω-methylsulfonylalkyl, and ω-alkenyl glucosinolates stop before the sulfoxides may be due in part to accidents of discovery. The natural existence of sulfide glucosinolates higher than (34) or sulfones past (24) would startle no one, although whether such compounds are artifacts—that is, products of nonenzymatic oxidation or reduction of sulfur—or not might not always be obvious. The enzymatically formed sulfoxide group does bear an inherent character pointing to its origin, namely chirality. All the sulfoxide glucosinolates of which derivatives have been isolated belong to the same stereochemical series (Klyne et al., 1960; Christensen and Kjær 1963b), in which the absolute configuration about sulfur can be uniformly designated as *R* (Cheung et al., 1965). This configuration of the sulfoxide group is opposite to that (Hine, 1962) of the *S*-methyl-L-cysteine sulfoxide formed enzymatically from *S*-methylcysteine in crucifers (Arnold and Thompson, 1962; Doney and Thompson, 1966), including plants that contain sulfoxide glucosinolates. The stereochemistry of the enzymatic oxidation of methionine in Cruciferae (Splittstoesser and Mazelis, 1967) has not been determined, although the absolute configurations of the methionine sulfoxides (Christensen and Kjær, 1965) are known.

The higher sulfoxide glucosinolates from 6-methylsulfinylhexyl to 9-methylsulfinylnonyl (35-38) are much more common in Cruciferae than published data would indicate. 7-Methylsulfinylheptylglucosinolate (36) has never been reported, but we have unequivocal chromatographic evidence for its presence in various mustard plants, and an attempt to isolate the glucosinolate or derivatives is now under way. For example, the compound occurs in small amount, along with more of its higher homologue (37) and still more of the well-known 2-phenylethylglucosinolate (49), in watercress seed. These sulfoxides are of course entirely overlooked in surveys that cover only volatile isothiocyanates.

The glucosinolates in group III derive without chain extension from phenylalanine (40), tyrosince (41-43) and its *meta* isomer (44-45), 3,4-dihydroxyphenylalanine (46), and tryptophan (47-48). None correspond yet to amino diacids with a carboxyl *meta* to the side chain, which occur in Cruciferae and Resedaceae (Larsen, 1967). The *O*-methylation or glycosylation observed is rare in phenolic amino acids. *O*-Methyl-L-tyrosine is a constituent of the *Streptomyces* antibiotic puromycin (Baker et al., 1955). The frequency with which the *m*- oxygenated derivatives (44 and (45) occur is of the same order as that for their *para* isomers (41-42), and contrasts not only with the rarity of the *meta* isomer of tyrosine but also with the unimportance in plants of compounds like *m*-hydroxybenzoic and *m*-hydroxycinnamic acids. *m*-Hydroxybenzaldehyde cyanohydrin glucoside is known from Rutaceae (Finnemore and Cooper, 1936). *m*-Hydroxyphenylalanine has recently been isolated from the latex of *Euphorbia myrsinites* L. (Euphorbiaceae)(Mothes et al., 1964); the latex of *E. lathyris* L. contains well over 1 percent of 3,4-dihydroxyphenylalanine, biosynthesized from tyrosine (Liss, 1961). The fungal synthesis of gliotoxin from phenylalanine begins with *meta* hydroxylation (Winstead and Suhadolnik, 1960). One microbiological route to *m*-oxygenated compounds is the reduction of 3,4-dioxygenated analogues (Booth and Williams, 1963). However, the fungal conversion of phenylalanine to volucrisporin (2,5-di-*m*-hydroxyphenyl-1,4-benzoquinone) appears to proceed by direct *meta* hydroxylation of the amino (or keto) acid, without passage through the tyrosine—3,4-dihydroxyphenylalanine series (Read et al., 1962; Chandra et al., 1966). The indole glucosinolates (47) and (48), with which we lack personal acquaintance, might be set off as a subgroup from the benzyl derivatives. The structure of the compound designated as 1-methoxy-3-indolylmethylglucosinolate (48) is probable but not proved. *N*-Oxygenated indoles are unfamiliar to most chemists, but 1-methoxyindole-2-carboxylic acid was synthesized by Reissert as long ago as 1896. Two *Gelsemium* (Loganiaceae) alkaloids have recently been shown to be *N*-methoxyoxindoles (Przybylska, 1962; Wenkert et al., 1962).

The two 2-phenylethyl glucosinolates (49) and (50) that make up group IV and can occur together originate by chain extension from phenylalanine as already discussed. Analogy to group II suggests that 3-phenylpropyl glucosinolates may be uncovered in the future. The compounds of group IV seldom if ever occur with benzylglucosinolate or any of its substituted derivatives. The absence to date of any chain-extended phenolic glucosinolates is also highly noticeable.

D. Occurrence

The distribution of glucosinolates in the plant kingdom, their exclusive locus, can now best be surveyed primarily on a family level. The botanical classification into families is one that nonbotanists can pretty well take at face value, and from this manner of considering the chemical results, a useful and interesting generalization emerges. The previous list (Kjær, 1960) of glucosinolate composition by genus and species could of course be brought up to date, but a more detailed systematic commentary on such a list would at present be largely anecdotal. Not only are the data yet sparse, but in a family like Cruciferae, where generic limits often are highly uncertain (cf. Rollins, 1959), growth in understanding of the chemical makeup and of the taxonomy of the plants may rightly go hand-in-hand. We are confident that insight will reward the active worker fitted with broad knowledge and clear, refined judgment. Our goal here is a small one.

The distribution of glucosinolates can hardly be considered without thought of the enzymes that act on them. The known modes of enzymatic attack on glucosinolates are two: removal of sulfate and cleavage of the glucosyl thioether link. The only animal enzyme found to affect glucosinolates is one of the molluscan sulfatases (Nagashima and Uchiyama, 1959; Takahashi, 1960). On the other hand, the claim that mustard seeds contain a sulfatase active against glucosinolates is dubious (cf. Calderon et al., 1966; Tsuruo et al., 1967). The typical action of a plant enzyme on glucosinolates, liberating isothiocyanates (equation 1, p. 89), is that of a thioglucoside hydrolase (Ettlinger et al., 1961). The presumable thioglucosidases do not show major specificity toward the glucosinolate side chains. Enzymes able to cleave glucosinolates in this fashion are produced by bacteria, in particular some that inhabit mammalian intestinal tracts (Oginsky et al., 1965), and by fungi (*Aspergillus;* Reese et al., 1958). These microorganisms are not known ordinarily to encounter such substrates. The possible existence of glucosinolates and corresponding enzymes in mushrooms will be discussed later. In higher plants, the occurrence of glucosinolates and that of thioglucosidases, termed myrosinases, capable of hydrolyzing them efficiently are very closely correlated. Regular β-glucosidases do not affect glucosinolates. From time to time, the presence of myrosinases without glucosinolates has been reported in common plants like beans, carrots, and onions (Bokorny, 1900) or wheat and clover (Kutáček, 1964), but these claims do not appear substantial. The old suggestions of myrosinase in Flacourtiaceae and Violaceae will be taken up later. It appears very near the truth to say that the sets of flowering plants containing a myrosinase and containing glucosinolates are identical. The rule may have exceptions—a plant organ may bear one kind of constituent and not the other—and it is well to remember that in one cyanogenetic plant, white clover (*Trifolium repens* L., Leguminosae), individuals vary with regard to the presence of readily detectable amounts of glucosides and, independently, of hydrolytic enzyme, and these variations are governed by separately inherited factors (Corkill, 1942-3; Atwood and Sullivan, 1943). The glucosides in clover, linamarin and lotaustralin, are derivatives of tertiary alcohols and are only slowly cleaved by the ordinary β-glucosidase (cf. Butler et al., 1965). No similar variation has yet been observed in isothiocyanate-producing plants, although quantitative differences in glucosinolate level are under study (see Appelqvist, 1962). A *Brassica* oilseed void of glucosinolates would have massive commercial value.

Observations that we accept as certain of glucosinolates in plants relate to eleven families of dicotyledonous angiosperms: Capparidaceae, Cruciferae, Resedaceae, Tovariaceae, Moringaceae; Limnanthaceae, Tropaeolaceae; Caricaceae; Euphorbiaceae; Gyrostemonaceae; and Salvadoraceae. One large family, Cruciferae, contains glucosinolates throughout. In the other large family listed, Euphorbiaceae, glucosinolates are known from only two species and appear to be absent, along with myrosinase, from many more. Capparidaceae, a medium-sized family, has glucosinolates consistently in major subfamilies. All the other families are small, numbering from one to half a dozen in genera.

The first five families listed, Capparidaceae through Moringaceae, are generally agreed to form a phyletic unit (see Kjær, 1963a, for summary and references), conveniently assembled in volume 17b of the second edition of "Die Natürlichen Pflanzenfamilien." *Koeberlinia*, which there is included in Capparidaceae but is frequently placed in a family Koeberliniaceae of variously assigned relationship, seems from our observations not to contain myrosinase or glucosinolates. The Papaveraceae (including Fumariaceae), which are linked in the "Pflanzenfamilien" with Capparidaceae and Cruciferae, differ chemically by the absence of myrosinase and glucosinolates and presence of the well-known characteristic alkaloids. Ordinal separation is now backed by botanical, including morphological, evidence (Merxmüller and Leins, 1967). The monotypic family Bretschneideraceae (*Bretschneidera sinensis* Hemsl.) is treated at the end of the capparidaceous alliance in the cited volume of the "Pflanzenfamilien," but it has also been put in or near Sapindaceae. Radlkofer (1908) observed cells in *B. sinensis* that looked to him just like the myrosinase-containing cells of *Moringa*, but no chemical tests of the plant have been reported.

Of the families that we have listed as glucosinolate-bearing after Moringaceae, two, Limnanthaceae and Tropaeolaceae, are usually taken to be related and are placed together in the order Geraniales. The other families are more scattered taxonomically: we do not mean that they constitute a random sample of dicotyledons, but simply that the divisions of the list separated by semicolons fall into different orders in any current scheme of classification. The Euphorbiaceae are commonly spoken of as polyphyletic. One suggestion of isothiocyanates in Euphorbiaceae, the bare remark that the essential oil, smelling like onions from the latex of *Jatropha multifida* L. possessed "a sulphurous component, probably benzyl-mustard oil" (Freise, 1935b), has no claim to recognition. Seed of *J. multifida*, according to our experiments, contained neither myrosinase nor glucosinolates. However, within the family glucosinolates and myrosinase do occur in *Putranjiva roxburghii* Wall. and in *Drypetes gossweileri* S. Moore (*D. armoracia* Pax and K. Hoffm.; see Brenan, 1952). Besides *D. gossweileri*, one or two other West African species of *Drypetes*, including *D. pellegrinii* Léandri, have a pungent bark (Cooper and Record, 1931; Léandri, 1934). Hurusawa (1954) proposed to merge *Putranjiva* as a subgenus into *Drypetes*, but in any event there is as yet no evidence that glucosinolates are usual in the more than 150 species of the latter genus. Our own experience indicates that glucosinolates and myrosinase may be confined in Euphorbiaceae to a small minority of genera. The occurrence of glucosinolates in Gyrostemonaceae, a group of Australian endemics placed next to, if not included in, Phytolaccaceae, also calls for brief systematic comment. The presence of betanin-type pigments, major chemical markers of Phytolaccaceae and other Centrospermae, has not yet been established in Gyrostemonaceae (see Mabry, 1966). There is no doubt of the existence (cf. Peckolt, 1895) of volatile sulfur compounds in several genera (*Agdestis*, *Gallesia*, *Petiveria*, *Seguieria*) of Phytolaccaceae, or of Agdestidaceae and Petiveriaceae if all the tribes of Phytolaccaceae in the old sense are raised to family rank, but the nature of these compounds appears unknown. Pietschmann (1924) treated distillates from root and stem of *Petiveria alliacea* L. with ammonia and phenylhydrazine and observed microscopically slight crystalline deposits that he took to be thioureas, but this attempted evidence is meaningless. Our own tests of *P. alliacea* seed, which has a powerful, garlicky taste, for myrosinase and glucosinolates (cf. Kjær et al., 1953) have given negative results. Further chemical investigation of the sulfur compounds of Phytolaccaceae would, however, certainly possess great interest.

TABLE 3

Familial Distribution of Grouped Glucosinolates (Table 2)

Capparidaceae	Ia (1, 3, 6, 7, 11, 12) Ib, III (47, 48)	Kjær and Thomsen (1963a); Schraudolf (1965)
Cruciferae	Ia [especially (3, 4, 6, 7)] Ic, II, III, IV	Kjær (1960)
Resedaceae	III (47), IV	Bertram and Walbaum (1894); Kjær and Gmelin (1958); Schraudolf (1965)
Tovariaceae	Ia (3), III (47, 48)	Schraudolf (1965); Kjær (1967)
Moringaceae	III (40, 43)	Kurup and Rao (1954); Badgett (1964)
Limnanthaceae	III (40, 44, 45)	Miller et al. (1964)
Tropaeolaceae	Ia (3, 6, 7), III (40, 41, 42, 44)	Kjær et al. (1953)
Caricaceae	III (40)	Ettlinger and Hodgkins (1956)
Euphorbiaceae	Ia (3, 4, 7, 10, 11), III (41)	Kjær and Friis (1962)
Gyrostemonaceae	Ia (7), III (40)	Bottomley and White (1950)
Salvadoraceae	III (40)	Patel et al. (1926)

The known occurrence of the several types of glucosinolate side chains is set forth in the family-wide summations collected in Table 3. In addition to the results in the major references given, the table incorporates unpublished data of our own. The work cited in Table 2 on the most recently discovered individual glucosinolates has also been taken into account. For all the families except Capparidaceae and Cruciferae, the listed data come from study of one to three species apiece. The statements about the first families in Table 3 merit brief comment. Methylglucosinolate (1) in Capparidaceae is the closest we have to a family-specific glucosinolate both in Capparidoideae and Cleomoideae, but is very rare outside Capparidaceae. On the other hand, the family peculiarly lacks benzylglucosinolate and its oxygenated derivatives; the compound occurs in the majority of species investigated. One general and important question which cannot now be answered is whether the broadly based chain extension by one carbon atom, utilizing acetate, that distinguishes glucosinolate production in Cruciferae and Resedaceae takes place also in Capparidaceae. The biogenesis of the methylpentyl skeletons of compounds (11) and (12) evidently is closely related to the formation of leucine from valine (see previous discussion in section IV C), and the pathway leading to the glucosinolates in group Ib, found so far only in two species of *Capparis*, is unknown. The possibility that glucosinolates of types II and IV may yet be encountered in Capparidaceae remains open. The family Cruciferae is characterized by its varied wealth of glucosinolates and in particular by the frequent presence of methionine-derived, extended-chain compounds of group II. Allyl isothiocyanate, which chemists think of as the typical natural mustard oil, is restricted, as far as we know, to this one family. The Resedaceae contain other glucosinolates besides the three listed, but their nature has not yet been determined.

From studies on the occurrence of the indole glucosinolates (47) and (48) in seedlings, Schraudolf (1967) concluded that these compounds are characteristic of Capparidaceae (namely *Cleome*) and its relatives Cruciferae, Resedaceae, and Tovariaceae. The compounds were absent from *Moringa oleifera* Lam. (Moringaceae) and from *Tropaeolum majus*. Whatever change in biochemistry of growth accompanies the indole glucosinolates is unknown; their distribution in the families of Table 3 has been only partly determined. The power of converting tryptophan to glucosinolates tentatively marks the group of the first four families listed.

A new generation that plainly results from the data in Table 3 is that in every family but Capparidaceae, Cruciferae, and Resedaceae the known glucosinolates all belong to the comparatively narrow structural range of groups I a and III and consist of plain or oxygenated immediate derivatives of valine, leucine, isoleucine, phenylalanine, tyrosine, or tryptophan. Group II stands restricted to Cruciferae and group IV to Cruciferae and Resedaceae. The homologization that leads in Euphorbiaceae to the 3-methylpentyl skeleton of (10) and (11) can reasonably be construed as an incidental accompaniment of the biosynthesis of leucine. From a broad distributional standpoint, the isopropyl (3) or 2-butyl (7) and the benzyl (40) compounds well illustrate glucosinolates. Another way to put the matter is that outside Capparidaceae—Cruciferae—Resedaceae the side chains of glucosinolates are equivalent in principle to those of the standard cyanohydrin glycosides. Departures from the usual pattern of direct biosynthesis from protein amino acids can occur for both classes of natural products. The cyanohydrin glucoside gynocardin, for example, was recognized as a biochemical maverick (Rosenthaler, 1923) long before its structure was determined (Coburn and Long, 1966). What the biogenetic relationship is between gynocardin and the likewise cyclopentenoid fatty acids with which its distribution in Flacourtiaceae is closely correlated (see Hegnauer, 1966) remains to be determined; meanwhile, the status of gynocardin may be slightly reminiscent of that of the glucosinolates in group I b. Further discoveries are certain to enlarge our knowledge but probably will not obscure the regularities now discernible.

The parallel in formation and occurrence between the glucosinolates and the cyanohydrin glycosides is exceedingly instructive when taken in a broad sense. Both kinds of compounds can be said, once the Cruciferae recede into perspective, to result ordinarily from simple transformations of the same protein amino acids. The cyanohydrin glycosides are widely distributed; the glucosinolates, though not found throughout higher plants, need not be confined to a unique group. Curiosity about many aspects of detailed relationships between these types of natural products, however, must be left unsatisfied for now. Cyanogenesis is frequent in Cruciferae (Honeyman, 1956), but as is so often the case, the source of the cyanide is not known. Familiar cyanohydrin glucosides occur in some genera of Euphorbiaceae (*Hevea*, *Manihot*, *Phyllanthus*), but the genetic relationships between these and the glucosinolate-producing genera are indefinite. Likewise, the usually assumed closeness between the

Caricaceae and the cyanogenetic Passifloraceae has yet to be made deeply significant.

Glucosinolates may occur outside the families listed in Table 3 and have indeed been reported to do so on the grounds of apparently solid evidence that we have felt bound to question. Sometimes, though, so flimsy an indication as a common name of a plant is a clue to glucosinolates worth being followed up. The attempt to form a comprehensive, open-minded, and critical opinion of the chemical, botanical, and pharmaceutical literature bearing on possible scattered occurrences of glucosinolates furnishes an occupation akin to wrestling with alligators. Nevertheless, we shall briefly enter on this region (which we have come to know) in order to guide readers among the existing uncertainties and point out opportunities for useful work. It will be understood that the possibilities of discovering glucosinolates in new sources are not to be bounded by speculation or precedent.

The pungent taste and odor of isothiocyanates can render the presence of glucosinolates in plants conspicuous. A variety of factors, however, confuses the sensory detection. Isothiocyanates often go unnoticed during botanical description or chemical analysis, partly because nonvolatile ones do not stand out, partly because the compounds are not formed in dry material or material in which the myrosinase is no longer active and are, once liberated, lost on storage. Furthermore, although the aglucones produced by enzymatic hydrolysis of glucosinolates decompose spontaneously in neutral solution to isothiocyanates and sulfate with half-lives of about a minute near room temperature, the reaction sequence can be diverted to other organic products (scheme 2; Miller, 1965; Ettlinger and Miller, 1966). In acidic medium or, very likely, by the action of certain

$$R{-}\underset{\underset{\large N-OSO_3^-}{\|}}{C}{-}S{-}Gl \xrightarrow{\text{myrosinase}}$$

$$R{-}\underset{\underset{\large N-OSO_3^-}{\|}}{C}{-}S^- \xrightarrow{-SO_4^=} \begin{cases} \xrightarrow{\text{enzyme}} R{-}S{-}C{\equiv}N \\ \longrightarrow R{-}N{=}C{=}S \\ \xrightarrow[\text{(or } Fe^{++} \text{ or enzyme)}]{H^+} R{-}C{\equiv}N + S \end{cases} \qquad (2)$$

other catalysts, the aglucones give nitriles, which are not easily picked out by the senses. One occurrence of benzyl cyanide (Sabetay et al., 1938), in flowers of *Leptactinia senegambica* Hook. f. (Rubiaceae), is not, so far as is known, connected with benzylglucosinolate. Special enzymes can cause a unique rearrangement of glucosinolate aglucones to thiocyanates, which impart to plants a taste and odor of garlic or, probably as a result of cleavage to thiols, a fetid character. Natural thiocyanates have been observed only as hydrolysis products of glucosinolates.

Examples of a garlic smell or other, unpleasant smells of often unidentified sulfur compounds in plants are manifold and will not be reviewed here. The characteristic sulfur compounds of *Allium* (Liliaceae) come from *S*-substituted cysteine sulfoxides; we know no basis for the statement (Clappaz and Esculier, 1963) that the odor by which *A. ursinum* L. can be recognized is due to allyl isothiocyanate. The old report of divinyl sulfide in *A. ursinum* is probably in error also (Ruigh and Erickson, 1939), but the plant contains *S*-allyl cysteine sulfoxide (Stoll and Seebeck, 1948). A sulfur-bearing constituent of the seed of an Indian *Entada*, possibly *E. pursaetha* DC. [Leguminosae-Mimosoideae; on the name *E. scandens* (L.) Benth., cf. Brenan, 1955], has been claimed to be a glycoside (Rangaswami and Rao, 1954), though not to be a glucosinolate. When aqueous or alcoholic extracts of the defatted seed were warmed with aqueous acid, an offensive, volatile compound with reactive sulfur was evolved. Treatment of the extracts with emulsin or myrosinase had no such effect. The precursor of the unidentified volatile substance was not isolated, and the lability of the precursor in acid seems to have been the only experimental reason for the supposition that it was a glycoside. That other possibilities exist is illustrated by the cleavage of *S*-methylcysteine sulfoxide in dilute acid with formation of dimethyl disulfide (Ostermayer and Tarbell, 1960). The Mimosoideae are known (Kjær, 1963a) to hold interesting cysteine derivatives.

Pungency or irritant properties in plants may be caused by a large variety of compounds other than glucosinolates. Vernacular names that refer to horseradish usually do indicate the production of some volatile isothiocyanate. Cress-type names may have the same meaning, but Para cress (*Spilanthes oleracea* Jacq., Compositae) owes its pungency to a *N*-isobutyldecatrienamide, spilanthol (affinin). In the old literature on materia medica, isothiocyanate-containing preparations appear to have been cited loosely to give a standard of likeness for smells with penetrating, aromatic qualities. An enigmatic instance concerns *Centaurium umbellatum* Gilib. [*C. minus* Moench, *Erythraea centaurium* (L.) Pers.; Gentianaceae]. Distillates of the plant (flowers and green parts) were recognized as pungent in the early days of plant analysis (Hermbstaedt, 1796) and compared to crucifer distillates (Deyeux, 1805; Méhu, 1862), but the agent seems never to have been identified.

The following survey takes up fungi and then dicotyledons, arranged alphabetically by families.

Basidiomycetes. Freise (1935a) claimed that evaporation of aqueous extracts of Brazilian mushrooms furnished crude glucosides, which on hydrolysis by enzymes prepared from the organisms gave very high yields (30 to 50 percent by weight) of essential oils. With no detail whatever, he stated that the oils consisted largely (75 to 90 percent) of isothiocyanates, some oils of the benzyl, others of the 2-phenylethyl derivative. The article appears to be a flight of fancy, but it is conceivable that higher fungi contain glucosinolates. The agaric described as *Lepiota globularis* Quél. (cf. Quélet, 1964) had the odor of cress. *Pholiota heteroclita* (Fr.) Quél. is agreed to smell like horseradish (Fries 1874, Lange, 1938). Mushrooms with a radish smell or flavor are common (Herrmann, 1918a,b; Gilbert, 1932), and though most of them probably owe their properties to traces of methanethiol or to a simple methyl sulfide, some may produce isothiocyanates (cf. Quélet, 1886).

Akaniaceae. The only species, the Australian *Akania lucens* (F. Muell.) Airy-Shaw (*A. hillii* Hook. f.), is called turnipwood or horseradish tree (Harms, 1940).

Annonaceae. The leaves of the Indian tree *Sageraea laurifolia* (Grah.) Blatter (*Bocagea dalzellii* Hook. f. and Thoms.; cf. Fries, 1959) have been described as follows (Dymock et al.,1890) "The aqueous extract contains a ferment which produces a pungent alliaceous odour.... It is precipitated from its aqueous solution by alcohol.... The distillate is oily, with a pungent odour and taste.... The leaves also contain a crystalline body extracted by boiling alcohol from the marc left by ether and cold alcohol exhaustion; it is probably the body related to sinigrin of mustard seed...."

Aquifoliaceae. According to a summary of a thesis (Smith, 1887), which has been noticed in subsequent literature, a petroleum ether extract of leaves of American holly, *Ilex opaca* Ait., contained a volatile fraction with "an acrid mustard-like odor." Our own experiments with the shrub gave no evidence of glucosinolates or myrosinase.

Bixaceae. Hegnauer (1964), from the irritant properties of leaves and unripe fruits of *Bixa orellana* L., has conjectured the possible presence of glucosinolates.

Flacourtiaceae. Enzyme preparations that hydrolyzed allylglucosinolate to the isothiocyanate have been reported from seeds of *Hydnocarpus kurzii* (King) Warb. (*Taraktogenos kurzii* King) (Power and Gornall, 1904) and *H. wightiana* Blume (Power and Barrowcliff, 1905). Crude enzymes from *H. anthelmintica* Pierre (Power and Barrowcliff, 1905) and *Gynocardia odorata* R. Br. (Power and Lees, 1905) had no myrosinase activity. All the seeds except those of *H. wightiana* were cyanogenetic, and none showed obvious signs of glucosinolates. Further study would be desirable.

Meliaceae. The fresh leaves of *Azadirachta indica* A. Juss. (*Melia azadirachta* L.) on steam distillation have been said to give a condensate containing "a distinct allyl or onion-smelling compound" (Hooper, 1903–4; the quotation in Watt, 1908, is more accessible). A news report ([Hooper,] 1903) presumably based on the same experiments carries the statement about "an allyl compound" with the addition that the "leaves crushed with water also give evidence of a mustard-like oil." The possible construction of "allyl compound" may be clarified by reference to Hooper's remark (1913) about papaya seeds that "they have a pungent mustard-like odour, and yield an allyl compound when distilled with water"; the substance in this case is benzyl isothiocyanate. Unquestionably *A. indica* and also species of *Cedrela* and *Dysoxylum* in Meliaceae possess characteristic sulfur compounds (on *A. indica* seed, cf. Watson et al., 1923), but what the substances are is unknown.

Plantaginaceae. Procházka (1959) reported an observation of 4-methylsulfinyl-3-butenyl isothiocyanate, corresponding to the glucosinolate (26), in plantain juice, presumably from an unidentified species of *Plantago*. The presence of a glucosinolate, especially one of group II, in a family systematically remote from any previously known to contain these glucosides would be notable. However, we have not been able to detect myrosinase or any glucosinolates in collections of *Plantago*, and there is no evidence of mustard oils in the family to be drawn from other literature. Since a single, chromatographically examined extract may include an apparently genuine constituent that came by mistake from another than the supposed source, we feel that the claim of isothiocyanates in Plantaginaceae needs to be verified.

Violaceae. Spatzier's report (1893) of myrosinase in *Viola* (including *V. odorata* L.) seed has been quoted without mention of the fact that it was contradicted by an outstandingly competent worker immediately afterwards (Guignard, 1893). Our own tests of *Viola* for myrosinase agree with Guignard's.

Note. The treatment in sections IV B and IV D has been reinforced, after its completion by discoveries that aldoximes (RR′ CH-CH=NOH) appear to be general biosynthetic intermediates both between amino acids (XXX) and cyanohydrin glucosides (XXXI; Tapper et al., 1967; Tapper and Butler, 1967) and between amino acids and glucosinolates (XXIX; Tapper and Butler, 1967; Underhill, 1967). Community between the biosynthesis of XXIX and XXXI hence truly cuts deep. The immediate precursors of aldoximes formed from amino acids (Underhill, 1967) may be α-nitroso acids (aqueous decarboxylation of α-oximino acids gives nitriles; Ahmed and Spenser, 1961), and the oximes may retain α-hydrogen of the amino acids: experiment can decide.

The intermediacy of α-nitro acids on the way to XXIX now seems less likely, but the plain nitro compounds—at least their *aci* tautomers (RR′ CH-CH=N^+ (OH)-O^-), which are the oxime *N*-oxides—might well be involved in biosynthesis of XXIX or XXXI. Use of O-labeled aldoximes may clarify the natural sequences. 1-Nitro-2-phenylethane (XXXV) in alkali gives benzaldehyde and cyanide (Gottlieb et al., 1962); a transformation of *aci*-nitro compound to α-hydroxy aldoxime (RR′ C(OH)-CH=NOH) can be written by prototropic shift of the double bond followed by migration (or loss and gain) of hydroxyl (the shift should also lead to a α,β-unsaturated oximes and so to β-hydroxylated products). The nitrile group of XXXI could be generated from an aldoxime-*O*-sulfonate.

ACKNOWLEDGMENT

We sincerely thank the Research Corporation for a grant that enabled this paper to be written.

REFERENCES

Achmatowicz, O., and Z. Bellen. 1962. Tetrahedron Lett., 1121.
_____, and J. T. Wróbel. 1964. Tetrahedron Lett., 129.
_____, H. Banaszek, G. Spiteller, and J. T. Wróbel. 1964. Tetrahedron Lett., 927.
Ahmad, A., and I. D. Spenser. 1961. Canad. J. Chem., 39:1340.
Alston, R. E., and J. Simmons. 1962. Nature (London), 195: 825.
_____, H. Rösler, K. Naifeh, and T. J. Mabry. 1965. Proc. Nat. Acad. Sci. U. S. A., 54: 1458.
Appelqvist, L.-Å.1962. Acta Chem. Scand., 16: 1284.

Arnold, W. N., and J. F. Thompson. 1962. Biochim. Biophys. Acta, 57: 604.
Atkinson, R. E., and R. F. Curtis. 1964. *In* Atkinson, R. E., R. F. Curtis, and G. T. Phillips. Chem. Industr. (London), 2101.
_____, and R. F. Curtis. 1965. Tetrahedron Lett., 297.
_____, R. F. Curtis, and G. T. Phillips. 1964. Tetrahedron Lett., 3159.
_____, R. F. Curtis, and G. T. Phillips. 1965. J. Chem. Soc., 7109.
_____, R. F. Curtis, and G. T. Phillips. 1966. J. Chem. Soc. C, 1101.
Atwood, S. S., and J. T. Sullivan. 1943. J. Hered., 34: 311.
Bachelard, H. S., M. T. McQuillan, and V. M. Trikojus. 1963. Aust. J. Biol. Sci., 16: 177.
Badgett, B. L. 1964. Ph.D. Thesis, Rice University, Houston, Texas. Dissertation Abstr., 25: 1556.
Baker, B. R., J. P. Joseph, and J. H. Williams. 1955. J. Amer. Chem. Soc., 77: 1.
Balenović, K., A. Deljac, I. Monković, and Z. Štefanac. 1966. Tetrahedron, 22: 2139.
Barclay, A. S., H. S. Gentry, and Q. Jones. 1962. Econ. Bot., 16: 95.
Barger, G., and F. P. Coyne. 1928. Biochem. J., 22: 1417.
Barothy, J., and H. Neukom. 1965. Chem. Industr. (London), 308.
Bell, E. A. 1964. Biochem. J., 91: 358.
Benkert, K. 1966. Naturwissenschaften, 53: 200.
Benn, M. H. 1962. Chem. Industr. (London), 1907.
_____, and D. Meakin. 1965. Canad. J. Chem., 43: 1874.
Bertram, J., and H. Walbaum. 1894. J. Prakt. Chem., [2] 50: 555.
Birnbaum, G. I. 1965. Tetrahedron Lett., 4149.
Bohlmann, F. 1966a. Fortschr. Chem. Forsch., 6: 65.
_____. 1966b. Symposium on Naturally Occurring Sulfur Compounds, Hellebæk, Denmark, June 23.
_____, and C. Arndt. 1966. Chem. Ber., 99: 135.
_____, and E. Berger. 1965. Chem. Ber., 98: 883.
_____, and E. Bresinsky. 1967. Chem. Ber., 100: 107.
_____, and P. Herbst. 1962. Chem. Ber., 95: 2945.
_____, and U. Hinz. 1965. Chem. Ber., 98: 876.
_____, and R. Jente. 1966. Chem. Ber., 99: 995.
_____, and K.-M. Kleine. 1963. Chem. Ber., 96: 1229.

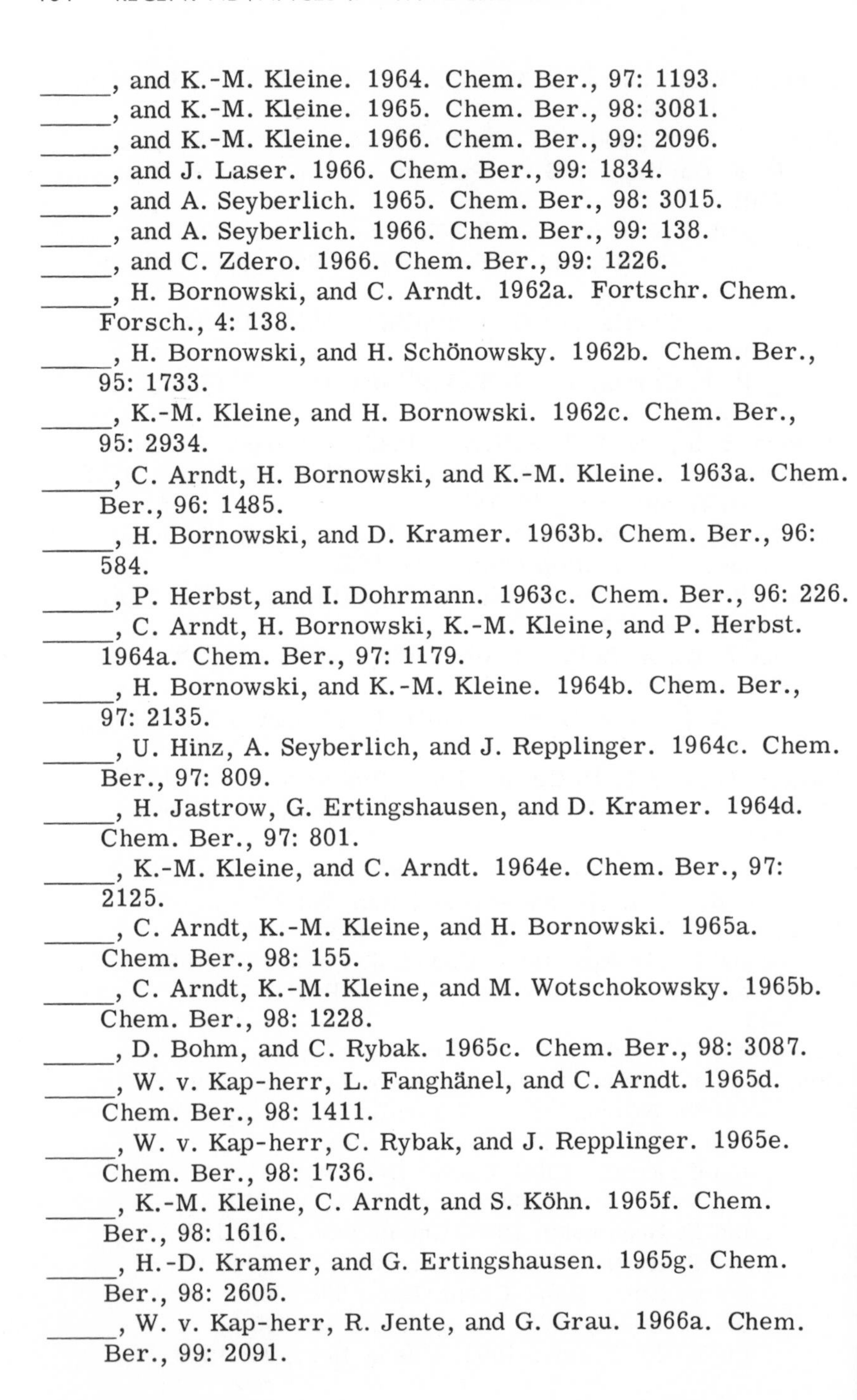

_____, and K.-M. Kleine. 1964. Chem. Ber., 97: 1193.
_____, and K.-M. Kleine. 1965. Chem. Ber., 98: 3081.
_____, and K.-M. Kleine. 1966. Chem. Ber., 99: 2096.
_____, and J. Laser. 1966. Chem. Ber., 99: 1834.
_____, and A. Seyberlich. 1965. Chem. Ber., 98: 3015.
_____, and A. Seyberlich. 1966. Chem. Ber., 99: 138.
_____, and C. Zdero. 1966. Chem. Ber., 99: 1226.
_____, H. Bornowski, and C. Arndt. 1962a. Fortschr. Chem. Forsch., 4: 138.
_____, H. Bornowski, and H. Schönowsky. 1962b. Chem. Ber., 95: 1733.
_____, K.-M. Kleine, and H. Bornowski. 1962c. Chem. Ber., 95: 2934.
_____, C. Arndt, H. Bornowski, and K.-M. Kleine. 1963a. Chem. Ber., 96: 1485.
_____, H. Bornowski, and D. Kramer. 1963b. Chem. Ber., 96: 584.
_____, P. Herbst, and I. Dohrmann. 1963c. Chem. Ber., 96: 226.
_____, C. Arndt, H. Bornowski, K.-M. Kleine, and P. Herbst. 1964a. Chem. Ber., 97: 1179.
_____, H. Bornowski, and K.-M. Kleine. 1964b. Chem. Ber., 97: 2135.
_____, U. Hinz, A. Seyberlich, and J. Repplinger. 1964c. Chem. Ber., 97: 809.
_____, H. Jastrow, G. Ertingshausen, and D. Kramer. 1964d. Chem. Ber., 97: 801.
_____, K.-M. Kleine, and C. Arndt. 1964e. Chem. Ber., 97: 2125.
_____, C. Arndt, K.-M. Kleine, and H. Bornowski. 1965a. Chem. Ber., 98: 155.
_____, C. Arndt, K.-M. Kleine, and M. Wotschokowsky. 1965b. Chem. Ber., 98: 1228.
_____, D. Bohm, and C. Rybak. 1965c. Chem. Ber., 98: 3087.
_____, W. v. Kap-herr, L. Fanghänel, and C. Arndt. 1965d. Chem. Ber., 98: 1411.
_____, W. v. Kap-herr, C. Rybak, and J. Repplinger. 1965e. Chem. Ber., 98: 1736.
_____, K.-M. Kleine, C. Arndt, and S. Köhn. 1965f. Chem. Ber., 98: 1616.
_____, H.-D. Kramer, and G. Ertingshausen. 1965g. Chem. Ber., 98: 2605.
_____, W. v. Kap-herr, R. Jente, and G. Grau. 1966a. Chem. Ber., 99: 2091.

_____, K.-M. Kleine, and C. Arndt. 1966b. Chem. Ber., 99: 1642.
_____, K.-M. Kleine, and C. Arndt. 1966c. Liebigs Ann. Chem., 694: 149
_____, M. Wotschokowsky, U. Hinz, and W. Lucas. 1966d. Chem. Ber., 99: 984.
_____, K.-M. Rode, and C. Zdero. 1967. Chem. Ber., 100: 537.
Bokorny, T. 1900. Chemiker-Ztg., 24: 771.
Booth, A. N., and R. T. Williams. 1963. Nature (London), 198: 684.
Bottomley, W., and D. E. White. 1950. Roy. Aust. Chem. Inst. J. and Proc., 17: 31.
Brenan, J. P. M. 1952. Kew Bull., 441.
_____. 1955. Kew Bull., 161.
Brieskorn, C. H., and J. Glasz. 1964. Naturwissenschaften, 51: 216.
_____, and J. Glasz. 1965. Pharmazie (Berlin), 20: 382.
_____, and J. Glasz. 1966. Arch. Pharm. (Weinheim), 299: 67.
Bu'Lock, J. D. 1964. Progr. Org. Chem., 6: 86.
_____. 1966. *In* Swain, T., ed., Comparative Phytochemistry, chap. 5, London, Academic Press.
_____, and G. N. Smith. 1967. J. Chem. Soc. C, 332.
Butler, G. W., and B. G. Butler. 1960. Nature (London), 187: 780.
_____, and E. E. Conn. 1964. J. Biol. Chem., 239: 1674.
_____, and L. Shen. 1963. Biochim. Biophys. Acta, 71: 456.
_____, R. W. Bailey, and L. D. Kennedy. 1965. Phytochemistry, 4: 369.
Calderon, P., C. S. Pederson, and L. R. Mattick. 1966. J. Agr. Food Chem., 14: 665.
Calvo, J. M., C. M. Stevens, M. G. Kalyanpur, and H. E. Umbarger. 1964. Biochemistry (Wash.), 3: 2024.
Canovas, J. L., M. Ruiz-Amil, and M. Losada. 1965. Arch. Mikrobiol., 50: 164.
Challenger, F. 1959. Aspects of the Organic Chemistry of Sulphur, London, Butterworth.
_____, and P. T. Charlton. 1947. J. Chem. Soc., 424.
_____, and J. L. Holmes. 1953. J. Chem. Soc., 1837.
Chandra, P., G. Read, and L. C. Vining. 1966. Canad. J. Biochem., 44: 403.
Cheung, K. K., A. Kjær, and G. A. Sim. 1965. Chem. Commun., 100.
Chisholm, M. D., and L. R. Wetter. 1964. Canad. J. Biochem., 42: 1033.

_____, and L. R. Wetter. 1966. Canad. J. Biochem., 44: 1625.
Christensen, B. W., and A. Kjær. 1963a. Acta Chem. Scand., 17: 279.
_____, and A. Kjær. 1963b. Acta Chem. Scand., 17: 846.
_____, and A. Kjær. 1965. Chem. Commun., 225.
Clapp, R. C., L. Long, Jr., G. P. Dateo, F. H. Bissett, and T. Hasselstrom. 1959. J. Amer. Chem. Soc., 81: 6278.
Clappaz, J.-P., and R. Esculier. 1963. Bull. Mens. Soc. Linnéenne Lyon, 32: 229.
Coburn, R. A., and L. Long, Jr. 1966. J. Org. Chem., 31: 4312.
Cooks, R. G., F. L. Warren, and D. H. Williams. 1967. J. Chem. Soc. C, 286.
Cooper, G. P., and S. F. Record. 1931. Bull. Yale Univ. School Forestry, no. 31, pp. 51-2.
Copenhaver, J. W. 1957. U. S. Pat. 2786865. Chem. Abstr., 51: 13920.
Corkill, L. 1942-3. New Zeal. J. Sci. Technol., 23B: 178.
Dannenberg, H., H. Stickl, and F. Wenzel. 1956. Z. Physiol. Chem., 303: 248.
Das, B. A., P. A. Kurup, and P. L. N. Rao. 1957. Indian J. Med. Res., 45: 191.
Dateo, G. P., Jr. 1961. Ph. D. Thesis, Rice University, Houston, Texas.
David, W. A. L., and B. O. C. Gardiner. 1966. Entom. Exp. Appl., 9: 247.
Daxenbichler, M. E., C. H. VanEtten, and I. A. Wolff. 1961. J. Org. Chem., 26: 4168.
_____, C. H. VanEtten, F. S. Brown, and Q. Jones. 1964. J. Agr. Food Chem., 12: 127.
_____, C. H. VanEtten, and I. A. Wolff. 1965. Biochemistry (Wash.), 4: 318.
Delaveau, P. G., and A. Kjær. 1963. Acta Chem. Scand., 17: 2562.
Deyeux, [N.] 1805. J. Médecine, 11: 141; Ann. Chim., 56: 316.
Doney, R. C., and J. F. Thompson. 1966. Biochim. Biophys. Acta, 124: 39.
Döpke, W. 1965. Naturwissenschaften, 52: 133.
Dymock, W., C. J. H. Warden, and D. Hooper. 1890. Pharmacographia Indica, vol. I, p. 46, London, Kegan Paul, Trench, Trübner and Co.
Emery, T. F. 1966. Biochemistry (Wash.), 5: 3694.
Ettlinger, M. G., and G. P. Dateo, Jr. 1961. Studies of Mustard Oil Glucosides, Final Report Contract DA19-129-QM-1059, U. S. Army Natick Laboratories, Natick, Massachusetts.

_____, and J. E. Hodgkins. 1956. J. Org. Chem., 21: 204.
_____, and H. E. Miller. 1966. *In* The Chemistry of Natural Products, Abstract Book, p. 154, International Union of Pure and Applied Chemistry, 4th International Symposium, Stockholm, Sweden.
_____, and C. P. Thompson. 1962. Studies of Mustard Oil Glucosides (II), AD no. 290747, Office of Technical Services, U. S. Department of Commerce, Washington, D. C.
_____, G. P. Dateo, Jr., B. W. Harrison, T. J. Mabry, and C. P. Thompson. 1961. Proc. Nat. Acad. Sci. U. S. A., 47: 1875.
_____, A. Kjær, C. P. Thompson, and M. Wagnières. 1966. Acta Chem. Scand., 20: 1778.
Finkbeiner, H. L., and M. Stiles. 1963. J. Amer. Chem. Soc., 85: 616.
Finnegan, R. A., and W. H. Mueller. 1965. J. Pharm. Sci., 54: 1136.
Finnemore, H., and J. M. Cooper. 1936. J. Proc. Roy. Soc. N. S. Wales, 70: 175.
Folkers, K., F. Koniuszy, and J. Shavel, Jr. 1944. J. Amer. Chem. Soc., 66: 1083.
Freise, F. W. 1935a. Perfumery Essent. Oil Record, 26: 91.
_____. 1935b. Perfumery Essent. Oil Record, 26: 219.
Fries, E. 1874. Hymenomycetes Europaei, p. 220, Upsala, E. Berling.
Fries, R. E. 1959. *In* Die Natürlichen Pflanzenfamilien (Engler-Prantl), 2d ed., vol. 17aII, pp. 72, 158, Berlin, Duncker and Humblot.
Friis, P., and A. Kjær. 1963. Acta Chem. Scand., 17: 1515.
_____, and A. Kjær. 1966. Acta Chem. Scand., 20: 698.
Gander, J. E. 1960. Plant Physiol., 35: 767.
_____. 1962. J. Biol. Chem., 237: 3229.
Gilbert, E.-J. 1932. Bull. Soc. Mycol. France, 48: 241.
Gmelin, R. 1963. Arzneimittelforschung, 13: 771.
_____, and J. B. s. Bredenberg. 1966. Arzneimittelforschung, 16: 123.
_____, and A. I. Virtanen. 1959. Acta Chem. Scand., 13: 1474.
_____, and A. I. Virtanen. 1961. Ann. Acad. Sci. Fenni. Ser. AII, no. 107.
_____, and A. I. Virtanen. 1962. Acta Chem. Scand., 16: 1378.
Gottlieb, O. R., and M. T. Magalhães. 1959. J. Org. Chem., 24: 2070.
_____, M. T. Magalhães, and W. B. Mors. 1961. Anais Acad. Brasil Cienc., 33: 301.

_____, I. S. de Souza, and M. T. Magalhães. 1962. Tetrahedron, 18: 1137.
Gottschalk, G., and H. A. Barker. 1966. Biochemistry (Wash.), 5: 1125.
Greer, M. A. 1962a. Recent Progr. Hormone Res., 18: 187.
_____. 1962b. Arch. Biochem. Biophys., 99: 369.
Guddal, E., and N. A. Sörensen. 1959. Acta Chem. Scand., 13: 1185.
Guignard, L. 1893. J. de Botan., 7: 345, 444.
Hamer, J., and A. Macaluso. 1964. Chem. Rev., 64: 473.
Harms, H. 1940. *In* Die Natürlichen Pflanzenfamilien (Engler-Prantl), 2d ed., vol. 19bI, p. 175, Leipzig, W. Engelmann.
Hart, N. K., and J. A. Lamberton. 1966. Aust. J. Chem., 19: 1259.
Hatanaka, S.-I., and A. I. Virtanen. 1962. Acta Chem. Scand., 16: 514.
Hegnauer, R. 1964. Chemotaxonomie der Pflanzen, vol. 3, pp. 282-4, Basel, Birkhäuser Verlag.
_____. 1966. Chemotaxonomie der Pflanzen, vol. 4, pp. 155-168, Basel, Birkhäuser Verlag.
Hermbstaedt, [S. F.] 1796. Berlinisches Jahrb. Pharm., 2: 146, 161.
Herrmann, E. 1918a. Pharm. Zentralhalle, 59: 7.
_____. 1918b. Pharm. Zentralhalle, 59: 105.
Heywood, V. H. 1966a. *In* Swain, T., ed., Comparative Phytochemistry, Chap. 1, London, Academic Press.
_____. 1966b. Chem. Plant Taxonomy Newsletter no. 7, p. 7.
Hine, R. 1962. Acta Crystalloger. (Copenhagen) 15: 635.
Hitchcock, C., and A. T. James. 1966. Biochim. Biophys. Acta, 116: 413.
Honeyman, J. M. 1956. Taxon, 5: 33.
[Hooper, D.] 1903. Pharm. J., 70: 755.
Hooper, D. 1903-4. Rept. Industr. Sect., Indian Museum (Calcutta), 30-31.
_____. 1913. Pharm. J., 91: 369.
Hurusawa, I. 1954. J. Fac. Sci. Univ. Tokyo, Sect. III (Botany), 6: 209, 334.
Ingraham, J. L., J. F. Guymon, and E. A. Crowell. 1961. Arch. Biochem. Biophys., 95: 169.
Kindl, H. 1964. Monatsh. Chem., 95: 439.
_____. 1965. Monatsh. Chem., 96: 527.
Kjær, A. 1958. *In* Ruhland, W., ed., Handbuch der Pflanzenphysiologie, vol. IX, p. 64, Berlin, Springer-Verlag.

_____. 1960. Fortschr. Chem. Org. Naturst., 18: 122.
_____. 1961. *In* Kharasch, N., ed., Organic Sulfur Compounds, vol. 1, chap. 34, New York, Pergamon Press.
_____. 1963a. *In* Swain, T., ed., Chemical Plant Taxonomy, chap. 16, London, Academic Press.
_____. 1963b. Pure Appl. Chem., 7: 229.
_____. 1966. *In* Swain, T., ed., Comparative Phytochemistry, chap. 11, London, Academic Press.
_____. 1967. Phytochemistry, in press.
_____, and B. W. Christensen. 1961. Acta Chem. Scand., 15: 1477.
_____, and B. W. Christensen. 1962a. Acta Chem. Scand., 16: 71.
_____, and B. W. Christensen. 1962b. Acta Chem. Scand., 16: 83.
_____, and P. Friis. 1962. Acta Chem. Scand., 16: 936.
_____, and R. Gmelin. 1956. Acta Chem. Scand., 10: 1100.
_____, and R. Gmelin. 1958. Acta Chem. Scand., 12: 1693.
_____, and S. E. Hansen. 1958. Botan. Tidsskrift, 54: 374.
_____, and K. Rubinstein. 1953. Acta Chem. Scand., 7: 528.
_____, and H. Thomsen. 1962a. Acta Chem. Scand., 16: 591.
_____, and H. Thomsen. 1962b. Acta Chem. Scand., 16: 2065.
_____, and H. Thomsen. 1963a. Phytochemistry, 2: 29.
_____, and H. Thomsen. 1963b. Acta Chem. Scand., 17: 561.
_____, and M. Wagnières. 1965. Acta Chem. Scand., 19: 1989.
_____, J. Conti, and I. Larsen. 1953. Acta Chem. Scand., 7: 1276.
_____, H. Thomsen, and S. E. Hansen. 1960. Acta Chem. Scand., 14: 1226.
Klyne, W., J. Day, and A. Kjær. 1960. Acta Chem. Scand., 14: 215.
Koukol, J., P. Miljanich, and E. E. Conn. 1962. J. Biol. Chem., 237: 3223.
Kowalewski, Z. 1960. Helv. Chim. Acta, 43: 1314.
Krishnaswamy, N. R., T. R. Seshadri, and B. R. Sharma. 1966. Tetrahedron Lett., 4227.
Krusius, F.-E., and P. Peltola. 1966. Acta Endocr. (Copenhagen), 53: 342.
Kurup, P. A., and P. L. N. Rao. 1954. Indian J. Med. Res., 42: 85.
Kutáček, M. 1964. Biol. Plant. (Praha), 6: 88.
_____, Ž. Procházka, D. Grünberger, and R. Stajkova. 1962a. Collection Czech. Chem. Commun., 27: 1278.

Kutáček, M., Ž. Procházka, and K. Vereš. 1962b. Nature (London), 194: 393.
Lancini, G. C., D. Kluepfel, E. Lazzari, and G. Sartori. 1966. Biochim. Biophys. Acta, 130: 37.
Lange, J. E. 1938. Flora Agaricina Danica, Vol. III, p. 58. Copenhagen, Recato A/S.
Langer, P. 1966a. Physiol. Bohemoslov., 15: 162.
_____. 1966b. Endocrinology, 79: 1117.
Larsen, P. O. 1965a. Acta Chem. Scand., 19: 1071.
_____. 1965b. Biochim. Biophys. Acta, 107: 134.
_____, 1967 Biochem. Biophys. Acta, 141: 27.
Léandri, J. 1934. Bull. Soc. Botan. France, 81: 458.
Leete, E., and J. B. Murrill. 1967. Tetrahedron Lett., 1727.
Liaaen Jensen, S., and N. A. Sørensen. 1961. Acta Chem. Scand., 15: 1885.
Lichtenstein, E. P., D. G. Morgan, and C. H. Mueller. 1964. J. Agr. Food Chem., 12: 158.
Liss, I. 1961. Flora (Jena), 151: 351.
Loder, J. W., and G. B. Russell. 1966. Tetrahedron Lett., 6327.
Loftfield, R. B., and E. A. Eigner. 1966. Biochim. Biophys. Acta, 130: 426.
Mabry, T. J. 1966. *In* Swain, T. ed., Comparative Phytochemistry, chap. 14, London, Academic Press.
Matsuo, M., and M. Yamazaki. 1963. Chem. Pharm. Bull., 11: 545.
_____, and M. Yamazaki. 1964. Chem. Pharm. Bull, 12: 1388.
_____, and M. Yamazaki. 1966. Biochem. Biophys. Res. Commun., 24: 786.
Meakin, D. 1967. Experentia, 23: 174.
Mechoulam, R., and A. Hirshfeld. 1967. Tetrahedron, 23: 239.
_____, F. Sondheimer, A. Melera, and F. A. Kincl. 1961. J. Amer. Chem. Soc., 83: 2022.
Méhu, [C.] 1862. Répert. Chim. Appl. (Bull. Soc. Chim. France A.), 4: 457.
Mentzer, C., J. Favre-Bonvin, and M. Massias. 1963. Bull. Soc. Chim. Biol. (Paris), 45: 749.
Merxmüller, H., and P. Leins. 1967. Botan. Jahrb., 86: 113.
Mikolajczak, K. L., C. R. Smith, Jr., M. O. Bagby, and I. A. Wolff. 1964. J. Org. Chem., 29: 318.
Miller, H. E. 1965. M. A. Thesis, Rice University, Houston, Texas.
Miller, R. W., M. E. Daxenbichler, F. R. Earle, and H. S. Gentry. 1964. J. Amer. Oil Chem. Soc., 41: 167.

Mitscher, L. A., W. McCrae, and S. E. DeVoe. 1965. Tetrahedron, 21: 267.
Mizuno, K. 1961. Bull. Chem. Soc. Jap., 34: 1633.
Morris, L. J., and S. W. Hall. 1967. Chem. Industr.(London), 32.
Mortensen, J. T., J. S. Sørensen. and N. A. Sørensen, 1964. Acta Chem. Scand., 18: 2392.
Mothes, K., H. R. Schütte, P. Müller, M. v. Ardenne, and R. Tümmler. 1964. Z. Naturforsch. [B], 19: 1161.
Nagashima, Z., and M. Uchiyama. 1959. Nippon Nogeikagaku Kaishi, 33: 1068.
Nelson, E. R., and J. R. Price. 1952. Aust. J. Sci. Res., 5A: 768.
Novotný, L., V. Herout, and F. Šorm. 1964. Collection Czech. Chem. Commun., 29: 2182.
Oginsky, E. L., A. E. Stein, and M. A. Greer. 1965. Proc. Soc. Exp. Biol. Med., 119: 360.
Ostermayer, F., and D. S. Tarbell. 1960. J. Amer. Chem. Soc., 82: 3752.
Pailer, M. 1960. Fortschr. Chem. Org. Naturst., 18: 55.
Patel, C. K., S. N. Iyer, J. J. Sudborough, and H. E. Watson. 1926. J. Indian Inst. Sci., 9A: 117.
Peckolt, T. 1895. Pharm. Rundschau (New York), 13: 215.
Pedersen, K. J. 1934. J. Phys. Chem., 38: 559.
_____. 1949. Acta Chem. Scand., 3: 676.
Pfaender, P., and T. Wieland. 1966. Liebigs Ann. Chem., 700:126.
Pietschmann, A. 1924. Mikrochemie, 2: 33.
Podoba, J., and P. Langer, ed. 1964. Naturally Occurring Goitrogens and Thyroid Function, Bratislava, Slovak Academy of Sciences.
Pollard, J. K., E. Sondheimer, and F. C. Steward. 1958. Nature (London), 182: 1356.
Power, F. B., and M. Barrowcliff. 1905. J. Chem. Soc., 87: 884.
_____, and F. H. Gornall. 1904. J. Chem. Soc., 85: 838.
_____, and F. H. Lees. 1905. J. Chem. Soc., 87: 349.
Procházka, Ž. 1959. Naturwissenschaften, 46: 426.
Przybylska, M. 1962. Acta Crystallogr. (Copenhagen), 15: 301.
Quélet, L. 1886. Bull. Soc. Mycol. France, 2: 82.
_____. 1964. Les Champignons du Jura et des Vosges. repr. ed., suppl. 19, Amsterdam, A. Asher.
Radlkofer, L. 1908. *In* Die Natürlichen Pflanzenfamilien (Engler-Prantl), 1st ed., Nachtr. III, pp. 202, 208, Leipzig, W. Engelmann.
Rangaswami, S., and V. S. Rao. 1954. Indian J. Pharm., 16: 152.

Read, G., L. C. Vining, and R. H. Haskins. 1962. Canad. J. Chem., 40: 2357.

Reese, E. T., R. C. Clapp, and M. Mandels. 1958. Arch. Biochem. Biophys., 75: 228.

Rimington, C. 1935. Onderstepoort J. Vet. Sci., 5: 445; S. Afr. J. Sci., 32: 154.

Rollins, R. C. 1959. Rhodora, 61: 253.

Rosenthaler, L. 1923. Biochem. Z., 134: 215.

Ruigh, W. L., and A. E. Erickson. 1939. J. Amer. Chem. Soc., 61: 915.

Rybak, C. 1965. Dissertation; in Bohlmann, et al. (1966a).

Sabetay, S., L. Palfray, and L. Trabaud. 1938. C. R. Acad. Sci. (Paris), 207: 540.

Saghir, A. R., L. K. Mann, and M. Yamaguchi. 1965. Plant Physiol., 40: 681.

Schraudolf, H. 1965. Experientia, 21: 520.

_____. 1966a. Phytochemistry, 5: 83.

_____. 1966b. Symposium on Naturally Occurring Sulfur Compounds, Hellebæk, Denmark, June 23 (unpublished).

_____, 1967. Experentia, 23: 103.

_____, and F. Bergmann. 1965. Planta (Berlin), 67: 75.

Schroth, W., F. Billig, and G. Reinhold. 1967. Angew Chem., 79: 685.

Schuetz, R. D., T. B. Waggoner, and R. U. Byerrum. 1965. Biochemistry (Wash.), 4: 436.

Schulte, K. E., and S. Foerster. 1966. Tetrahedron Lett., 773.

_____, J. Reisch, and L. Hörner. 1962. Chem. Ber., 95: 1943.

_____, J. Reisch, W. Herrmann, and G. Bohn. 1963. Arch. Pharm. (Weinheim), 296: 456.

_____, G. Rücker, W. Meinders, and W. Herrmann. 1966. Phytochemistry, 5: 949.

Shaw, P. D. 1967. Biochemistry (Wash.), 6: 2253.

Smith, W. A. 1887. Amer. J. Pharm., 59: 228.

Sørensen, J. S., and N. A. Sørensen. 1958. Acta Chem. Scand., 12: 771.

_____, T. Bruun, D. Holme, and N. A. Sørensen. 1954. Acta Chem. Scand., 8: 26.

_____, J. T. Mortensen, and N. A. Sørensen. 1964. Acta Chem. Scand., 18: 2182.

Sørensen, N. A. 1961. Proc. Chem. Soc., 98.

_____. 1963. *In* Swain, T., ed., Chemical Plant Taxonomy, chap. 9, London, Academic Press.

Spatzier, W. 1893. Jahrb. Wiss. Botan., 25: 39.

Splittstoesser, W. E., and M. Mazelis. 1967. Phytochemistry, 6: 39.
Staron, T., C. Allard, and N. D. Xuong. 1965. C. R. Acad. Sci. (Paris), 260: 3502.
Stern, J. R., C. S. Hegre, and G. Bambers. 1966. Biochemistry (Wash.), 5: 1119.
Stoll, A., and E. Seebeck. 1948. Helv. Chim. Acta, 31: 189.
Strassman, M., and L. N. Ceci. 1966. J. Biol. Chem., 241: 5401.
Sugii, M., Y. Suketa, and T. Suzuki. 1964. Chem. Pharm. Bull., 12: 1115.
Suzuki, T., M. Sugii, and T. Kakimoto. 1962. Chem. Pharm. Bull., 10: 328.
Takahashi, N. 1960. J. Biochem. (Tokyo), 47: 230.
Tapper, B. A., and G. W. Butler. 1967. Arch. Biochem. Biophys., 120: 719.
_____, and D. B. MacGibbon. 1967. Phytochemistry, 6: 749.
_____, E. E. Conn, and G. W. Butler. 1967. Arch. Biochem. Biophys., 119: 593.
Thompson, J. F., C. J. Morris, and G. E. Hunt. 1964. J. Biol. Chem., 239: 1122.
Tookey, H. L., C. H. VanEtten, J. E. Peters, and I. A. Wolff. 1965. Cereal Chem., 42: 507.
Towers, G. H. N., A. G. McInnes, and A. C. Neish. 1964. Tetrahedron, 20: 71.
Tsuruo, I., M. Yoshida, and T. Hata. 1967. Agr. Biol. Chem., 31: 18.
Uhlenbroek, J. H., and J. D. Bijloo. 1959. Rec. Trav. Chim., 78: 382.
Underhill, E. W. 1965a. Canad. J. Biochem., 43: 179.
_____. 1965b. Canad. J. Biochem., 43: 189.
_____, 1967. Eur. J. Biochem., 2: 61.
_____, and M. D. Chisholm. 1964. Biochem. Biophys. Res. Commun., 14: 425.
_____, M. D. Chisholm, and L. R. Wetter. 1962. Canad. J. Biochem. Physiol., 40: 1505.
Uribe, E. G., and E. E. Conn. 1966. J. Biol. Chem., 241: 92.
Vaughan, J. G., and A. Waite. 1967. J. Exp. Botany, 18: 100.
_____, J. S. Hemingway, and H. J. Schofield. 1963. J. Linnean Soc., Botany, 58: 435.
Virtanen, A. I. 1961. Experientia, 17: 241.
_____. 1962. Arch. Biochem. Biophys., suppl. 1, p. 200.
_____. 1963. *In* Virtanen, A. I., ed., Investigations on the Alleged Goitrogenic Properties of Milk, p. 1, Helsinki, Biochemical Institute.

_____. 1965. Phytochemistry, 4: 207.
Waser, J., and W. H. Watson. 1963. Nature (London), 198: 1297.
Watson, E. R., N. G. Chatterjee, and K. C. Mukerjee. 1923. J. Soc. Chem. Industr., 42: T387.
Watt, Sir G. 1908. The Commercial Products of India, p. 780, London, John Murray.
Webster, R. E., and S. R. Gross. 1965. Biochemistry (Wash.) 4: 2309.
Wenkert, E., J. C. Orr, S. Garratt, J. H. Hansen, B. Wickberg, and C. L. Leicht. 1962. J. Org. Chem., 27: 4123.
Wetter, L. R. 1964. Phytochemistry, 3: 57.
Wiley, P. F., R. R. Herr, F. A. MacKellar, and A. D. Argoudelis 1965. J. Org. Chem., 30: 2330.
Winstead, J. A., and R. J. Suhadolnik. 1960. J. Amer. Chem. Soc., 82: 1644.
Winterfeldt, E. 1963. Chem. Ber., 96: 3349.
Wright, W. G., and F. L. Warren. 1967a. J. Chem. Soc. C, 283.
_____, and F. L. Warren. 1967b. J. Chem. Soc. C, 284.
Zechmeister, L., and J. W. Sease. 1947. J. Amer. Chem. Soc., 69: 273.

4

THE BETALAINS

TOM J. MABRY
The Cell Research Institute and Department of Botany
The University of Texas, Austin

ANDRE S. DREIDING
The Organic Chemistry Institute
The University of Zurich, Switzerland

I. INTRODUCTION

The betacyanins and betaxanthins are red-violet and yellow pigments, respectively, which occur in flowering plants belonging to 10 families usually allied in the order Centrospermae. On the basis of structural and biogenetic considerations that are discussed later, we have introduced the name betalains in the title of this paper to describe the class of substances to which both betacyanins and betaxanthins belong. These pigments occur abundantly in many familiar plants, such as the red beet, cacti, cockscomb, pokeberry, and so on.

Like the more widespread anthocyanins, which have not yet been found in betalain-containing families, betacyanins and betaxanthins occur in the cell vacuoles of the flowers, fruits, and leaves. Moreover, betalains often accumulate in the stalk, and of course in the beet they are found in high concentration in underground parts. It has also been noted that these pigments frequently accumulate at wound or injury sites in the plants which synthesize them normally. Betalains are water soluble, occurring naturally as internal salts, that is, zwitterions. These pigments are readily recognized and characterized by their electrophoretic migration towards the anode in weakly acidic buffers while anthocyanins migrate toward the cathode under similar conditions. Protocyanins, the high molecular weight blue pigments of certain flowers including the cornflower, migrate in the same direction as the betalains during electrophoresis but are readily distinguished by their colors; furthermore, protocyanins yield anthocyanins on acidic treatment. (For a review of these complex pigments see Bayer, 1966.)

Before 1960, little was known about the chemical nature of the betalains; but now the absolute structures of several betacyanins and betaxanthins have been firmly established and evidence has been adduced concerning their biogenesis. Their possible evolutionary significance, which has long been appreciated (e.g., Reznik, 1955, 1957), has been recently reevaluated. Recent reviews by Dreiding (1961) and Mabry (1964, 1966) cover most of the publications pertaining to the betalains prior to 1965. Therefore this review will emphasize subsequent work.

II. THE CHROMOPHORE OF THE BETALAINS

The general formula (I) for the betalains represents formally the condensation product of a primary or secondary amine with substance II, which is called betalamic acid (Wyler et al., 1965). Betalamic acid has not been isolated from natural sources or prepared synthetically and may not be an intermediate in the biosynthesis of the betalains; however, it does represent a unifying structure for all known betacyanins and betaxanthins. So far as is known, the structures of all betalains are based on formula I, in which R′ may be an H or be cyclizied to R.

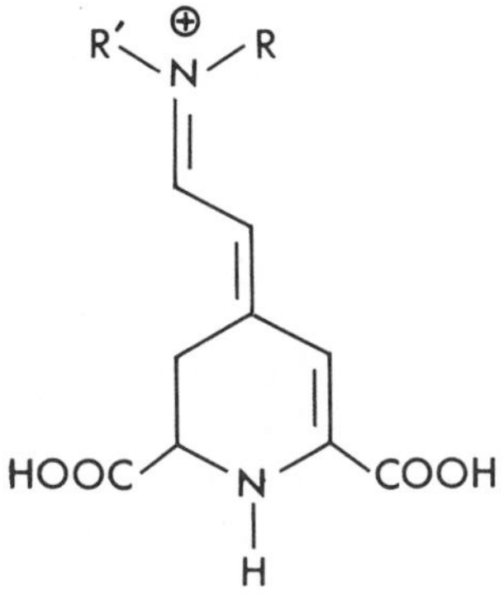

I 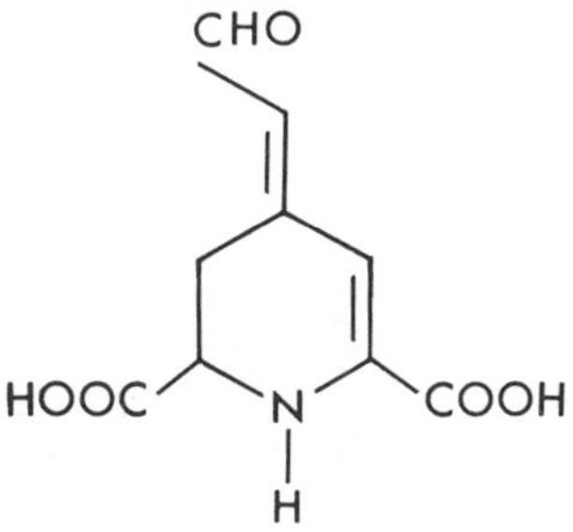 II

It should be noted that the chromophore responsible for the colors of all known betalains contains the cation III, shown in two important resonance structures. Therefore the betalains

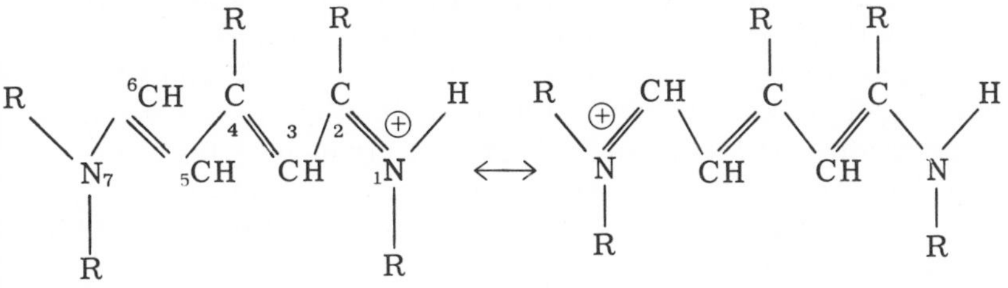

III

belong to a group of compounds for which König (1922, 1926; see also Staab, 1959) proposed the expression "polymethin pigments." Recently a modification of König's nomenclature was suggested in such a way as to include in the count of the methin units the terminal atoms of the conjugated system (Mabry et al., 1967). Thus the chromophore in the generalized formula for a betalain (I and III) would be described as a protonated 1,2,4,7,7-pentasubstituted 1,7-diazaheptamethin system.

When R and R′ do not extend the conjugation of the 1,7-diazaheptamethin system present in formula I, the substances are yellow (long-wavelength absorption near 480 nm) and are called betaxanthins. In the betacyanins, R′ is a substituted aromatic group that is conjugated to the 1,7-diazaheptamethin cation; this extension of conjugation in the chromophore produces a bathochromic shift to around 540 nm, which accounts for the red-violet color of the betacyanins.

III. NATURALLY OCCURRING BETACYANINS

Early in the twentieth century the pigment of the red beet (*Beta vulgaris*) was named betanin by Schudel (1918), a student of Willstätter, and the class of pigments to which betanin belonged was described by the term "betacyanins," presumably to express an undefined relationship to the more common anthocyanins. Other individual betacyanin-type pigments have been named for the original plant source, for instance, amarantin from the genus *Amaranthus*.

The hydrolysis of all betacyanins investigated so far leads to either betanidin, isobetanidin, or a mixture of the two isomeric aglycones (Wyler and Dreiding, 1961; Piattelli and Minale, 1964). The established structure for betanidin (Mabry et al., 1962; Wyler et al., 1963) is represented here by formula IV; isobetanidin is its C-15 epimer (Wilcox et al., 1965a), VII. Betanin (V) and isobetanin (VIII) are 5-O-β-D-glucosides of IV and VII, respectively (Piattelli et al., 1964a; Wilcox et al., 1965b). Both the betanidin and isobetanidin series of pigments have the S configuration at the C-2 asymmetric center, and have the S and R configurations, respectively, at C-15 (Wyler et al., 1963; Wilcox et al., 1965a).

Another pair of isomeric betacyanins, amarantin (VI) and isoamarantin (IX), were isolated from the leaves of *Amaranthus tricolor* (Piattelli et al., 1964b). Hydrolysis of these isomeric glycosides with β-glucuronidase followed by β-glucosidase transformed them into betanidin and isobetanidin, respectively. The *Amaranthus* pigments were subsequently formulated as the O-(β-D-glucopyranosyluronic acid)-5-O-β-D-glucopyranosides of betanidin and isobetanidin.

The first indication that the betacyanins contained a novel ring system resulted from the structural analysis of a group of primary conversion products of betanidin, which were called neobetanidin derivatives (Mabry et al., 1962, 1967). These substances have the 14,15-dehydrobetanidin structure; for example, one of these primary conversion products, which was obtained directly from the diazomethane methylation of either betanidin or isobetanidin, can be represented by formula X. The structures

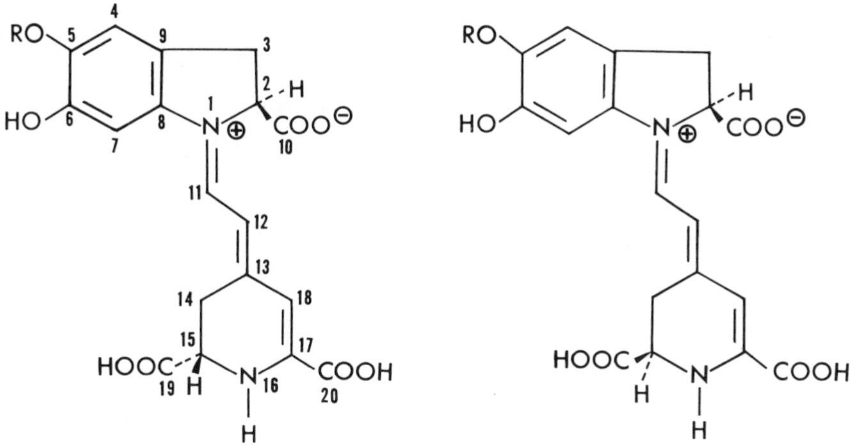

IV. R = H, Betanidin
V. R = β-D-Glucosyl, betanin
VI. R = O-(β-D-Glucosyluronic acid)-β-D-glucosyl, amarantin
VII. R = H, isobetanidin
VIII. R = β-D-Glucosyl, isobetanin
IX. R = O-(β-D-Glucosyluronic acid)-β-D-glucosyl, isoamarantin

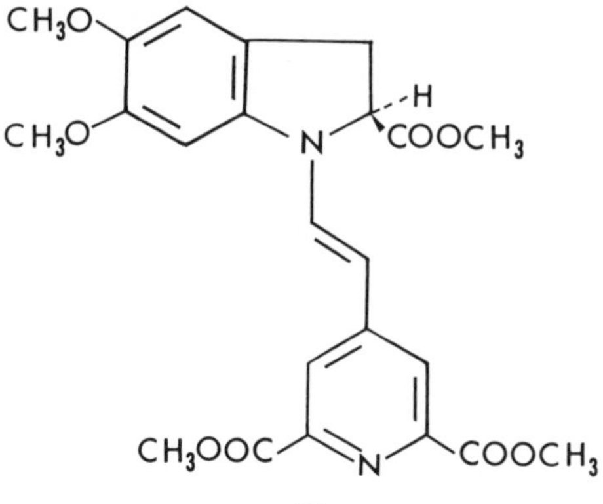

X

determined for the neobetanidins confirmed the fact that the betacyanins could not be structurally related to the flavylium salts, and the misleading expression "nitrogenous anthocyanins," which had been applied to the betacyanins, was finally discarded. All the neobetanidin derivatives exhibit a shift to longer wavelengths in their visible absorption spectra upon protonation, a phenomenon typical for the diazaheptamethin system. The spectral shift is from near 400 nm in methanol to about 500 to 520 nm in acidic methanol.

In addition to betanin, isobetanin, amarantin, and isoamarantin, a number of other betacyanin glycosides have been detected by chromatography, spectral methods, or electrophoresis (Reznik, 1955, 1957; Wyler et al., 1959; Wyler and Dreiding, 1961; Piattelli and Minale, 1964). The most comprehensive of these reports described 42 different betacyanin glycosides (Table 1) from 37 species, representing seven Centrospermae families (Piattelli and Minale, 1964). The betacyanin glycosides were isolated by polyamide chromatography and characterized by visible spectroscopy and electrophoresis. All the pigments showed an absorption maximum between 534 and 552 nm and electrophoretic mobilities of about 0.3 to 1.78 relative to betanin. Twenty-nine of the 42 betacyanins were hydrolyzed, and the hydrolyzates were subsequently examined by electrophoresis; only isobetanidin or a mixture of isobetanidin and betanidin was detected.

TABLE 1

Betacyanins[a]

Pigment	λ_{max} Water nm	E_b[b]	
		pH 4.5	pH 2.4
1. Amarantin	536	1.16	1.00
2. Isoamarantin	536	1.16	1.00
3. Betanin	538	1.00	1.00
4. Isobetanin	538	0.93	0.93
5. Iresinin-I	538	1.17	1.00
6. Iresinin-II	538	1.17	1.00
7. Phyllocactin	538	1.25	1.00
8. Celosianin	544-546	1.26	1.11
9. Isophyllocactin	538	1.14	0.93
10. Isocelosianin	542-544	1.26	1.11
11. Bougainvillein-I	538-540	0.96	0.94
12. Bougainvillein-II	540-542	0.90	0.92
13. Gomphrenin-I	535-537	1.00	0.96
14. Betanidin	542-546	1.00	0.70
15. Bougainvillein-III	540-542	1.00	0.94
16. Bougainvillein-IV	540-542	0.96	0.81
17. Isobetanidin	542-546	0.87	0.70
18. Gomphrenin-II	536-538	0.91	0.89
19. Oleracin-I	534-536	0.80	0.72

TABLE 1 (continued)

Pigment	λ_{max} Water nm	E_b[b] pH 4.5	E_b[b] pH 2.4
20. Oleracin-II	534-536	0.80	0.72
21. Gomphrenin-III	536-538	0.91	0.89
22. Bougainvillein-V	540-542	0.84	0.81
23. Bougainvillein-VI	544-546	0.84	0.81
24. Bougainvillein-VII	544-546	0.84	0.81
25. Gomphrenin-IV	540-542	0.80	0.72
26. Prebetanin	540-542	1.34	1.78
27. Bougainvillein-VIII	544-546	0.84	0.81
28. Isoprebetanin	540-542	1.21	1.78
29. Bougainvillein-IX	544-546	0.84	0.81
30. Gomphrenin-V	542-544	0.80	0.72
31. Bougainvillein-X	544-546	0.84	0.81
32. Gomphrenin-VI	542-544	0.80	0.72
33. Rivinianin	541-543	1.34	1.78
34. Bougainvillein-XI	544-546	0.84	0.81
35. Gomphrenin-VII	542-544	0.80	0.72
36. Bougainvillein-XII	548-550	0.66	0.52
37. Gomphrenin-VIII	540-542	0.80	0.72
38. Bougainvillein-XIII	549-551	0.66	0.52
39. Bougainvillein-XIV	550-552	0.66	0.52
40. Bougainvillein-XV	544-546	0.51	0.37
41. Bougainvillein-XVI	544-546	0.51	0.37
42. Mesembryanthemin-I	540-542	0.48	0.28
43. Mesembryanthemin-II	540-542	0.48	0.28
44. Mesembryanthemin-III	540-542	0.48	0.28

[a]Adapted from Piattelli and Minale (1964).

[b]Paper electrophoretic mobilities relative to betanin in a 0.05 *M* pyridine formate buffer, pH 4.5 and 0.1 *M* formic acid, pH 2.4.

It is therefore likely that members of the betacyanins differ only in stereochemistry at C-15, in the nature of the sugar moiety, and in esterification of the carboxyl and sugar groups.

IV. NATURALLY OCCURRING BETAXANTHINS

The yellow betalain pigments, which are structurally related to the betacyanins, are called betaxanthins (Greek: *xanthos*, yellow). The nomenclature is based on an analogous situation in flavonoid chemistry; flavonoids, other than anthocyanins, are often referred to as anthoxanthins since many of these other flavonoids are yellow in color. The betaxanthins have electrophoretic properties similar to betacyanins but show visible absorption maxima in the range 474 to 486 nm.

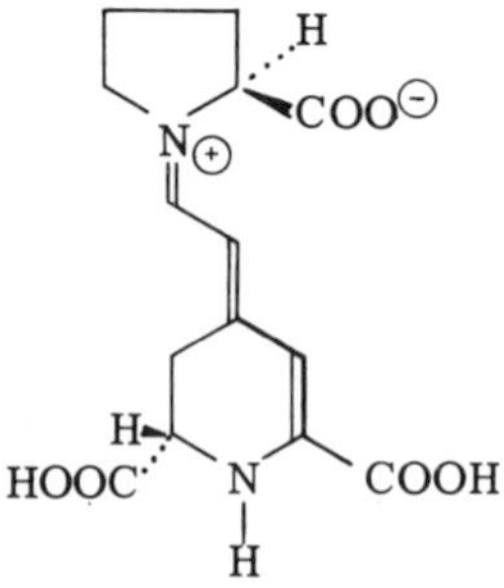

XI. Indicaxanthin

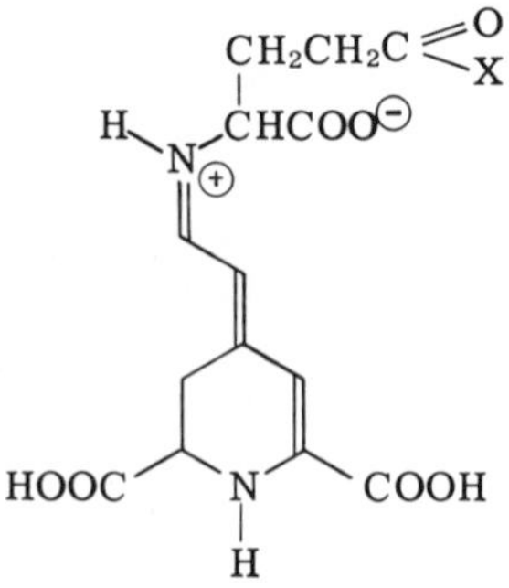

XII. Vulgaxanthin-I, X = OH
XIII. Vulgaxanthin-II, X = NH_2

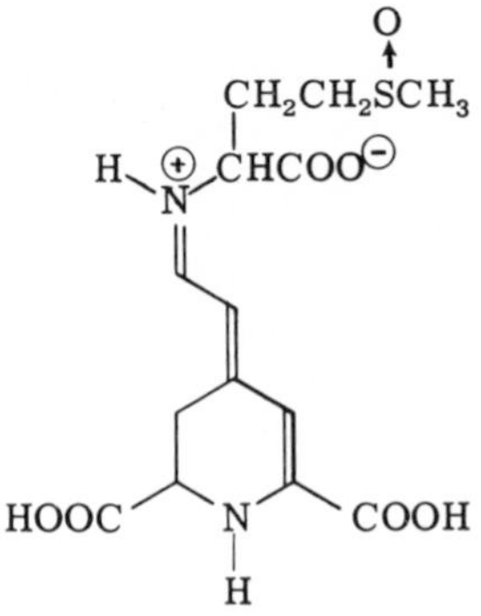

XIV. Miraxanthin-I

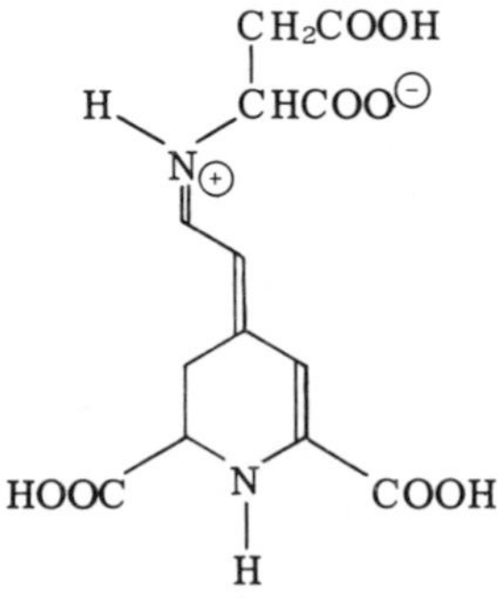

XV. Miraxanthin-II

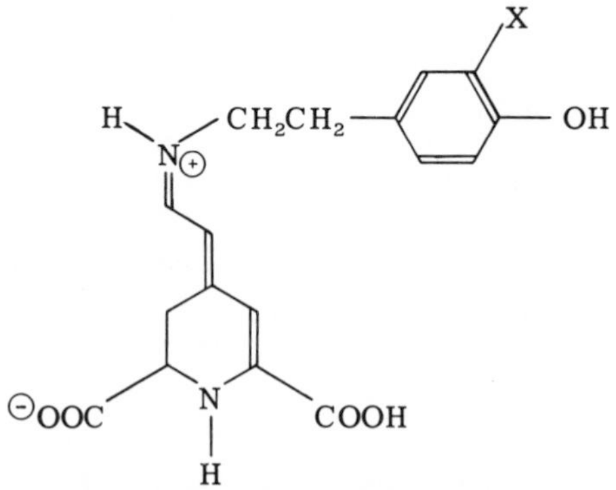

XVI. Miraxanthin-III, X = H
XVII. Miraxanthin-V, X = OH

Detailed structural information is available for only one betaxanthin, indicaxanthin (XI). This yellow pigment (λ_{max} = 483 nm) was isolated from the fruits of a cactus, *Opuntia ficus-indica*, and its structure was determined by a number degradation reactions which yielded compounds such as L-proline and 4-methylpyridine-2,6-dicarboxylic acid (XIX) (Piattelli et al., 1964c). A partial synthesis of indicaxanthin was effected by a novel reaction: indicaxanthin was formed by the treatment of betanin (V) with L-proline (Wyler et al., 1965). Furthermore indicaxanthin could be transformed into betanidin by allowing it to react with 5,6-dihydroxy-2,3-dihydroindole-2-carboxylic acid. The mechanism of these imino acid exchange reactions with the dihydropyridine component of the betalains has not been clarified; however, it has been suggested (Wyler et al., 1965) that these exchange reactions could proceed by hydrolysis of the pigments to yield betalamic acid (II), which could then condense with available amino acids or amines to produce betalains. Alternatively, the reaction might proceed via an intermediate substance such as XVIII.

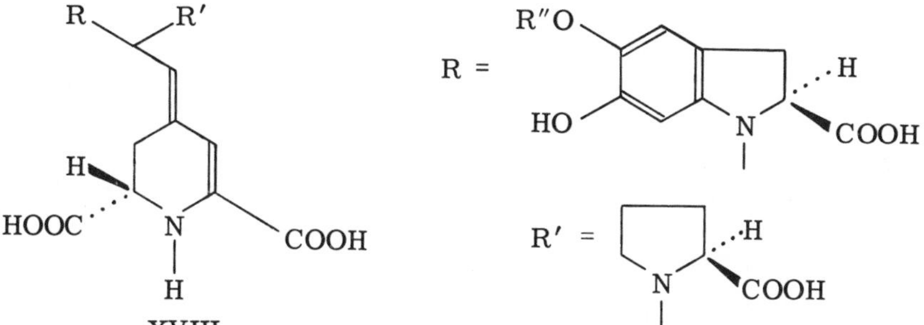

R R′ H HOOC N COOH H

XVIII

R = R″O HO N H COOH

R′ = N H COOH

Small quantities of two betaxanthins, vulgaxanthin-I (λ_{max} = 477 nm) and vulgaxanthin-II (λ_{max} = 479 nm), were obtained from the red beet (Piattelli et al., 1965b). On the basis of alkali and acid degradation reactions, structures XII and XIII were proposed for the two yellow *Beta vulgaris* pigments.

In addition to indicaxanthin and vulgaxanthin-I, six other betaxanthins were isolated from the flowers of the "four o'clock," *Mirabilis jalapa* (Piattelli et al., 1965a). The six new betaxanthins were named miraxanthin-I (XIV), -II (XV), -III (XVI), -IV, -V (XVII), and -VI; structures were proposed and partial syntheses provided for all except miraxanthin-IV and -VI. All six of the miraxanthin pigments yielded 4-methylpyridine-2,6-dicarboxylic acid (XIX) on fusion with alkali, which indicated the presence of a dihydropyridine moiety in each of the pigments. On acid hydrolysis miraxanthin-I afforded methionine sulfoxide, and miraxanthin-II gave aspartic acid. Miraxanthin-III and -V are noteworthy in that they are the first naturally

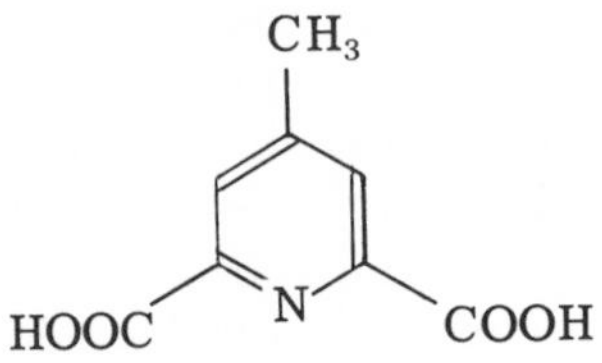

XIX

occurring betalains that do not combine an α-amino acid with the dihydropyridine moiety. Acid hydrolysis of miraxanthin-IV yielded an unidentified basic substance which coupled with diazotized sulfanilic acid and showed chromatographic and electrophoretic behavior similar to tyramine. Miraxanthin-VI, which appeared to contain a dihydroxyphenol group, also yielded on hydrolysis a basic substance which coupled with diazotized sulfanilic acid.

Miraxanthin-I, -II, -III, and -V were synthesized by dissolving betanin in water, saturating the solution with sulfur dioxide, and concentrating the solution in vacuo. The appropriate amino acid dissolved in water was then added to the concentrate, and, on workup, the reaction mixture yielded the expected betaxanthin. It was observed that methylation of the betaxanthins with diazomethane gave methylated derivatives in which the dihydropyridine ring had been oxidized to a pyridine ring. These substances can be called neobetaxanthins since they correspond to the previously described neobetanidin derivatives of betanidin. The methylated neobetaxanthins showed absorption maxima between 340 and 360 nm in neutral solvents and displayed a bathochromic shift of 80 to 100 nm in acidic solutions.

It seems reasonable that combinations of other amines or amino acids with the dihydropyridine moiety common to all betalains will account for additional betaxanthins.

V. BIOGENESIS OF THE BETALAINS

A biosynthetic pathway for the betalains can be formulated on the basis of L-dopa (L-5,6-dihydroxyphenylalanine), XX, undergoing an oxidative cleavage and subsequent cyclization to betalamic acid (II). The latter substance could then condense with available amines or amino acids to produce betalains. Condensation of II with cyclodopa or a derivative of cyclodopa (XXI) would yield betacyanins while other amino acids or amines would afford betaxanthins.

XXI

→ Betacyanins

XX. L-Dopa

II. Betalamic acid

→ Betaxanthins

Amines or amino acids
(other than XXI)

Beet seedlings grown on a medium containing D,L-dopa-2-^{14}C were reported to synthesize labeled betanin (Hörhammer et al., 1964). The conclusion was based on the radioactivity measured for small quantities of noncrystalline electrophoretically purified betanin. However, it is now known that under the incorporation and isolation procedures used by these workers a radioactive impurity accompanies the labeled betanin during electrophoresis (Wilcox, Wyler, Wohlpart, Rösler, Dreiding, and Mabry, unpublished results).

Other workers (Minale et al., 1965) reported that when the homogenized pulp of *Opuntis ficus-indica* fruits was incubated with L-proline-^{14}C (randomly labeled), indicaxanthin was formed that contained label only in the proline moiety. Furthermore, a similar experiment using D,L-dopa-2-^{14}C produced labeled betanin and labeled indicaxanthin. The data for the latter compound suggest that dopa might serve as a precursor for the dihydropyridine portion of indicaxanthin and probably for betanin as well. However, the results are inconclusive since a labeled compound derived from the dihydropyridine ring of the betalains was not isolated and the experiments were performed in vitro.

TABLE 2

Betacyanin-Betaxanthin Containing Genera[a,b]

Chenopodiaceae (over		Portulacaceae (19 genera,		Cactaceae (about 200	
100 genera, 1500 species)		500 species)		genera, 2000 species)	
Atriplex	(150:5)	*Anacampseros*	(70:1)	*Ariocarpus*	(8:5)
Beta[c]	(12:1)	*Calandrinia*	(150:1)	*Aylostera*	(10:1)
Chenopodium	(200:7)	*Claytonia*	(20:3)	*Cereus*[c]	(40:3)
Corispermum	(50:2)	*Montia*	(50:1)	*Chamaecerceus*[c]	(11:1)
Cycloloma	(1:1)	*Portulaca*[c]	(100:3)	*Cleistocactus*[c]	(35:1)
Enchylaena[d]	(1:1)	*Spraguea*	(5:1)	*Gymmocalycium*	(70:3)
Kochia	(80:2)	*Talinum*[c]	(50:—)	*Hariota*[c]	(2:1)
Rhagodia[d]	(12:2)	Basellaceae (4 or 5		*Hylocereus*	(20:1)
Salicornia	(30:1)	genera, 20 species)		*Lobivia*[c]	(75:2)
Salsola	(100:1)	*Anredera*	(2:1)	*Mammillaria*[c]	(300:7)
Spinacia	(3:1)	*Basella*	(5:2)	*Metrocactus*	(35:1)
Suaeda	(100:2)	Stegnospermaceae (one		*Monvillea*	(5:1)
Amaranthaceae (over 60		genus, 3 species)		*Neoporteria*[c]	(40:1)
genera, 900 species)		*Stegnosperma*	(3:1)	*Nopalxochia*	(2:1)
Achyranthes	(5:1)	Phytolaccaceae (about 17		*Notocactus*	(20:1)
Aerva	(10:1)	genera, 120 species)		*Opuntia*[c]	(200:6)
Alternanthera[c]	(170:6)	*Phytolacca*	(35:3)	*Parodia*[c]	(35:3)
Amaranthus[c]	(50:13)	*Rivina*[c]	(3:2)	*Pereskia*[c]	(20:1)
Celosia[c]	(60:4)	*Trichostigma*	(3:1)	*Phyllocactus*	(20:1)
Froelichia	(10:1)	Aizoaceae (Ficoidaceae;		*Rebutia*[c]	(40:3)
Gomphrena	(100:4)	130 genera, 2500 species)		*Selinocereus*	(23:1)
Iresine	(70:2)	*Aptenia*[c]	(2:1)	*Thelocactus*	(20:1)

Mogiphanes	(12:1)	*Bergeranthus*[c]	(15:—)	*Zygocactus*[c]	(2:1)
Tidestromia	(6:1)	*Conophytum*[c]	(300:17)		
Didieraceae (4 genera,		*Dorotheanthus*	(6:1)		
11 species)		*Faucaria*[c]	(35:—)		
Alluandia	(6:3)	*Fenestraria*[c]	(2:1)		
Alluandiopsis	(2:2)	*Gibbaeum*	(30:2)		
Decarya	(1:1)	*Glottiphyllum*[c]	(60:—)		
Didierea	(2:2)	*Lampranthus*[c]	(200:2)		
Nyctaginaceae (30 genera,		*Lithops*[c]	(70:1)		
300 species)		*Malephora*	(1:1)		
Abronia	(20:1)	*Mesembryanthemum*	(45:4)		
Boerhaavia	(20:1)	*Pleiospilos*[c]	(30:2)		
Bougainvillea[c]	(14:2)	*Rabiaea*[c]	(5:—)		
Cryptocarpus	(2:1)	*Rhombophyllum*[c]	(3:—)		
Cyphomeris	(2:1)	*Sesuvium*	(8:1)		
Mirabilis[c]	(60:3)	*Tetragonia*	(60:1)		
Nyctaginia	(1:1)	*Trianthema*	(15:1)		
Oxybaphus	(30:1)	*Trichodiadema*	(26:1)		

[a]Reproduced from Mabry (1966).

[b]Numbers in parentheses following generic names represent the estimated number of species in the genus relative to the number of species known to contain betacyanins.

[c]Contain betaxanthins (Reznik, 1955, 1957; Mabry and Turner, 1964; Mabry and Wohlpart, unpublished results).

[d]A. Wohlpart, unpublished results.

More recently it was observed that L-tyrosine was a good precursor of amarantin in a cultivar, Molten Fire, of *Amaranthus* species (Garay and Towers, 1966). However, again the conclusion was based on noncrystalline material that had been purified, in this instance, by paper chromatography. These workers also reported that the biosynthesis of amaranthin from tyrosine or dopa required light. It is known, however, that light is not a general requirement for betacyanin and betaxanthin synthesis. Indeed, of six Centrospermae species investigated, four readily produced betalains, albeit in reduced amounts, in the absence of light (Wohlpart and Mabry, 1968).

It appears that the biogenetic results reported to date do not provide sufficient evidence to confirm the biogenetic scheme for the betalains outlined above.

VI. PHYLOGENETIC SIGNIFICANCE OF THE BETALAINS

Since the phylogenetic implications of the distribution of betalains have been reviewed recently (Mabry, 1966), only a brief treatment of this topic will be presented here. The betalains are restricted to 10 families belonging to the order Centrospermae: Chenopodiaceae, Amaranthaceae, Portulacaceae, Nyctaginaceae, Phytolaccaceae, Stegnospermaceae, Aizoaceae, Basellaceae, Cactaceae, and Didieraceae. All of these families are known to contain betacyanins and, with the exceptions of the Basellaceae, Didieraceae, and Stegnospermaceae, betaxanthins as well. More extensive surveys for betaxanthins will probably reveal their presence in all the betacyanin-containing families.

Although most species in the betalain-containing families remain uninvestigated (Table 2) the data suggest, nevertheless, that plants containing the betalains are phylogenetically closely related. Therefore the previous proposal (Mabry et al., 1963), which recognized on chemical and morphological considerations a Centrospermae order containing the 10 betalain families, is retained here.

Order Centrospermae

Chenopodiaceae
Amaranthaceae
Nyctaginaceae
Portulacaceae
Phytolaccaceae (including the Petiveriaceae)
Stegnospermaceae
Aizoaceae
Basellaceae
Cactaceae
Didieraceae

Meeuse (1963; see also Mabry, 1966) has proposed a tentative phylogeny which shows the Centrospermae originating at least 150 million years ago along with several other major parallel lines in the Angiosperms. Meeuse arrived at this conclusion from purely morphological considerations, but his proposal for the Centrospermae, as recognized here, is supported by the available chemical evidence.

ADDENDUM

Recently, eight acylated betacyanins from three species were reported (Minale *et al.*, 1966): phyllocactin and isophyllocactin from *Phyllocactus hybridus* (Cactaceae); celosianin and isocelosianin from a cultivar of *Celosia cristata* (Amarantceceae); and iresinin-I, isoiresinin-I, iresinin-III and iresinin-IV from *Iresine herbstii* (Amaranthaceae). Phyllocactin and isophyllocactin have an O-malonyl group, $O{=}\overset{|}{C}CH_2COOH$, attached to the C-6 position of the glucose moiety in betanin and isobetanin, respectively. The hydrolytic results for celosianin and its C-15 epimer, isocelosianin, were in accord with their formulation as p-coumaroyl-feruloyl derivatives of amarantin and isoamarantin, respectively. The evidence favored the suggestion that the p-coumaric acid is attached to the glucose moiety and the ferulic acid to the glucuronic acid unit.

Of the four acylated betacyanins from *Iresine herbstii* only one, iresinin-I, was obtained crystalline. Based primarily on hydrolytic data, iresinin-I and isoiresinin-I were postulated to be 5-0-[2-0-(β-D-glucopyranosyluronic acid)-6-0-(3-hydroxy-3-methylglutaryl)-β-D-glucopyranosyl] derivatives of betanin and isobetanin, respectively. It was proposed that in iresinin-III a hydroxycinnamic acid is esterified to one of the carboxyl groups of betanidin or isobetanidin. Although iresinin-IV was only partially characterized it was reported that on alkaline hydrolysis it gave, in addition to amarantin and isosmarantin, trans-ferulic and trans-sinapic acids.

REFERENCES

Bayer, E. 1966. Angew. Chem., 78: 834.

Dreiding, A. S. 1961. *In* Ollis, W. D., ed., Recent Developments in the Chemistry of Natural Phenolic Compounds, 194, Oxford, Pergamon.

Garay, A. S., and G. H. N. Towers. 1966. Canad. J. Bot., 44: 231.
Hörhammer, L., H. Wagner, and W. Fritzsche. 1964. Biochem. Z., 399: 398.
König, W. 1922. Chem. Ber., 55: 3297.
_____. 1926. J. prakt. Chem. [2], 112: 1.
Mabry, T. J. 1964. *In* Leone, C. A., ed., Taxonomic Biochemistry and Serology, 239, New York, Ronald Press.
_____. 1966. *In* Swain, T., ed., Comparative Phtochemistry, 231, London, Academic Press.
_____, and B. L. Turner. 1964. Taxon, 13: 197.
_____, H. Wyler, G. Sassu, M. Mercier, I. Parikh, and A. S. Dreiding. 1962. Helv. Chim. Acta, 45: 640.
_____, A. Taylor, and B. L. Turner. 1963. Phytochemistry, 2: 61.
_____, H. Wyler, I. Parikh, and A. S. Dreiding. 1967. Tetrahedron, 23.
Meeuse, A. D. J. 1963. Acta Biotheor, 16 (III): 9.
Minale, L., M. Piattelli, and R. A. Nicolaus. 1965. Phytochemistry, 4: 593.
_____, M. Piattelli, S. De Stefano, and R. A. Nicolaus. 1966. Phytochemistry, 5: 1037.
Piattelli, M., and L. Minale. 1964. Phytochemistry, 3: 547.
_____, L. Minale, and G. Prota. 1964a. Ann. Chim., 54: 955.
_____, L. Minale, and G. Prota. 1964b. Ann. Chim., 54: 963.
_____, L. Minale, and G. Prota. 1964c. Tetrahedron, 20: 2325.
_____, L. Minale, and R. A. Nicolaus. 1965a. Phytochemistry, 4: 817.
_____, L. Minale, and G. Prota. 1965b. Phytochemistry, 4: 121.
Reznik, H. 1955. Z. Bot., 43: 499.
_____. 1957. Planta, 49: 406.
Schudel, G. 1918. Ph.D. Dissertation. Zurich-ETH.
Staab, H. A. 1959. Einführung in die theoretische organische Chemie, 323ff., Weinheim, Verlag Chemie.
Wilcox, M. E., H. Wyler, and A. S. Dreiding. 1965a. Helv. Chim. Acta, 48: 1134.
_____, H. Wyler, T. J. Mabry, and A. S. Dreiding. 1965b. Helv. Chim. Acta, 48: 252.
Wyler, H., and A. S. Dreiding. 1961. Experientia, 17: 23.
_____, G. Vincenti, M. Mercier, G. Sassu, and A. S. Dreiding. 1959. Helv. Chim. Acta, 42: 1696.
_____, T. J. Mabry, and A. S. Dreiding. 1963. Helv. Chim. Acta, 46: 1745.
_____, M. E. Wilcox, and A. S. Dreiding. 1965. Helv. Chim. Acta, 48: 361.
Wohlpart, A. and T. J. Mabry. 1968. Plant Physiol.

5

ALKALOID CHEMISTRY AND THE SYSTEMATICS OF *PAPAVER* AND *ARGEMONE*

FRANK R. STERMITZ
Department of Chemistry
Colorado State University, Fort Collins

I. INTRODUCTION

In recent reviews (Alston and Turner, 1963; Hegnauer, 1963) on the use of alkaloid content in biochemical systematics or chemotaxonomy, considerable emphasis was placed on interfamily and intergeneric relationships. The value of viewing alkaloid content in terms of relatively large groupings based on homologous biosynthetic pathways (e.g., synthesis from tyrosine, or tryptophan, or anthranilic acid) was stressed. One example was given (Alston and Turner, 1963) of the use of alkaloid content to compare species of a single genus, *Veratrum*.

In the present paper, an evaluation will be made of detailed alkaloid content as a mark of *intrageneric* relationships in two genera of poppies, *Papaver* and *Argemone*. Because biosynthetic studies on the alkaloids of the Papaveraceae have been proceeding at a rapid rate in a number of laboratories, it is possible in some instances to pinpoint the possible enzymatic or biochemical differences that are responsible for the observed variations in alkaloid content. The biosynthetic relationships among the alkaloids of *Papaver* have recently been summarized (Kühn and Pfeifer, 1965a). Many of the same results are presumably applicable to the alkaloids of *Argemone*, although the origin of the pavine types of alkaloids, which are of importance for systematics, is as yet unknown.

Although only two genera will be discussed here, other examples of the value of comparing alkaloid content among species of a single genus are available in the recent literature. For example, a study of the quinolizidine alkaloid (sparteine, retamine, etc.) content of *Genista* species showed some interesting correlations among the total alkaloid contents of species from various sections of this genus (Bernasconi et al., 1965). The establishment of identical alkaloid content in what were regarded as two species of *Heimia* (Douglas et al., 1964) gave added weight to the opinion of some taxonomists that *Heimia* is a monotypic genus.

II. THE GENUS *PAPAVER*

A. Introduction

Perhaps no other genus of higher plants has been investigated chemically so extensively as has *Papaver*. A comprehensive review (Kühn and Pfeifer, 1963) listed the results of alkaloid determinations in 33 species, and many additional species have been studied since that time. Many workers have made contributions in the past, and groups under Pfeifer in Germany and Slavik and Santavy in Czechoslovakia are continuing to investigate the genus intensively.

A standard botanical reference for the genus is that of Fedde (1909), who recognized 99 species, which were divided into nine sections (Table 1). Although some earlier attempts have been made to correlate alkaloid content with taxonomic classification in the genus (see, for example, Santavy, 1962), the most complete review that combined both chemical and taxonomic data was that of Kühn and Pfeifer (1963). The discussion here will draw heavily on the latter paper. Thus, following Kühn and Pfeifer (1963), the

TABLE 1

Sections of the Genus *Papaver* (Fedde)

Section 1. *Orthorhoeades*

P. rhoeas L.
P. rapiferum Fedde
P. caespitosum Fedde
P. tenerifae Fedde
P. subpiriforme Fedde
P. hirto-dubium Fedde
P. rhopalothece Stapf.
P. bipinnatum C. A. Mey.
P. polytrichum Boiss. & Kotschy.
P. pseudo-haussknechtii Fedde
P. integrifolium Vig.
P. syriacum Boiss. & Blanche
P. humile Fedde
P. roubiaei Vig.
P. subadpressiusculo-setosum Fedde
P. schweinfurthii Fedde
P. thaumasiosepalum Fedde
P. ameristophyllum Fedde
P. trilobum Wallr.
P. strigosum (Bönningh.) Schur.
P. chelidoniifolium Boiss. & Buhse
P. commutatum Fisch. & Mey.
P. tenuissimum (Heldr.) Fedde
P. californicum A. Gray
P. lemmonii Greene
P. dubium L.
P. apicigemmatum Fedde
P. arenarium Marsch.-Bieb.
P. pinnatifidum Moris
P. tunetanum Fedde
P. simoni Foucaud
P. modestum Jord.
P. erosulum Jord.
P. stipitatum Fedde
P. tenuifolium Boiss. & Hohen.
P. robertianella Fedde
P. posti Fedde
P. humifusum Fedde
P. stylatum Boiss. & Bal.
P. subumbilicatum Fedde
P. clavatum Boiss. & Hausskn.
P. gürlekense Stapf
P. umbonatum Boiss.
P. cassandrinum Charrel
P. exspectatum Fedde

TABLE I (continued)

Section 2. *Argemonorhoeades*

P. argemone L.
P. virchowii Aschers. & Sint.
P. belangeri Boiss.
P. apulum Ten.
P. hybridum L.
P. pavonium Fisch. & Mey.

Section 3. *Carinatae*

P. macrostomum Boiss. & Huet
P. piptostigma Bienert
P. tubuliferum Fedde
P. dalechianum Fedde
P. kurdistanicum Fedde
P. divergens Fedde & Bornmüller
P. bornmülleri Fedde

Section 4. *Mecones*

P. somniferum L.
P. setigerum DC.
P. glaucum Boiss. & Hausskn.
P. gracile Auch.
P. decaisnei Hochst. & Steud.

Section 5. *Miltantha*

P. tauricolum Boiss.
P. bartuschianum Fedde
P. flahaultii Fedde
P. persicum Lindl.
P. caucasicum Marsch.-Bieb.
P. hyoscyamifolium Boiss. & Hausskn.
P. floribundum Desf.
P. acrochaetum Bornmüller
P. fugax Poir.
P. urbanianum Fedde
P. armeniacum (L.) DC.
P. triniaefolium Boiss.
P. libanoticum Boiss.
P. polychaetum Schott & Kotschy

Section 6. *Pilosa*

P. spicatum Boiss. & Bal.
P. pilosum Sibth. & Smith
P. apokrinomenon Fedde
P. heldreichii Boiss.
P. strictum Boiss. & Bal.
P. pseudostrictum Fedde
P. lateritium C. Koch
P. oreophilum F. J. Rupr.
P. monanthum Trautv.
P. ramosissimum Fedde
P. rupifragum Boiss. & Reut.
P. atlanticum Ball

Section 7. *Macrantha*

P. orientale L.
P. bracteatum Lindl.
P. lasiothrix Fedde
P. paucifoliatum (Trautv.) Fedde

Section 8. *Scapiflora*

P. suaveolens Lap.
P. pyrenaicum (L.) A. Kerner
P. alpinum L.
P. nudicaule L.
P. anomalum Fedde

Section 9. *Horrida*

P. aculeatum Thunb.

genus will be discussed from the viewpoint of inter- and intra-sectional similarities and differences in alkaloid content.

B. *Orthorhoeades*

This section is preeminently that of species which accumulate rhoeadine (I) and its congeners and the papaverrubines, or *N*-desmethylrhoeadines. A number of workers have proposed structure I for rhoeadine (Santavy et al., 1965; Slavik et al., 1965a; Pfeifer, 1966 and references contained therein). The above alkaloids are the major alkaloids of this section, although rhoeadine types of alkaloids are widespread throughout the entire genus and papaverrubines were noticed in all 30 species tested (Pfeifer and Banerjee, 1962). No biosynthetic work has as yet been reported, as might be expected during a period of structure revision and establishment.

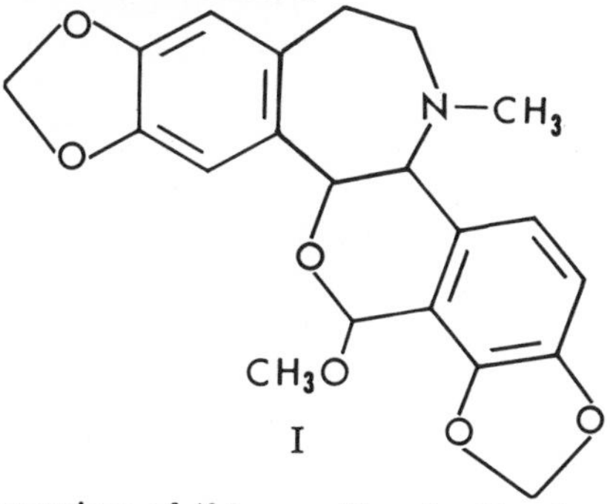

$N-CH_3$

CH_3O

I

The typical species of this section is *P. rhoeas* L., and this species has undergone considerable investigation. Most recently, Pfeifer and Hanus (1965) have established the presence of (–)-sinactine (II), which was reported to be the first tetrahydroprotoberberine alkaloid found in the genus *Papaver*. However, tetrahydroprotoberberines are generally thought to be precursors of the protopine, berberine, and sanguinarine types of alkaloids,

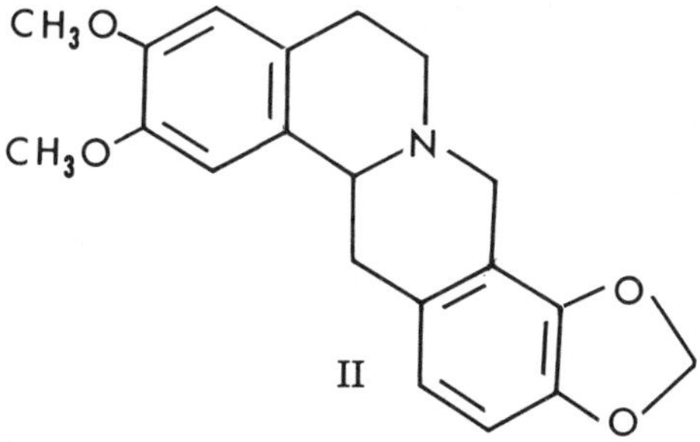

CH_3O

CH_3O

II

all of which abound in the genus. The finding of II is therefore of biosynthetic interest but is of less systematic value. In general, the most valuable alkaloids in a systematic sense are those that can be considered end products[1] in a particular species. Some tetrahydroprotoberberines and tetrahydroisoquinolines (e.g., reticuline) apparently play central roles in alkaloid biosynthesis in *Papaver*. Therefore these alkaloids may prove to be ubiquitous, given sufficient plant material and appropriate analytical techniques. Actually, among the tetrahydroprotoberberines it is the usual failure to isolate these alkaloids in *Papaver* that is probably of most systematic interest. The genus apparently converts these alkaloids rapidly to protopine, berberine, and so on and this ability serves to distinguish *Papaver* from *Corydalis*, *Dicentra*, and *Chelidonium*, all of which accumulate significant amounts of the tetrahydroprotoberberines.

Difficulties in the section *Orthorhoeades* are presented by the finding of thebaine (III) in *P. strigosum* Schur. and *P. intermedium* (Becker) O. Ktze. (Santavy et al., 1960), and papaverine (IV) and isocorydine (V) in *P. commutatum* Fisch. et Mey (Mnatsakanyan and Yunusov, 1961; Slavik et al., 1965b). Morphologically, these species stand quite close to *P. rhoeas* L., yet

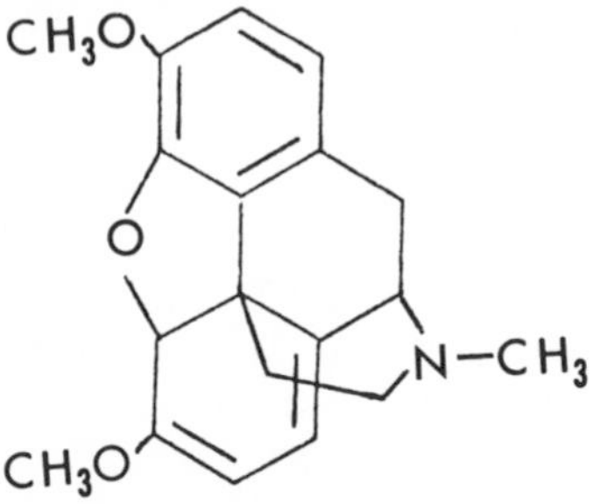

III

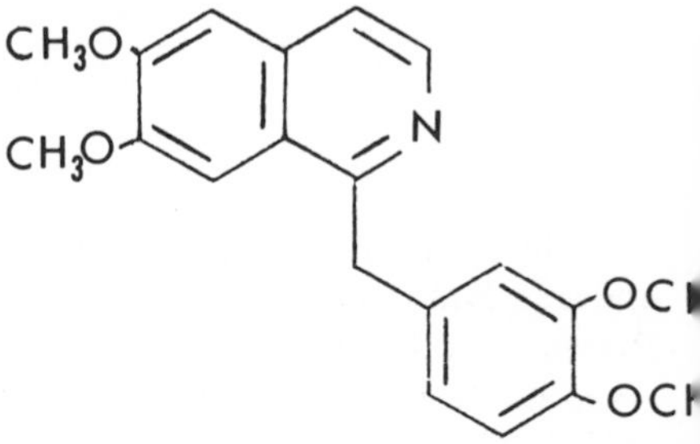

IV

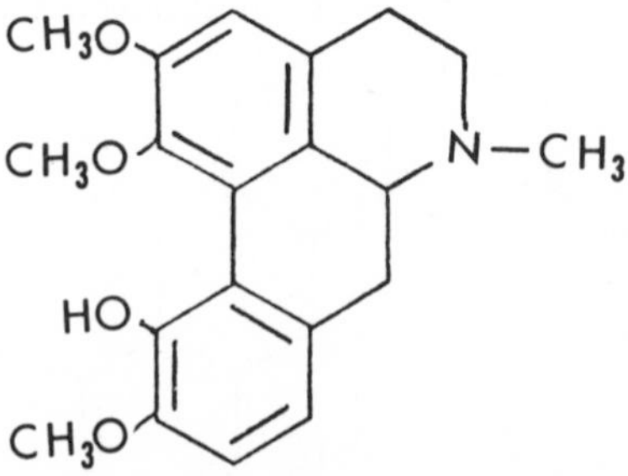

V

[1]By end products is meant those alkaloids that are not converted to other alkaloids. However, there is some evidence (Fairbairn and Wassel, 1964) that even a compound such as morphine can be converted to nonalkaloid materials.

chemically they have a resemblance to *P. somniferum* L. and *P. setigerum* DC. of the section Mecones. The latter two species accumulate thebaine, codeine, and morphine, and also are the only other species in which papaverine has been found. Perhaps phylogenetically *P. strigosum*, *P. intermedium*, and *P. commutatum* are intermediate between species of the section *Mecones* and the other members of *Orthorhoeades*. On the other hand, it was reported (Maturova et al., 1966) that traces of thebaine were found in *P. oreophilum* F. J. Rupr. of the section Pilosa. Perhaps the appearance of traces of thebaine will prove to be common and hence of little value systematically.

C. *Argemonorhoeades*, *Carinatae*, and *Horrida*

There have been no striking developments in the chemistry of these relatively small sections and the reader is referred to the earlier literature (Kühn and Pfeifer, 1963).

D. *Mecones*

This small section is best known from the extensively studied species *P. somniferum* L. and *P. setigerum* DC. These are the only plant species so far found to contain morphine (VI). They also accumulate thebaine and codeine (the precursors of morphine) and various benzylisoquinolines. It has been suggested (Santavy, 1962; Kühn and Pfeifer, 1963) that the above two species should be set apart from two other members of the section, *P. glaucum* Boiss. et Hauskn. and *P. gracile* Auch. These last two species have been found to contain no morphine or benzylisoquinoline alkaloids and instead accumulate the rhoeadine and papaverrubine alkaloids. Since there are a number of morphological similarities between these two species and most of the section *Orthorhoeades*, it seems reasonable that they should be transferred to that section as has been suggested (Santavy, 1962; Kühn and Pfeifer, 1963). No chemical data have as yet been reported on the fifth member of the section, *P. decaisnei* Hochst.

The opium poppy, *P. somniferum* L., continues to be intensively studied especially for its trace alkaloid content (Brochmann-Hanssen et al., 1965 and references therein) and is also widely used for biosynthetic studies. Good recent reviews (Dyke, 1965; Kühn and Pfeifer, 1965a) are available on the chemistry of this species.

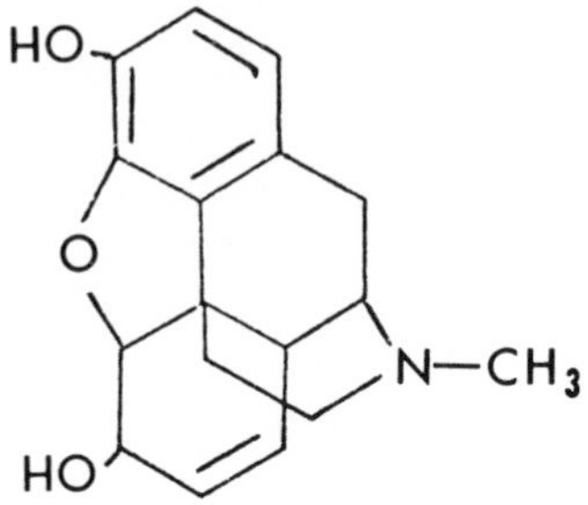

VI

A number of workers have been engaged in studying various races of *P. somniferum*. For example, Tetenyi and Vagujfalvi (1965) made an extensive study of the relationships among genetics, climate, and alkaloid content of numerous strains of this species.

E. *Miltantha*

Fedde (1909) included 14 species in this section. It was noted (Kühn and Pfeifer, 1963) that five species had been investigated up to 1963, with numerous alkaloids having been isolated. However, the structure of only one compound, armepavine (VII), was

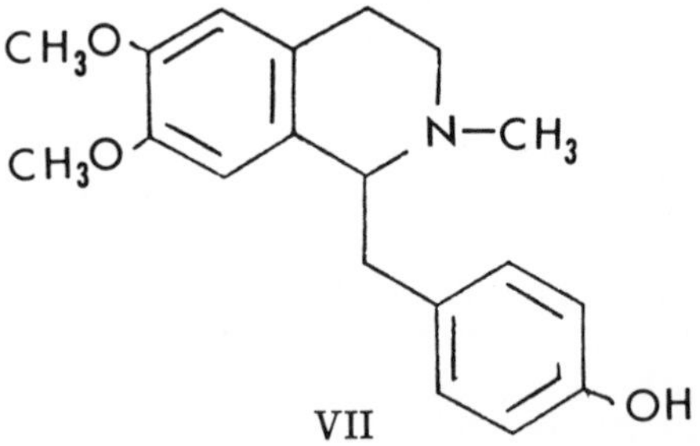

VII

known for certain. The section has received extensive work, and many of the previously isolated alkaloids have had a structure assigned. Only one additional species has been recently investigated in this section. The characteristic alkaloids of the section are armepavine (VII), aporphines (such as roemerine, VIII), and the immediate biosynthetic precursors of the latter, the proaporphines (such as pronuciferine, IX).

The widespread occurrence of aporphine alkaloids throughout a number of angiosperm families has been well documented, and structure work on the proaporphines, which are precursors of aporphine alkaloids, has recently been accomplished indepen-

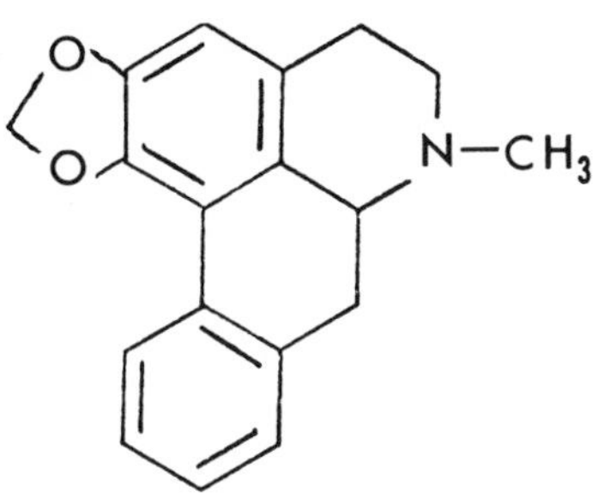

VIII 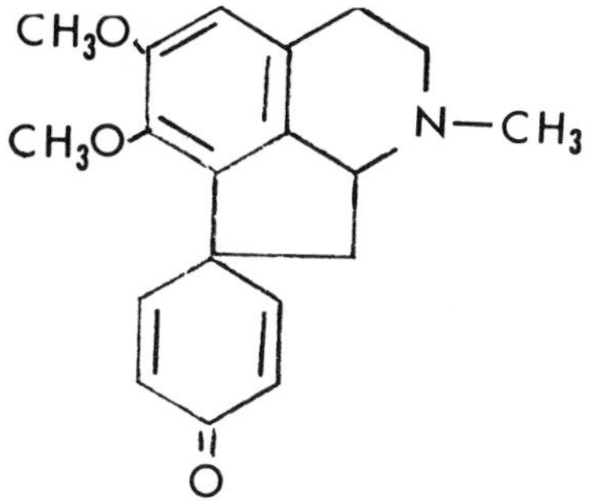 IX

dently for several of the families. Refined isolation techniques and a knowledge of the chemistry of the proaporphines have now simplified their identification. An excellent comprehensive review (Kühn and Pfeifer, 1965a) is now available covering this area of alkaloid chemistry.

The characteristic alkaloids of the section *Miltantha* are not accumulated by *Papaver* species of any other section. This chemical feature was noted by Kühn and Pfeifer (1965b), who investigated the alkaloids of four species of the section Miltantha. According to Kühn and Pfeifer, "These results point out the close relationships among the individual members of the section and their demarcation from other sections." On the basis of data acquired from the investigation of six of the fourteen species recognized in this section, the group appears to be quite homogeneous in its alkaloid chemistry.

F. *Pilosa*

Half of the 12 species in this section have been investigated. The members of this section generally accumulate rhoeadine and the protopine types of alkaloids and are therefore similar in their alkaloid chemistry to the *Orthorhoeades*. However, one alkaloid, latericine, has been reported from three species and may be unique to this section. Its structure has as yet not been reported.

G. *Macrantha*

This section is also well defined chemically and morphologically. Of the four species present, two (*P. orientale* L. and *P. bracteatum* Lindl.) have been examined a number of times for alkaloid content. These two species are closely related morphologically. Chemically, they are distinguished from the species

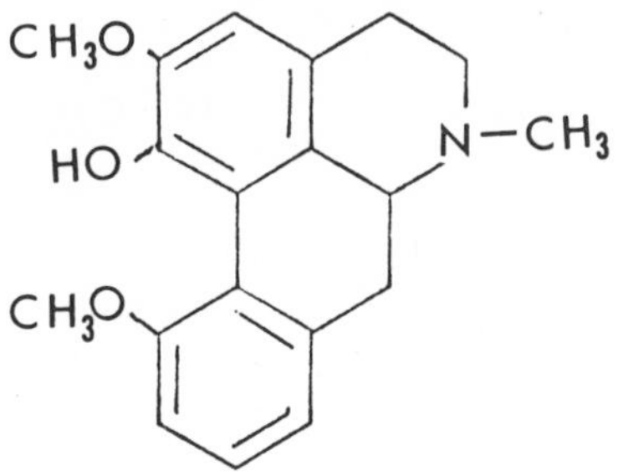

X 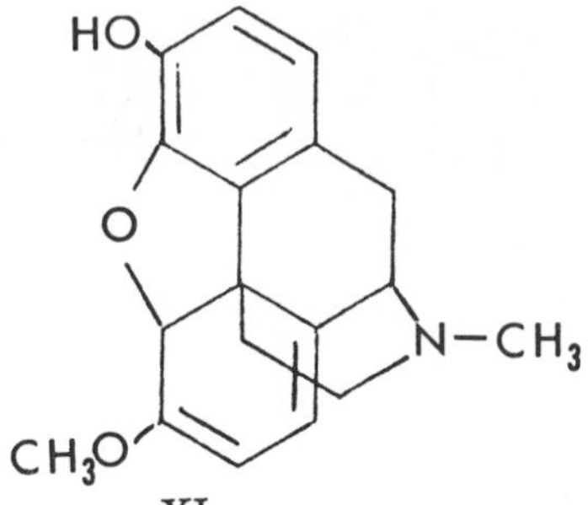 XI

of other sections by the accumulation of thebaine (III) and isothebaine (X). The presence of thebaine indicates a closeness in alkaloid chemistry to Mecones as does the occurrence of oripavine (XI) in *P. orientale* L. The (presumed) demethylation of thebaine to yield oripavine is paralleled only by the similar demethylation in the section *Mecones* which produces morphine from codeine.

An interesting example of chemical variation within a species is provided by *P. bracteatum*. In two investigations (Kiselev and Konovalova, 1948; Heydenreich and Pfeifer, 1965) isothebaine was found to be the main alkaloid. However, a chemical variety of *P. bracteatum* has been developed (Neubauer and Mothes, 1963; Neubauer, 1965) in which thebaine represents 95 percent of the total alkaloid content. Although there are a number of species in the Papaveraceae that show strong tendencies toward the development of such chemical varieties, this is perhaps one of the most striking examples. Such reports emphasize the dangers inherent in basing systematics on alkaloid content alone. One may question the idea that alkaloid content could have any fundamental significance in plant biochemistry if barely distinguishable varieties of the same species are found to exhibit such marked chemical differences. However, even though the final structures of thebaine (III) and isothebaine (X) appear to be quite different, the basic phenol coupling reactions that lead to the elaboration of III and X could easily be mediated by the same enzyme system.

H. *Scapiflora*

Recent chemical studies have indicated that this section will be as unique chemically as it is in morphology and geographical range. In contrast to all other members of the genus (which occur mainly in temperate regions), species of the *Scapiflora* section occur principally in the high mountains of the temperate zone or in Iceland, Greenland, and other near-Arctic or Arctic regions.

Several varieties of *P. nudicaule* L. have been intensely investigated. The alkaloids of the section as a whole are grouped into three categories. Most of the species have been found to contain, along with protopine, muramine, which has recently been shown (Cross et al., 1965; Manske, 1966) to be XII. Elsewhere,

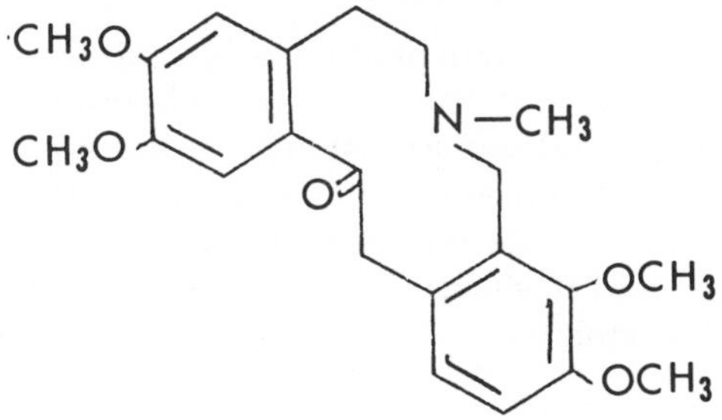

XII

muramine has only been reported in *P. pilosum* Sibth. et Smith and in *Argemone munita* Dur. et Hilg. subsp. *rotundata* (Rydb.) G. B. Ownbey (Stermitz and Seiber, 1966). The second group of structures is represented by amurine and nudaurine. These substances have been characterized as either morphinandienones (Flentje et al., 1965) or proaporphines (Maturova et al., 1965).

The last group of compounds is the most interesting. Compounds such as amurensine and amurensinine are representative. These alkaloids have been found in five members of the section and the basic isopavine structure XIII was assigned (Santavy et al., 1966). The evidence as presented did not seem to exclude

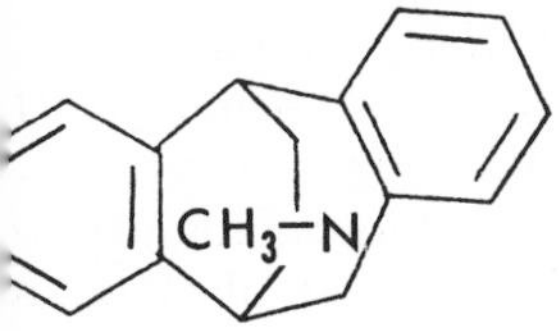

XIII 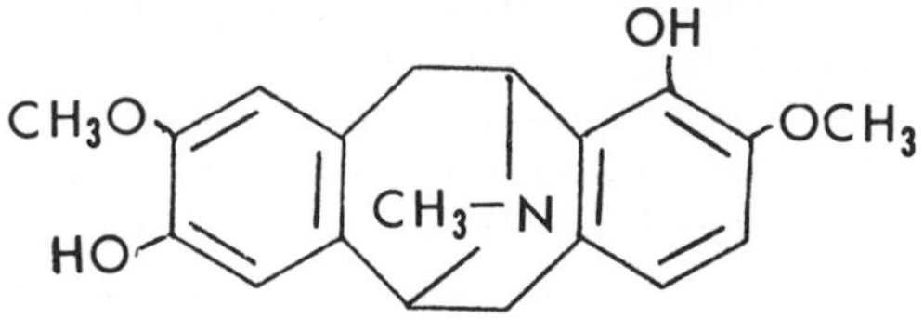 XIV

pavines with an unsymmetrical substitution pattern such as is present in munitagine (XIV) (Stermitz and Seiber, 1966). Isopavines have not otherwise been reported as natural products and pavines have been found only in *Argemone* (where they are common) and in *Eschscholtzia*. No other section of *Papaver* contains alkaloids of this type.

It is possible that the occurrence of muramine and isopavine (or pavine) alkaloids in the section Scapiflora indicates a closeness between these species and the American genera *Argemone* and *Eschscholtzia*. This idea will be amplified subsequently.

I. Summary

It is apparent that the morphologically distinct sections of the genus *Papaver* also have distinct alkaloid chemistry. In instance where the morphological demarcations are not strong, the alkaloid content is also less distinctive. Sometimes the alkaloid content may be able to provide information regarding the connection between sections, and it should thus be useful for determining phylogenetic relationships.

III. THE GENUS *ARGEMONE*

A. Introduction

Although the early classification of *Papaver* species by Fedde in 1909 has required only minor revision, this has not proved true of the genus *Argemone*. Fedde (1909) adopted a conservative approach and recognized only nine species of *Argemone*, although he included many varieties. A comprehensive reevaluation of the classification was recently accomplished (Ownbey, 1958, 1960); see Table 2. Altogether, 23 species in North America and the West Indies were described (Ownbey, 1958), and, of these, all but two were studied in the field. Many were cultivated, crossing experiments were conducted, and chromosome numbers were determined whenever possible. Subsequentl the taxa of South America and Hawaii were investigated, with four species from South America and one from Hawaii identified (Ownbey, 1960). The discrepancy between the number of species described by Fedde and by Ownbey perhaps overstates the difference between the two classifications. Ownbey added a number of species that were completely unknown to Fedde, but in many instances raised to species level taxa that were recognized as varieties by Fedde.

The availability of such a complete and recent review of the genus has greatly aided the comparison of morphological and

TABLE 2

The Genus *Argemone* (Ownbey)

1. *A. fruticosa* Thurb. ex Gray
2. *A. mexicana* L.
   - 2a. *A. mexicana* f. *leiocarpa* (Greene) Ownb.
3. *A. ochroleuca* Sweet
   - 3a. *A. ochroleuca* subsp. *ochroleuca*
   - 3b. *A. ochroleuca* subsp. *stenopetala* (Prain) Ownb.
4. *A. superba* Ownb.
5. *A. aenea* Ownb.
6. *A. aurantiaca* Ownb.
7. *A. echinata* Ownb.
8. *A. squarrosa* Greene
   - 8a. *A. squarrosa* subsp. *squarrosa*
   - 8b. *A. squarrosa* subsp. *glabrata* Ownb.
9. *A. hispida* Gray
10. *A. munita* Dur. & Hilg.
    - 10a. *A. munita* subsp. *robusta* Ownb.
    - 10b. *A. munita* subsp. *rotundata* (Rydb.) Ownb.
    - 10c. *A. munita* subsp. *munita*
    - 10d. *A. munita* subsp. *munita* X subsp. *rotundata*
    - 10e. *A. munita* subsp. *argentea* Ownb.
11. *A. subintegrifolia* Ownb.
12. *A. brevicornuta* Ownb.
13. *A. arizonica* Ownb.
14. *A. pleiacantha* Greene
    - 14a. *A. pleiacantha* subsp. *pleiacantha*
    - 14b. *A. pleiacantha* subsp. *pinnatisecta* Ownb.
    - 14c. *A. pleiacantha* subsp. *ambigua* Ownb.
15. *A. platyceras* L. & O.
16. *A. arida* Rose
17. *A. sanguinea* Greene
18. *A. chisosensis* Ownb.
19. *A. corymbosa* Greene
    - 19a. *A. corymbosa* subsp. *corymbosa*
    - 19b. *A. corymbosa* subsp. *arenicola* Ownb.
20. *A. gracilenta* Greene
21. *A. grandiflora* Sweet
    - 21a. *A. grandiflora* subsp. *grandiflora*
    - 21b. *A. grandiflora* subsp. *armata* Ownb.
22. *A. polyanthemos* (Fedde) Ownb.
23. *A. albiflora* Hornem.
    - 23a. *A. albiflora* subsp. *albiflora*
    - 23b. *A. albiflora* subsp. *texana* Ownb.

24. *A. rosea* Hook
25. *A. hunnemanni* Otto & Dietr.
26. *A. crassifolia* Ownb.
27. *A. subfusiformis* Ownb.
    27a. *A. subfusiformis* subsp. *subfusiformis*
    27b. *A. subfusiformis* subsp. *subinermis* (Fedde) Ownb.
28. *A. glauca* L. ex Pope
    28a. *A. glauca* var. *glauca*
    28b. *A. glauca* var. *decipiens* Ownb.
    28c. *A. glauca* var. *inermis* Deg. & Deg.

chemical data. Our own research group is fortunate in being located within easy access to many of the taxa collected and studied by Ownbey. We are in the process of investigating and reporting on ten species, all collected personally in the same locations as described in the monograph (Ownbey, 1958), and for each of which voucher specimens have been deposited in the Intermountain Herbarium at Utah State University and at Colorado State University.

B. Morphological Comparisons

The genus *Argemone* presents quite different problems from those of *Papaver*. As has been noted, in the latter genus quite well-defined sections have been established on the basis of morphology, and an analysis of alkaloid content has shown these sections, in the main, to be distinguishable chemically. In *Argemone*, on the other hand, the situation can best be summarized by a quotation from Ownbey (1958):

> Some effort has been made without success to discover groupings of species in *Argemone* which might form the basis for establishing sections within the genus. There is no evidence for the existence of groupings of this magnitude. Some poorly definable alliances among species may, however, be pointed out, but there are many species which cannot be assigned with any assurance to any alliance, nor do the alliances themselves appear to have any degree of reality.... It is probably unwise to attempt any taxonomic subdivision of *Argemone* above the rank of species.

It might be expected, in view of this statement, that little success would be forthcoming in attempting to set up sections of the genus by simply adding one more characteristic, that of alkaloid content, to the description of the various species. However, by 1963, at least partial data were available on the alkaloid content of seven species, and it was suggested (Slavik and Slavikova, 1963) that these species seemed to fall into two definite groups on the basis of their alkaloid content. It appeared evident to us that these two groups not only coincided with two of Ownbey's "poorly definable alliances," but that a similar division could be inferred from the earlier morphological classification (Fedde, 1909). Further discussion of these ideas will be deferred until after the following section, which will review published work and our own recent studies on seven species.

C. Alkaloid Content and Taxonomic Alliances

In Table 3 are listed the alkaloid structures found to be of importance for systematics in the species which have been investigated. Often, more than one investigation has been carried out on a single species, but reference is given only to the most recent (or most complete) study. The species are all named in accordance with the Ownbey (1958) monograph, even where this name differs from that in the reference. A pertinent example is *A. albiflora* Hornem, which is now used for *A. alba* Lestib. Alkaloids belonging to other structural classes have been found in various *Argemone* species, but these compounds have not yet proven to be of systematic value. For example, reticuline (XV) has been identified in *A. munita, A. hispida*, and *A. corymbosa* (Stermitz and Sieber, 1966), but, because it is presumably a precursor for almost all of the alkaloids of this genus, its presence in small amounts is not of significance for systematics.

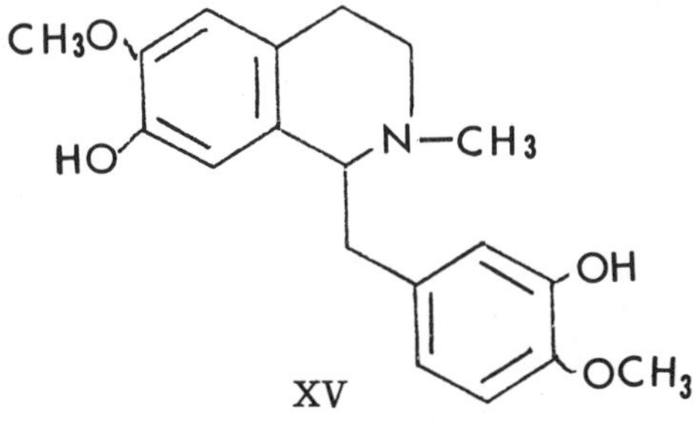

CH_3O / HO / $N-CH_3$ / OH / OCH_3

XV

TABLE 3

Distribution of Types of Alkaloids in Argemone

	Protopine Type	Berberine Type	Pavine Type
	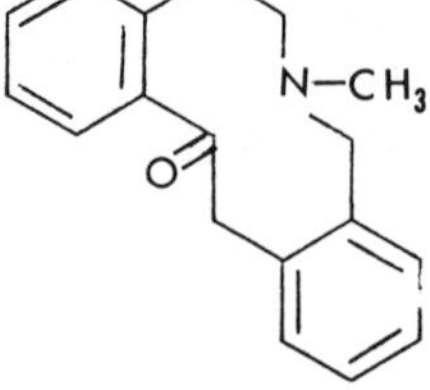	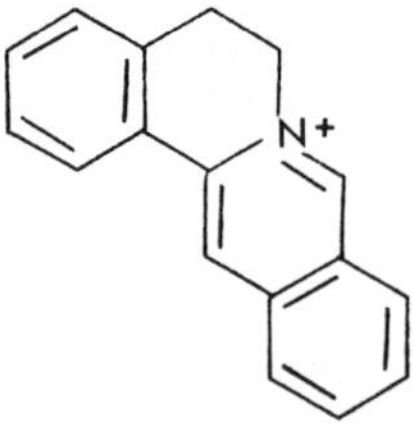	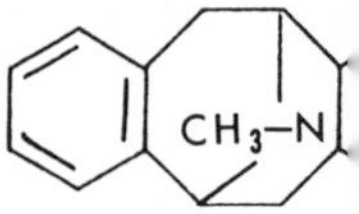
	XVI	XVII	XVIII
A. mexicana L.[a]	+	+	−
A. aenèa Ownb.[b]	+	+	−
A. albiflora Hornem.[c]	+	+	−
A. polyanthemos (Fedde) Ownb.[d]	+	+	−
A. corymbosa Greene subsp. *arenicola*[d]	+	+	−
A. ochroleuca Sweet[e]	+	?	−
A. squarrosa Greene subsp. *squarrosa*[f]	+	−	−
A. arizonica Ownb.[g]	+	?	−
A. platyceras L. & O.[h]	+	+	+
A. pleiacantha Greene subsp. *ambigua*[i]	+	+	+
A. hispida Gray[j]	−	−	+
A. gracilenta Greene[g]	−	−	+
A. munita Dur. & Hilg. subsp. *rotundata*[j]	Trace	−	+

[a]Slavikova and Slavik (1956); [b]Dominguez and Barragin (1965); [c]Slavikova et al. (1960); [d]Stermitz (1966); [e]Giral and Sotelo (1959); [f]Soine and Willette (1960); [g]Stermitz and McMurtrey (1965); [h]Slavik and Slavikova (1963); [i]Stermitz and Coomes (1965); [j]Stermitz and Seiber (1966).

Table 4 illustrates the possible alliances of species as suggested by Ownbey (1958). This treatment may be compared with the classification of Fedde (1909), who described under one species, *A. platyceras*, varieties now recognized as *A. pleiacan-*

Table 4

Possible Alliances in the Genus *Argemone* (Ownbey)

A	B	C
A. mexicana	*A. pleiacantha*	*A. albiflora*
aenea	*echinata*	*polyanthemos*
superba	*aurantiaca*	*grandiflora*
ochroleuca	*hispida*	*gracilenta*
	munita	
	squarrosa	

tha, *A. gracilenta*, *A. hispida*, and *A. munita*. Since Ownbey has stated (1958) that *A. platyceras* is probably more closely related to *A. pleiacantha* than to any other species, it could presumably be added to alliance B. All the above groupings have been based on characters such as morphology, cytology, and so on.

With respect to the chemical characteristics, the suggestion was made (Slavik and Slavikova, 1963) that the genus can be divided into two groups depending on the presence or absence of the pavine types of alkaloids (XVIII). This suggestion has certainly been strengthened by subsequent work as is seen from a comparison of Tables 3 and 4. Thus, in Table 4, if *A. gracilenta* is moved to alliance B and *A. squarrosa* removed from that alliance, then B contains all the species in which the pavine types of alkaloids (XVIII) have been found. Even more striking is the fact that in the classification of Fedde (1909), everyone of the taxa that he listed as varieties of *A. platyceras* (and which have been investigated chemically) has been found to contain pavine types of alkaloids, but in no instance have these alkaloids been detected in taxa outside this group. Although Fedde included *A. gracilenta* as a variety under *A. platyceras*, *A. squarrosa* was maintained by him as a species. It is interesting that when Ownbey raised Fedde's varieties of *A. platyceras* to the species level (which is apparently entirely justified), the apparent relationship among the plants of this group was no longer implied. Alkaloid analysis has now established the existence of a chemical relationship. Within this group of pavine-containing species, however, there is an additional striking demaraction. The species *A. platyceras* and *A. pleiacantha* and their several varieties have an intensely colored (usually bright yellow-orange) latex, the color of which is due to the presence of alkaloids having either

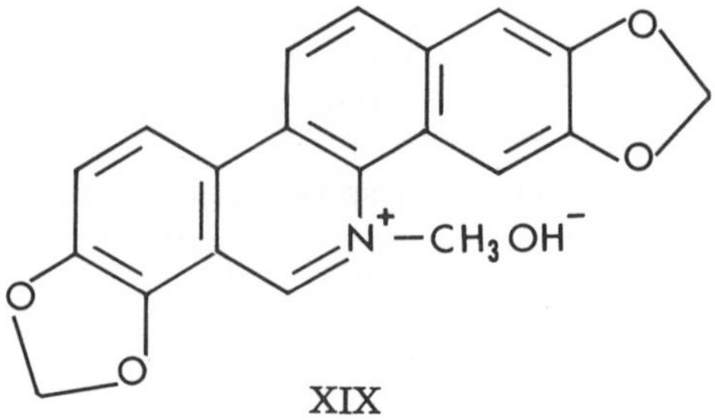

XIX

the berberine type of structure (XVII) or that of sanguinarine (XIX). The latter group are not indicated in Table 3 but have been found in a number of the species. These alkaloids are completely lacking in three members whose latex is of a pale cream color: *A. gracilenta*, *A. hispida*, and *A. munita*. Not all species with yellow-orange sap can be related to *A. platyceras* and *A. pleiacantha*, since most species outside alliance B have this characteristic, although they do not contain the pavine types of alkaloids.

Further study of Tables 3 and 4 indicates that at least two additional groupings can be made in the genus. The species *A. aenea*, *A. albiflora*, *A. mexicana*, *A. ochroleuca*, *A. polyanthemos*, and *A. corymbosa* contain alkaloids of the protopine type (XVI) and the berberine type (XVII), with no accompanying pavines (XVIII). This relationship is even closer than the gross structures of the alkaloids indicated, since it is found that the three major alkaloids of each species are exactly the same: protopine (XX), allocryptopine (XXI), and berberine (XXII).

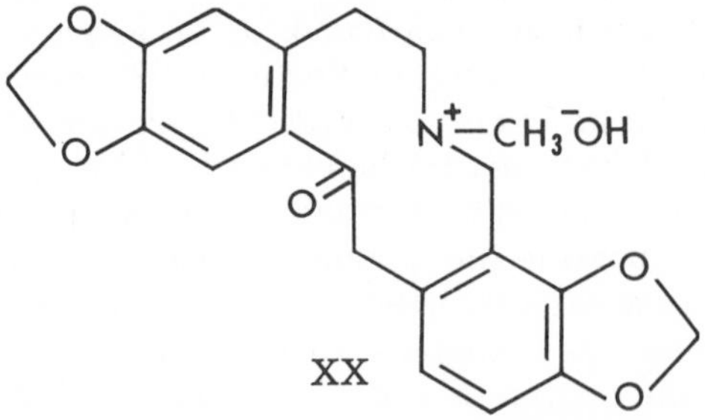

XX

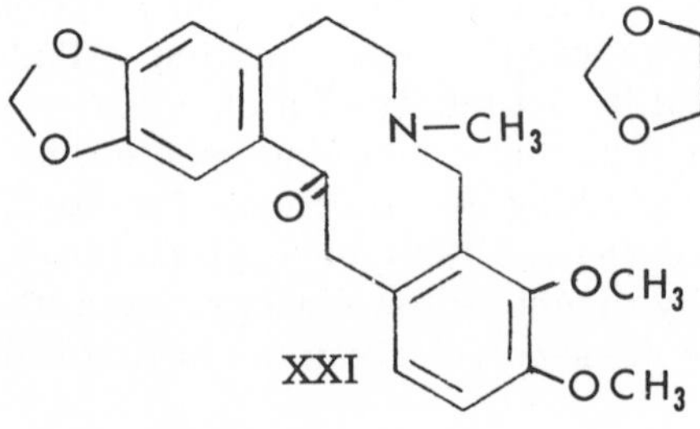

XXI

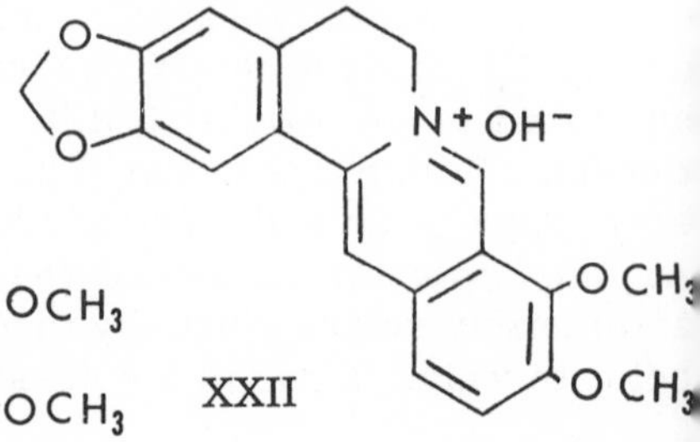

XXII

Table 3 indicates that traces of the protopine types of alkaloids have been found in *A. munita* subsp. *rotundata*. Although these traces could probably be ignored for the purposes of systematics, the protopine alkaloids are not those just described but rather the protopine types, cryptopine (**XXIII**) and muramine (**XXIV**). Thus, the trace alkaloids in this example simply provide further evidence for separating the taxa. *A. squarrosa* and *A. arizonica*

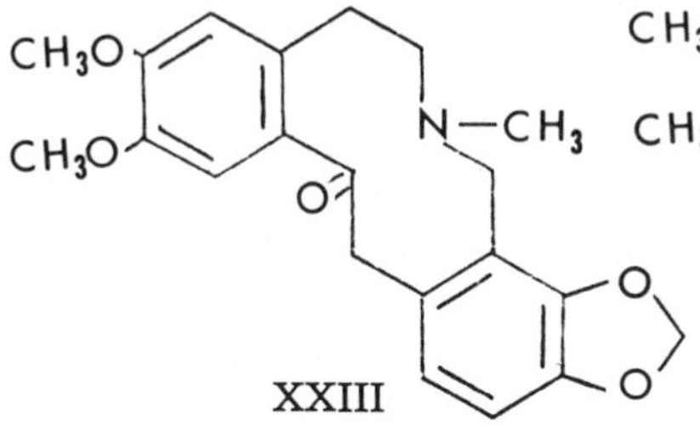

CH_3O CH_3O $N-CH_3$ O XXIII

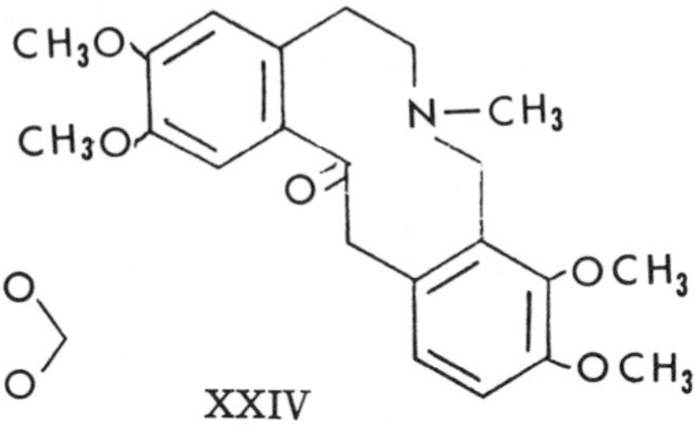

CH_3O CH_3O $N-CH_3$ O OCH_3 OCH_3 XXIV

have not been completely investigated, but there is evidence that these two species are set apart in their alkaloid chemistry from the groups so far discussed. Both species have a pale cream-colored latex, which indicates that alkaloids of the type **XVII** or **XIX** are probably absent. Some preliminary investigations (Stermitz and McMurtrey, 1965) indicate that there is a small amount of an alkaloid of the **XVII** group in *A. arizonica*, but protopine appears to be the only alkaloid present in significant amounts. It has been reported (Soine and Willette, 1960) that only allocryptopine (**XXI**) is to be found as a major alkaloid component in *A. squarrosa*. No close relationship between the two species has been suggested on morphological grounds, and further work and study would be necessary to place these two species.

By combining all present knowledge, including the alkaloid chemistry, regarding the well-studied species, we can arrive at a suggested division of the genus *Argemone* into the groups of Table 5. It would be possible, on the basis of comparative

Table 5

Revised Alliances in the Genus *Argemone*

I	II	III	IV
A. gracilenta	*A. platyceras*	*A. albiflora*	*A. mexicana*
hispida	*pleiacantha*	*polyanthemos*	*aenea*
munita		*corymbosa*	*ochroleuca*

morphology alone, to assign several additional species to one or the other of these groups. However, it is likely that this can be done with more assurance after the chemical evidence is added.

Some additional comments should be made in regard to Table 5. First, no attempt has been made to give the groupings any special name. For example, they have not been designated as sections. Such names have special taxonomic meanings, and it will be the task of the taxonomist to make any decision regarding whether or not the alliances described conform with the standard botanical use of a term such as section. Such a decision would, of course, be made with strong consideration of the various morphological characteristics, which have not been described to any extent here. Second, as has been pointed out (Alston and Turner, 1963), one must be aware of the possibility of chemical variations within a species. Some comment should be made on this aspect in the genus *Argemone*. In alliance I, we have studied *A. munita* subsp. *rotundata* from several locations throughout its range and have found little variation in alkaloid content. Our results on this species and that of *A. hispida* have also correlated well with the pioneering reports on these two species by Soine and Kier (1963). In alliances III and IV, the remarkable finding of the same three alkaloids representing the major ones of every species indicates that there is little likelihood that important chemical variants occur among these species. However, we have found (Stermitz and Coomes, 1965) considerable variation in *A. pleiacantha* subsp. *ambigua*, of alliance II. None of the variations so far found are important enough to have a bearing on the systematics discussed, but are of interest biosynthetically. Finally, it is evident that alliances III and IV have been divided without regard to the chemical content of the species included.

D. Possible Phylogenetic Relationships

It has been observed (Ownbey, 1958, 1960) that speciation in *Argemone* appears to have been the result primarily of geographical separation with ecological diversity as a second factor. Geographical relationships among the genera and species of Papaveraceae were discussed at some length by Fedde (1909, pp. 65–72). The following suggestion is especially interesting relevant to *Papaver* and *Argemone*. The family Papaveraceae was considered to be of northern origin, with some species having spread southward before the climate changes associated with the ice ages. Appearance of the ice presumably eliminated all the most northern species with the exception of several members of the section Scapiflora of the genus *Papaver*, which are found native today largely in high mountains or near-Arctic regions. The widespread and numerous species of the genus *Papaver* that occur throughout central Europe today are regarded as having developed chiefly from a northward spreading of those species which had originally migrated into the Mediterranean region and had thus escaped the ice.

As a result of such considerations, we could perhaps assume that the species of the Papaveraceae with the closest link to those of early northern origin would be species which had already reached southern regions prior to the advent of the ice age and species of the section *Scapiflora*, the ice age relics. It is in striking agreement with this hypothesis that there exists a unique chemical tie-up between the *Papaver* section *Scapiflora* and at least two genera occurring south of the ice age limit in the Americas. This is the occurrence of the pavine (or pavine-like) alkaloids in the section *Scapiflora* and also in *Eschscholtzia* and *Argemone*. If this is to be considered a primitive chemical characteristic, then it would follow that, among species of *Argemone*, *A. gracilenta*, *A. hispida*, and *A. munita* are among the most primitive. In an intermediate phylogenetic position would be the species *A. platyceras* and *A. pleiacantha*, which contain the pavine types of alkaloids and also numerous others. The placement of *A. platyceras* as an intermediate species in evolutionary development was suggested earlier (Slavik and Slavikova, 1963). The most recently evolved species would then be those of alliances III and IV, which have lost the capacity to accumulate the pavine types of alkaloids. These suggestions are also supported by some of the morphological characteristics. For example, in general, perennials are considered (Hutchinson, 1964) to be more primitive than biennials, with annuals representing the most advanced plants. In *Papaver*, the section *Scapiflora* is composed only of perennials, while the possibly more recently evolved species of other sections (e.g., *P. somniferum* or *P. rhoeas*) are all annuals. In *Argemone*, *A. hispida*, *A. munita*, and *A. gracilenta* are all perennials, while those species of alliance IV are annuals (although some can, in addition, exhibit perennial character).

These tentative phylogenetic hypotheses are put forth in full knowledge of the difficulties involved in proving or disproving such suggestions. It is of course extremely difficult to evaluate the direction of phylogeny even where strong circumstantial evidence exists for assuming close relationships. It is hoped that, at the very least, the hypotheses will arouse sufficient interest to provide the impetus for further research in this area.

ACKNOWLEDGMENT

The author is indebted to the National Institute of General Medical Sciences, United States Public Health Service, for a Research Career Development Award which has allowed him time to prepare this paper, and for Grant GM-09300, under which the experimental work was done at Utah State University.

REFERENCES

Alston, R. E., and B. L. Turner. 1963. Biochemical Systematics, 155-179, Englewood Cliffs, New Jersey, Prentice-Hall.

Barton, D. H. R., R. H. Hesse, and G. W. Kirby. 1965. J. Chem. Soc.: 6379.

Bernasconi, R., St. Gill, and E. Steinegger. 1965. Pharm. Acta Helv., 40: 275.

Brochmann-Hanssen, E., B. Nielsen, and G. E. Utzinger. 1965. J. Pharm. Sci., 54: 1531.

Cross, A. D., L. Dolejs, V. Hanus, M. Maturova, and F. Santavy. 1965. Collection Czech. Chem. Commun., 30: 1335.

Douglas, B., J. L. Kirkpatrick, R. F. Raffauf, O. Ribeiro, and J. A. Weisbach. 1964. Lloydia, 27: 25.

Dominguez, X. A., and V. Barragin. 1965. J. Org. Chem., 30: 2049.

Dyke, S. F. 1965. Lect. Roy. Inst. Chem., No. 1.

Fairbairn, J. W., and G. Wassel. 1964. Phytochemistry, 3: 583.

Fedde, F. 1909. *In* Das Pflanzenreich, Engler, A., ed., vol. 40, Leipzig, Wilhelm Englemann.

Flentje, H., W. Doepke, and P. W. Jeffs. 1965. Naturwissenschaften, 52: 259.

Giral, F., and A. Sotelo. 1959. Ciencia, 19: 67.

Hegnauer, R. 1963. *In* Chemical Plant Taxonomy, Swain, T., ed., 389-427, New York, Academic Press.

Heydenreich, K., and S. Pfeifer. 1965. Pharmazie, 20: 521.

Hutchinson, J. 1964. The Genera of Flowering Plants, vol. I, 10-11, London, Oxford Univ. Press.

Kiselev, V. V., and R. A. Konovalova. 1948. Zh. Obsch. Khim., 18: 142.
Kühn, L., and S. Pfeifer. 1963. Pharmazie, 18: 819.
_____, and S. Pfeifer. 1965a. Pharmazie, 20: 659.
_____, and S. Pfeifer. 1965b. Pharmazie, 20: 520.
Manske, R. H. F. 1966. Private communication.
Maturova, M., L. Hruban, F. Santavy, and W. Wiegrebe. 1965. Arch. Pharm. (Weinheim), 298: 209.
_____, D. Pavlaskova, and F. Santavy. 1966. Planta Medica, 14: 22.
Mnatsakanyan, V. A., and S. Yu. Yunusov. 1961. Dokl. Akad. Nauk Uz. S. S. R., 3: 34.
Neubauer, D. 1965. Arch. Pharm. (Weinheim), 295: 737.
_____, and K. Mothes. 1963. Planta Medica, 11: 387.
Ownbey, G. B. 1958. Memoirs Torrey Botanical Club, 21, No. 1.
_____. 1960. Brittonia, 13: 91.
Pfeifer, S. 1966. J. Pharm. Pharmacol., 18: 133.
_____, and S. K. Banerjee. 1962. Pharmazie, 19: 286.
_____, and V. Hanus. 1965. Pharmazie, 20: 394.
Santavy, F. 1962. Collection Czech. Chem. Commun., 27: 1717.
_____, M. Maturova, A. Nemeckova, and M. Horak. 1960. Planta Medica (Stuttgart), 8: 167.
_____, J. L. Kaul, L. Hruban, L. Dolejs, V. Hanus, K. Blaha, and A. D. Cross. 1965. Collection Czech. Chem. Commun., 30: 335.
_____, M. Maturvoa, and L. Hruban. 1966. Chem. Commun., 36: 144.
Slavik, J., and L. Slavikova. 1956. Collection Czech. Chem. Commun., 21: 211.
_____, and L. Slavikova. 1963. Collection Czech. Chem. Commun., 28: 1728.
_____, V. Hanus, K. Vokac, and L. Dolejs. 1965a. Collection Czech. Chem. Commun, 30: 2464.
_____, J. Appelt, and L. Slavikova. 1965b. Collection Czech. Chem. Commun., 30: 3961.
Slavikova, L., and J. Slavik. 1960. Collection Czech. Chem. Commun., 25: 756.
Soine, T. O., and L. B. Kier. 1963. J. Pharm. Sci., 52: 1013.
_____, and R. E. Willette. 1960. J. Amer. Pharm. Ass., 49: 368.
Stermitz, F. R. 1966. Unpublished results.
_____, and R. M. Coomes. 1965. Unpublished results.
_____, and K. McMurtrey. 1965. Unpublished results.
_____, and J. N. Seiber. 1966. J. Org. Chem., 31: 2925.
Tetenyi, P., and D. Vagujfalvi. 1965. Pharmazie, 20: 731.

part III

ACETATE- AND MEVALONATE-DERIVED COMPOUNDS

6

THE TAXONOMIC SIGNIFICANCE OF ACETYLENIC COMPOUNDS

N. A. SØRENSEN
Organic Chemistry Laboratories
Norway Institute of Technology, Trondheim

I. INTRODUCTION

Today many taxonomists are interested in the distributions of secondary compounds of plants. Numerous examples exist where problems concerning the position of a genus or a species are not capable of solution by classical morphological methods.

Plant chemistry provides one possible additional method in taxonomy.

To solve these problems the taxonomist has frequently introduced new methods such as embryology or palynology, and more recently chemical data. In fact, bitter principles, colors, odors, and latex have played a minor part in systematics since pre-Linnaean days. In a more precise sense, chemistry has been of great importance in certain instances, for example, lichen systematics. Nevertheless, there is still a widespread, skeptical attitude among some botanists toward chemical taxonomy. Most natural products were isolated only because they happened to be abundant—or in some way obvious, like the perfume plant or the fetid plant. However, acetylenic compounds seldom occur in large quantities in the plants in which they are found, although some plants may have a very high content of acetylenic fatty acids. Most botanists are poorly informed of the extent to which modern natural products chemistry is dependent on instrumentation. From a physicist's point of view nothing important has happened in ultraviolet (UV) or infrared (IR) spectroscopy since Victor Henri wrote his "Etudes de Photochimie" in 1919. But in our technological age the chemist had to wait for 40 years to obtain convenient and reliable commercial instruments.

In Fig. 1, the UV spectra of two widely distributed acetylenes are given, dehydromatricaria ester and trideca-pent-ynene. The spacing of the fine structure immediately suggests that the compounds are polyacetylenic. The extinction coefficients are high so that often less than a milligram of material may be detected in the crude extracts of the plants. Obviously, plants where these types of acetylenes occur can be readily discovered.

The UV spectra of three other acetylenic compounds are given in Fig. 2. If these curves, without the attached formulas, were presented to an organic chemist, he would conclude that the spectra might represent any of at least a dozen different types of natural compounds; thus, there is a good chance that numerous fractions with such spectra have been discarded. If such acetylenic compounds happen to be the major components of an extract, an infrared spectrum would reveal their acetylenic nature although the stretching frequency of the triple bond is "forbidden" in IR. One of the consequences of this fact is that the intensity of this band is variable. For example, it practically disappears in some propargylic alcohols.

Table 1 lists some simple acetylenic structures that may have a very wide distribution in nature. In UV, these substances

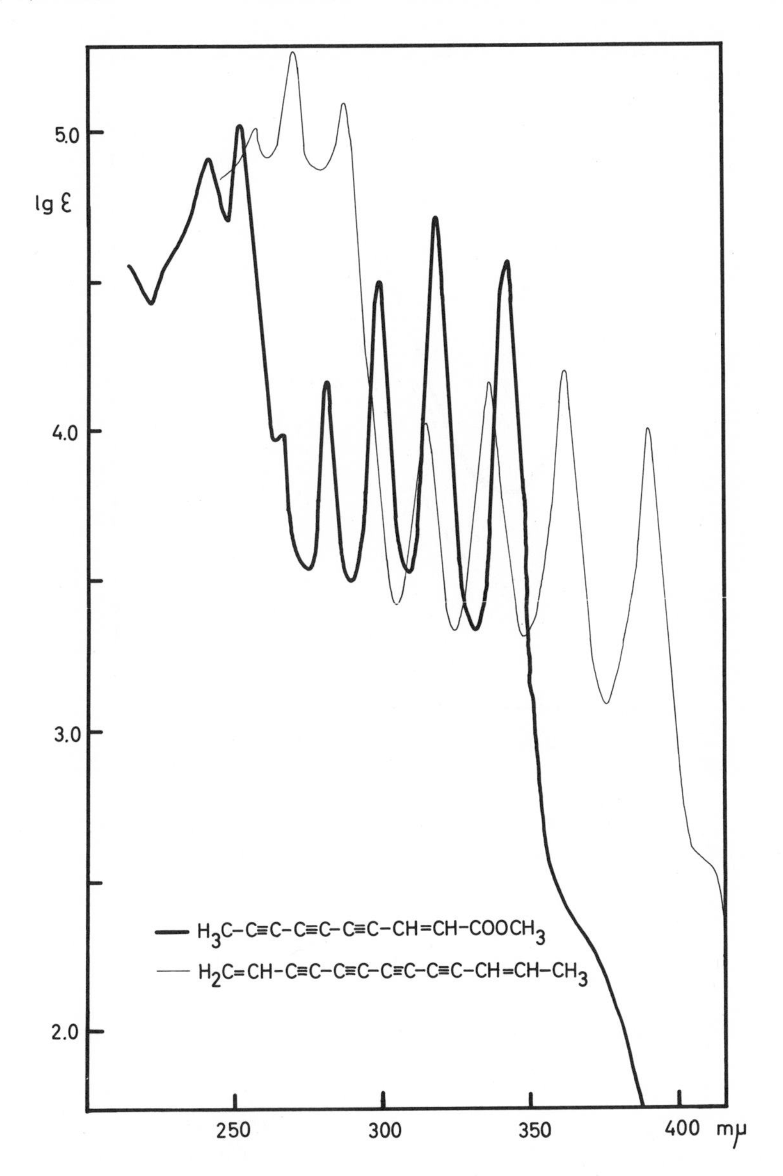

Fig. 1. Two examples of fine structure in the ultraviolet absorption spectra of polyacetylenic compounds.

are either not recognizable or they cannot, in crude extracts, be distinguished from certain di- or tri-enes, or unsaturated carbonyl compounds. Furthermore, as may be seen from the last

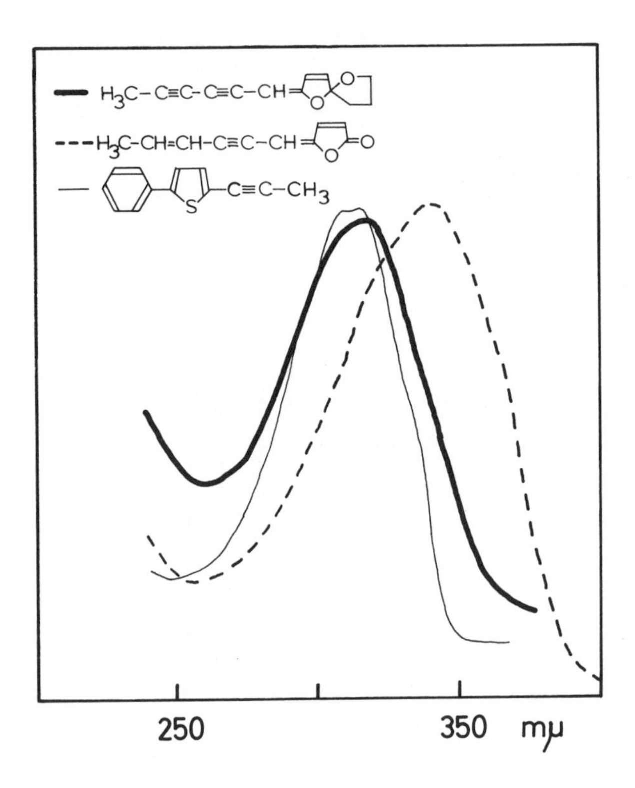

Fig. 2. Three examples of broad-banded ultraviolet absorption spectra of polyacetylenic compounds.

TABLE 1

Chromophore Types Where UV and IR Spectroscopy Do Not Indicate Acetylenic Structure

Structure	Ultraviolet Absorption, mμ	Infrared, $2200\ cm^{-1}$ Band
$R-(CH_2)_x-C\equiv C-(CH_2)_y-R$	178	Absent
$R-(CH_2)_x-CH=CH-C\equiv C-(CH_2)_y-R'$	223	Very weak
$-CH=CH-CH=CH-C\equiv C-$	267 (277)	Weak
$-CH=CH-C\equiv C-CH=CH-$	266 275	Very weak
$-CH=CH-CH=CH-CH=CH-$	253 263 274	—

column of the table, infrared spectroscopy also provides little assistance in detecting the triple bond in these compounds.

The triple bond stretching frequency is always strong in the Raman spectrum. When Sir C. V. Raman introduced his method 40 years ago, some 200 ml of a substance of extreme stability was required; therefore, the method was nearly useless in natural products research. In about 1955, scientific equipment was constructed that produced good Raman spectra on less than 1 gm of material in about 15 minutes. At the present time an instrument costing about $15,000 is available that only requires 50 mg of pure substance, and some work has been published with samples as small as 5–10 mg. Possibly, in a few years the natural products chemist may have a Raman spectrometer working on the scale of 1–2 mg. It is impossible to predict how Raman spectroscopy will change our picture of the distribution of acetylenic compounds. Personally, I expect the change to be tremendous. Thus, the more a botanist knows about the present situation, that the organic chemist of today has developed into an appendix to spectroscopic instruments, the more reserved will be his attitude toward premature applications of chemotaxonomy.

The acetylenic compounds also raise other problems. Acetylenes from fungi are mostly excreted into the nutritional medium. Sir Ewart Jones and his collaborators in a number of instances have extracted the washed mycelium and found it to be essentially devoid of acetylenic compounds. Recently, I had the pleasure of discussing these problems with Dr. Anchel at the New York Botanical Garden. Dr. Anchel informed me that the fruiting bodies of basidiomycetes do contain acetylenes, apparently the same ones that are known to be excreted into the culture medium. However, one should remember that the fruiting bodies of the basidiomycetes are so constructed that they are never extracted by rain or dew. An excreted metabolic product is of course a natural compound, but, as mycologists know very well, the compounds excreted from fungi are often quite dependent on culture conditions. In a recent paper, Jones et al. (1966) stated that *Fistulina hepatica* is a poor producer of acetylenes when grown in surface culture; but when the cultures are shaken the mold becomes an excellent source of polyacetylenes.

Another case with fungi may become a classic. The first acetylenic thiophene, junipal, was isolated by Birkinshaw and Chaplen (1955) from *Daedelia juniperina* grown under the standard conditions in Professor A. Raistrick's institute (cf. Table 2). While about 50 acetylenic thiophenes have since been isolated

TABLE 2

Metabolic Products from *Daedalea Juniperina*

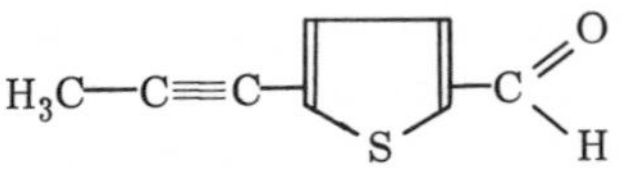 $H_3C{-}C{\equiv}C{-}(\text{2,5-thienyl}){-}CHO$	Junipal
$H_3C{-}C{\equiv}C{-}C{\equiv}C{-}C{\equiv}C{-}CH{=}CH{-}CH_2OH$	Dehydromatricarianol
$H_3C{-}C{\equiv}C{-}C{\equiv}C{-}CH{=}C{=}CH{-}CH_2{-}CH_2OH$	
$H_3C{-}CH{=}CH{-}CH{=}CH{-}CH{=}CH{-}CH{=}CH{-}CH_2OH$	2:4:6:8 Decatetraenol

from the Compositae, no other thiophenes have been isolated from fungi. In a recent reinvestigation of *Daedalea juniperina* (Bew et al., 1966) under various culture conditions, no trace of junipal was detected; instead, some new polyacetylenes and polyenes were found. While all four products in Table 2 are natural products synthesized by *Daedalea juniperina*, their importance from a taxonomic point of view is subject to question.

Higher plants are well known often to accumulate characteristic secondary plant products only in certain specific plant parts. Furthermore, and most importantly, some secondary plant products are produced only during a brief period in the life cycle. As an example, the cotyledons of *Xanthium* in Australia are ill-reputed for their toxic action, while the developed plants are quite harmless. Many young grasses are dangerous because of their high content of cyanogenetic glycosides, whereas the mature grass plant gives negative test for cyanogens. If such transient compounds are not conspicuous they are apt to be easily overlooked. It is well known that the concentrations and relative proportions of acetylenes may change in many of the Compositae during the growing season. But most interesting observations of this type were made by Bu'Lock and Smith (1963) with the fatty acids from *Eucaria acuminata*. Hatt and Wailes at CSIRO, Melbourne (Wailes et al., 1960) had demonstrated that unsaturation in the fatty acids of the Santalaceae always increased from the ripe seed fat, through the unripe seed, the leaves, and stems and reached the maximum number of acetylenic bonds in the root. Bu'Lock demonstrated that the sprouting seed and the seedling of *Eucaria acuminata* synthesized transiently acetylenic fatty acids still more unsaturated than those known from the root glycerides.

In the related family, the Loranthaceae, acetylenic fatty acids are found occasionally in the mature plant. What should be done, of course, is to look for a transient occurrence of acetylenes in sprouting seeds of those Loranthaceae which reputedly lack these substances. Classical morphology is a science of the entire plant during all stages of life, and so, of course, must be natural products chemistry if it is to play a part in taxonomy.

Finally, I should like to mention another restriction on the use of chemical compounds in taxonomy, that is, that the compound isolated should be a regular metabolic product and not be due to infection, symbiosis, and so on. I shall consider an example involving couch-grass. In 1947, Treibs (1947) demonstrated that a sample of an essential oil from the root of couch-grass (*Agropyrum repens*), which was provided by Schimmel and Co., consisted of about 90 percent of an acetylenic compound, agropyrene. The correct constitution of agropyrene was shown by Craig (1959) to be

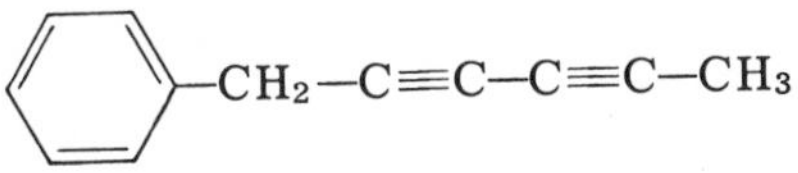

$$C_6H_5{-}CH_2{-}C{\equiv}C{-}C{\equiv}C{-}CH_3$$

As indicated by Craig, agropyrene might have originated from the *Artemisia dracunculus* roots present as an impurity in the starting material. Couch-grass roots and the roots of the closely related *Triticum* have been reinvestigated recently by Schulte et al. (1965). No indication of the occurrence of agropyrene was given. Instead minute amounts of some aliphatic poly-ynes were described. The composition of these polyacetylenes varied from sample to sample. Personally I do not accept that these results prove that grasses synthesize acetylenes. If compounds are not found consistently, fungal infections are suspect and reinvestigations on pure and healthy material are imperative.

Among higher plants numerous examples are known where an infection gives rise to the production of quite new compounds, sometimes in astonishing amounts. This is a very fascinating and important chapter in itself, but it has nothing to do directly with secondary plant products in relation to taxonomy.

It is unfortunate that one has to look for the transient occurrence of acetylenic compounds during short periods of the life cycle of a plant. The requirement that the acetylenic compound shall be a regular product of the "pure" plant itself when grown under physiologically normal conditions, however, has to be considered rigorous.

TABLE 3

Acetylenic Fatty Acids

Structure	Number of Atoms	Genus First Isolated	Family
$H_3C—(CH_2)_7—C{\equiv}C—(CH_2)_6—COOH$	17	*Hibiscus*	Malvaceae
$H_3C—(CH_2)_5—CH{=}CH—C{\equiv}C—(CH_2)_6—COOH$	17	*Acanthosyris*	Santalaceae
$H_3C—(CH_2)_5—CH{=}CH—C{\equiv}C—CHOH—(CH_2)_5—COOH$	17	*Acanthosyris*	Santalaceae
$H_2C{=}CH—(CH_2)_4—CH{=}CH—C{\equiv}C—(CH_2)_6—COOH$	17	*Acanthosyris*	Santalaceae
$CH_2{=}CH—(CH_2)_4—CH{=}CH—C{\equiv}C—CHOH—(CH_2)_5—COOH$	17	*Acanthosyris*	Santalaceae
$H_3C—(CH_2)_{10}—C{\equiv}C—(CH_2)_4—COOH$	18	*Picramnia*	Simarubaceae
$H_3C—(CH_2)_7—C{\equiv}C—(CH_2)_7—COOH$	18	[*Pyrularia*, *Sterculia*	[Santalaceae, Sterculiaceae
$H_2C{=}CH—(CH_2)_6—C{\equiv}C—(CH_2)_7—COOH$	18	*Acanthosyris*	Santalaceae
$CH_2{=}CH—(CH_2)_4—CH{=}CH—C{\equiv}C—(CH_2)_7—COOH$	18	*Acanthosyris*	Santalaceae
$CH_2{=}CH—(CH_2)_4—CH{=}CH—C{\equiv}C—CHOH—(CH_2)_6—COOH$	18	*Acanthosyris*	Santalaceae
$H_2C{=}CH—(CH_2)_4—C{\equiv}C—C{\equiv}C—(CH_2)_7—COOH$	18	*Ongokea*	Olacaceae
$H_2C{=}CH—(CH_2)_2—CH{=}CH—C{\equiv}C—C{\equiv}C—(CH_2)_7—COOH$	18	*Ongokea*	Olacaceae
$H_2C{=}CH—(CH_2)_4—C{\equiv}C—C{\equiv}C—CHOH—(CH_2)_6—COOH$ (?)	18	*Ongokea*	Olacaceae
$H_2C{=}CH—(CH_2)_2—CH{=}CH—C{\equiv}C—C{\equiv}C—CHOH—(CH_2)_6—COOH$ (?)	18	*Ongokea*	Olacaceae
$H_3C—(CH_2)_5—CH{=}CH—C{\equiv}C—CHOH—(CH_2)_6—COOH$	18	*Ximenia*	Olacaceae
$H_3C—(CH_2)_5—CH{=}CH—C{\equiv}C—(CH_2)_7—COOH$	18	[*Ximenia*, *Santalum*	[Olacaceae, Santalaceae
$H_3C—(CH_2)_3—CH{=}CH—C{\equiv}C—C{\equiv}C—(CH_2)_7—COOH$	18	*Exocarpus*	Santalaceae
$H_3C—(CH_2)_3—CH{=}CH—CH{=}CH—C{\equiv}C—(CH_2)_7—COOH$	18	*Eucarya*	Santalaceae
$H_3C—CH_2—CH{=}CH—CH{=}CH—C{\equiv}C—C{\equiv}C—(CH_2)_7—COOH$ (?)	18	*Leptomeria*	Santalaceae
$H_3C—CH_2—CH{=}CH—C{\equiv}C—C{\equiv}C—C{\equiv}C—(CH_2)_7—COOH$ (?)	18	*Leptomeria*	Santalaceae
$H_3C—(CH_2)_4—C{\equiv}C—CH{=}CH—CHOH—(CH_2)_7—COOH$	18	*Helichrysum*	Compositae
$H_3C—(CH_2)_4—C{\equiv}C—CH_2—CH{=}CH—(CH_2)_7—COOH$	18	*Crepis*	Compositae
$H_3C—(CH_2)_4—CH{=}CH—CH{=}CH—CHOH—(CH_2)_7—COOH$	18	*Tribe Arctotidae*	Compositae

II. ACETYLENIC FATTY ACIDS

The known acetylenic fatty acids are tabulated in the first two divisions of Table 3.

Owing to recent papers from the Northern Regional Research Laboratory (Powell and Smith, 1966; Powell et al., 1966) the number of acetylenic acids has increased appreciably. From the seed fat of the South American tree *Acanthosyris spinescens* (Mart. et Eich.) Grisebach, not less than seven new acetylenic fatty acids were isolated. The most remarkable point with this seed fat is that more than 50 percent of the acids had a chain length of 17 carbon atoms. The two monoacetylenic acids (stearolic acid and heptadec-8-ynoic acid) and the two acids from *Helichrysum* and *Crepis* (Table 3) have also only recently been described.

The family Santalaceae shows a remarkably consistent picture. About 20 species have been investigated and, so far, acetylenic fatty acids have been detected in all parts of all species of these plants. As noted earlier, unsaturation always increases from the ripe seed fat to the root glycerides. Outside the Santalaceae, only the seed fats have been studied, except for *Ximenia americana*, wherein Hatt and collaborators found the same general situation to prevail as in the Santalaceae.

It is obvious that further investigations of all parts of the plants and, of course, seedlings, are necessary before important taxonomic conclusions may be permitted concerning the order Santalales in general, especially the families Olacaceae and Loranthaceae, and the isolated genus *Picramnia* of the Simarubaceae.

The two new monoacetylenic acids deserve special attention. *Sterculia* and *Hibiscus* supplied the first members of the interesting cyclopropene acids. In the laboratory the only rational way to synthesize such compounds is by a carbene reaction on acetylenes. By reinvestigating *Sterculia* and *Hibiscus* for minor components Smith and Bu'Lock (1965) recently found monoacetylenic acids as companions of the cyclopropene acids.

The second division of Table 3 contains two new acetylenic acids reported from Compositae seed oils. A few seed oils from Compositae are world trade articles, for example, sunflower, safflower, and Guizotia oil from Abyssinia. These oils and many noncommercial oils from the same family have been thoroughly investigated and all of them appear to contain only normal ethylenic fatty acids; although many different acetylenes have been found in other parts of these Compositae.

The seed fats from two tribes of the Compositae are rather peculiar. In 1960 Wolff and co-workers (Earle et al., 1960) from Northern Regional Research Laboratory described 9-hydroxy-10,12-octadeca-dienoic acid ("morphecolic acid") from *Dimorphotheca*. Four years earlier McLean and Clark (1956) described 8,10,12-octadeca-trienoic acid from *Calendula* seeds. In 1964 Wolff and his co-workers (Earle et al., 1964) published the results of an additional investigation of 29 species from five genera of Calenduleae. When my wife and I were visiting CSIRO, Melbourne, in 1960 and 1961, we also investigated this situation and found that the conjugated tri-enes were restricted to the tribe Calenduleae in the narrower sense, morphecolic acid to the tribe Arctotidae, while both these tribes contained up to penta-acetylenes in their root parts.

I expected the seed fats of the Compositae to be devoid of acetylenes; thus it was somewhat of a surprise to me when Wolff and collaborators in 1964 (Mikolajczak et al., 1964) described "crepenynic acid" from *Crepis foetida*. It was a double surprise since *Crepis* belongs to the subfamily Liguliflorae, where acetylenes had never previously been found, despite the fact that the compounds had been found in members of all 12 tribes of the subfamily Tubuliflorae.

Another extremely interesting point is the fact that Bu'Lock (cf. Jones, 1966) demonstrated that the basidiomycete *Tricholoma grammopodium* converts labeled oleic acid into crepenynic acid, further into 14-dehydrocrepenynic acid, and finally to the shorter-chain acetylenic metabolites characteristic of a number of other basidiomycetes. One may conclude from these experiments by Bu'Lock that crepenynic acid may therefore represent a key compound in the chemistry of naturally occurring acetylenes. One may immediately recognize from its constitutional formula that neither UV nor IR spectroscopy is of much assistance in the search for crepenynic acid, and that frequently it may have been overlooked.

III. ACETYLENES FROM BASIDIOMYCETES

The polyacetylenes isolated from fungi have been summarized by Jones (1959) and by Bu'Lock (1961, 1964; Bu'Lock and Powell, 1965). In Figs. 3 to 7, I have tried to give a compact presentation of the compounds; the present discussion will be restricted to certain special cases.

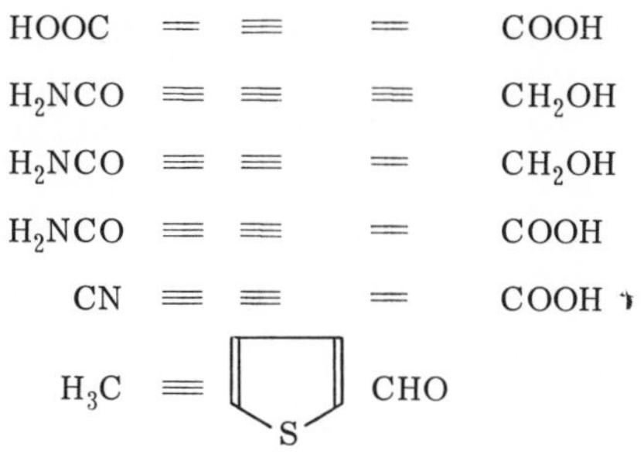

$HOOC \quad = \quad \equiv \quad = \quad COOH$

$H_2NCO \quad \equiv \quad \equiv \quad \equiv \quad CH_2OH$

$H_2NCO \quad \equiv \quad \equiv \quad = \quad CH_2OH$

$H_2NCO \quad \equiv \quad \equiv \quad = \quad COOH$

$CN \quad \equiv \quad \equiv \quad = \quad COOH$

$H_3C \quad \equiv \quad$ (thiophene ring, S) $\quad CHO$

Fig. 3. C_8-Polyacetylenes from basidiomycete fungi.

$H \quad \equiv \quad \equiv \quad \equiv \quad = \quad CH_2OH$

$H \quad \equiv \quad \equiv \quad \equiv \quad = \quad CHO$

$H \quad \equiv \quad \equiv \quad \equiv \quad CH(OH){-}CH(OH){-}CH_2OH$

$H \quad \equiv \quad \equiv \quad \equiv \quad CH{-}CH{-}CH_2OH$ (CH—CH bridged by O, epoxide)

$H \quad \equiv \quad \equiv \quad \equiv \quad = \quad COOH$

$H \quad \equiv \quad \equiv \quad = \quad CH_2CH_2CH_2OH$

$H \quad \equiv \quad \equiv \quad = \quad CH(OH){-}CH(OH){-}CH_2OH$

$H \quad \equiv \quad \equiv \quad = \quad CH_2CH_2COOH$

$H_3CCH_2{-}CHOH \quad \equiv \quad \equiv \quad CHOHCH_2OH$

$H_3CCH_2{-}CO \quad \equiv \quad \equiv \quad CHOHCH_2OH$

$HOCH_2{-}CH_2 \quad \equiv \quad \equiv \quad = \quad CH_2OH$

$HOOC{-}CH_2 \quad \equiv \quad \equiv \quad = \quad CH_2OH$

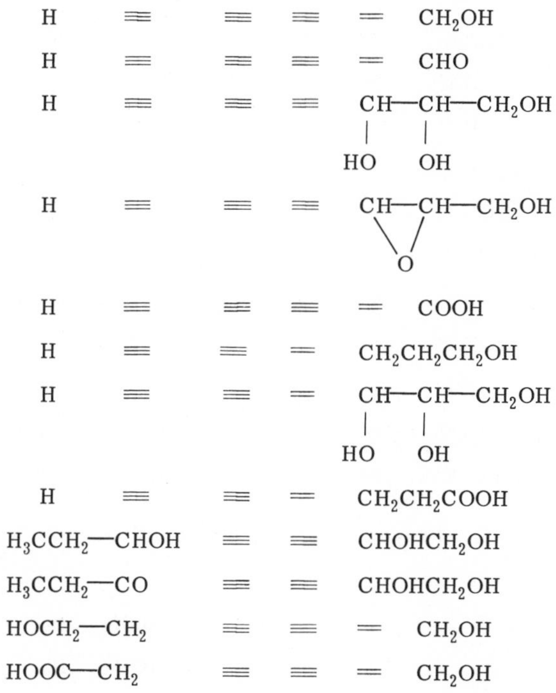

Fig. 4. C_9-Polyacetylenes from basidiomycete fungi.

It may be considered odd to include mycomycin under the present heading. This antibiotic substance was isolated by Celmer and Solomons (1952). The constitution was fully established by scientists of the Pfizer laboratories:

$$HC{\equiv}C{-}C{\equiv}C{-}CH{=}C{=}CH{-}CH{=}CH{-}CH{=}CH{-}CH_2{-}COOH$$

H_3C	≡	≡	≡	=	$CH_2OH\ (CH_3)$
H_3C	≡	≡	≡	=	CHO
H_3C	≡	≡	≡	=	$COOH$
HOH_2C	≡	≡	≡	=	CH_2OH
HOH_2C	=	≡	≡	≡	CHO
$(H_3C)\ HOOC$	=	≡	≡	≡	CH_2OH
$(H_3C)\ HOOC$	=	≡	≡	≡	$COOH$
HOH_2C	≡	≡	≡	$CH_2\ CHOHCOOH$	
$HOOC$	≡	≡	≡	$CH_2\ CH_2\ COOH$	
H_3C	=	≡	≡	=	CH_2OH
HOH_2C	=	≡	≡	=	CH_2OH
H_3C	=	≡	≡	=	$COOH\ (CH_3)$
HOH_2C	=	≡	≡	=	$COOCH_3$
$(H_3C)\ HOOC$	=	≡	≡	=	$COOH$
H_3C	=	≡	≡	$CH_2\ CH_2\ CH_2\ OH$	
HOH_2C	=	≡	≡	$CH_2\ CH_2\ CH_2\ OH$	
H_3C	=	≡	≡	CH_2CH_2COOH	
$(H_3C)\ HOOC$	=	≡	≡	$CH_2\ CH_2\ CH_2\ OH$	
$(H_3C)\ HOOC$	=	≡	≡	$CH_2\ CH_2\ COOH$	

Fig. 5. C_{10}-Polyacetylenes from basidiomycete fungi.

C_{13}							
	H_3C	≡	≡	≡	≡	$(CH_2)_3\ CH_3$	
	HOH_2C	=	≡	≡	≡	$(CH_2)_3\ CH_3$	
	H_3C	≡	≡	≡	≡	$(CHOH)_3\ CH_2OH$	
C_{14}							
	$HOOC$	≡	≡	≡	CH_2	=	$(CH_2)_3\ COOH$

Fig. 6. C_{13}- and C_{14}-Polyacetylenes from basidiomycete fungi.

The microorganism used was stated to be the actinomycete *Nocardia acidophilus* by the discoverers, Johnson and Burdon (1947). What we know today from the work of Jones's group

$\mathrm{H{-}C{\equiv}C{-}C{\equiv}C{-}CH{=}C{=}CH{-}CH{=}CH{-}CH{=}CH{-}CH_2{-}COOH}$	Mycomycin
$\mathrm{H{-}C{\equiv}C{-}C{\equiv}C{-}CH{=}C{=}CH{-}CHOH{-}CH_2{-}CH_2{-}COOH}$	Nemotinic acid
$\mathrm{H{-}C{\equiv}C{-}C{\equiv}C{-}CH{=}C{=}CH{-}\underset{H}{C}}$ [ring: $\mathrm{CH_2{-}CH_2}$, $\mathrm{C{=}O}$, O]	Nemotin
$\mathrm{H{-}C{\equiv}C{-}C{\equiv}C{-}CH{=}C{=}CH{-}CHOH{-}CH_2{-}CH_2{-}CO{-}O{-}xylose}$	Nemotinic acid xyloside
$\mathrm{H{-}C{\equiv}C{-}C{\equiv}C{-}CH{=}C{=}CH{-}CHOH{-}CH_2{-}CH_2{-}COOCH_3}$	
$\mathrm{H_3C{-}C{\equiv}C{-}C{\equiv}C{-}CH{=}C{=}CH{-}CHOH{-}CH_2{-}CH_2{-}COOH}$	Odyssic acid
$\mathrm{H_3C{-}C{\equiv}C{-}C{\equiv}C{-}CH{=}C{=}CH{-}\underset{H}{C}}$ [ring: $\mathrm{CH_2{-}CH_2}$, $\mathrm{C{=}O}$, O]	Odyssin
$\mathrm{H{-}C{\equiv}C{-}C{\equiv}C{-}CH{=}C{=}CH{-}CH_2{-}CH_2OH}$	± Marasin
$\mathrm{H{-}C{\equiv}C{-}C{\equiv}C{-}CH{=}C{=}CH{-}CH_2{-}COOH}$	R=H and $\mathrm{CH_3}$
$\mathrm{H_2C{=}C{=}CH{-}C{\equiv}C{-}C{\equiv}C{-}CH{=}CH{-}CH_2{-}COOH}$	Drosophilin D
$\mathrm{H{-}C{\equiv}C{-}C{\equiv}C{-}CH{=}C{=}CH{-}CH_2OH}$	(*Cortinellus*)
$\mathrm{H{-}C{\equiv}C{-}C{\equiv}C{-}CH{=}C{=}CH{-}CH_2{-}CH_2{-}CH_2OH}$	(*Odontia*)
$\mathrm{H{-}C{\equiv}C{-}C{\equiv}C{-}CH{=}C{=}CH{-}CH_2{-}CH_2{-}CH_2{-}CH_2OH}$	(*Odontia*)
$\mathrm{H{-}C{\equiv}C{-}C{\equiv}C{-}CH{=}C{=}CH{-}CHOH{-}CH_2{-}CH_2OH}$	(*Odontia*)
$\mathrm{H{-}C{\equiv}C{-}C{\equiv}C{-}CH{=}C{=}CH{-}CHOH{-}CH_2{-}CH_2{-}CH_2OH}$	(*Odontia*)
$\mathrm{H{-}C{\equiv}C{-}C{\equiv}C{-}CH{=}C{=}CH{-}CH_2{-}CHOH{-}CH_2{-}CH_2OH}$	(*Flammula*)
$\mathrm{H_3C{-}C{\equiv}C{-}C{\equiv}C{-}CH{=}C{=}CH{-}CH_2{-}CH_2OH}$	(*Daedalea*)

Fig. 7. Allene-acetylene compounds from basidiomycetes.

(Bew et al., 1966a) is that mycomycin is produced by some basidiomycete fungi. Quite a lot of work has been published on metabolic products from actinomycetes, and also about *Nocardia* species, but no acetylenes have otherwise turned up. Therefore, it is very satisfying to know that a check on the identity of the original strain of Johnson and Burdon is underway. Dr. Anchel has kindly allowed me to report briefly the results of this reinvestigation, which is still in progress. Dr. Burdon was able to deliver one living culture and Canham, Bistis, and Anchel have been able to prove that this organism contains discrete nuclei. So the organism is definitely not a *Nocardia*, nor even an actinomycete; no decision has yet been reached on the question of whether the organism actually belongs to any known genus of basidiomycetes. Regardless of the outcome of this reexamination, it is noteworthy that mycomycin is the first allenic compound found in nature. Jones (cf. Bew et al., 1966a) has recently summarized 18 acetylenic allenes from basidiomycetes, mycomycin included (Fig. 7). Allenes apparently are quite rare elsewhere in the plant kingdom, and the basidiomycetes are so far the most frequent producers of allenes known. But one should not forget that, as with most of the fungal acetylenes, these allenes are excretion products; and as far as I know, no one has investigated whether higher plants that produce acetylenes excrete allenes.

The length of the carbon chain in the acetylenic compounds from basidiomycetes varies from C_8 to C_{13}, with a dominance of C_9 and C_{10}. The end groups in the compounds are alcohols, aldehydes, acids, lactones, methyl esters, amides, nitriles, epoxides, and glycols up to tetra-ols. Except for lactones and epoxides, heterocyclics have not been found, nor have any aromatic rings been detected, with the single exception of the thiophene aldehyde junipal, which was mentioned earlier.

The most remarkable difference between acetylenes from these fungi and those of Compositae is the rare occurrence of hydrocarbons in the fungal species. The first to be found, trideca-2,4,6,8-tetra-yne, was described only a few months ago (Jones et al., 1966).

The discovery of the fungal acetylenes is due to the antibiotic activity shown by a number of them. This antibiotic property was described by Robbins et al. (1947). The disclosure of the acetylenic nature of these compounds was due to Anchel (1952) and to her cooperation with Sir Ewart Jones. From the point of view of chemotaxonomy, the presence of antibiotic substances is quite advantageous since they nearly always initiate the publication of a survey of all related species whether they contain antibiotics or not. Although none of these acetylenic antibiotics has found use in medicine, so many new and interesting compounds were discovered that both Anchel and Jones extended their

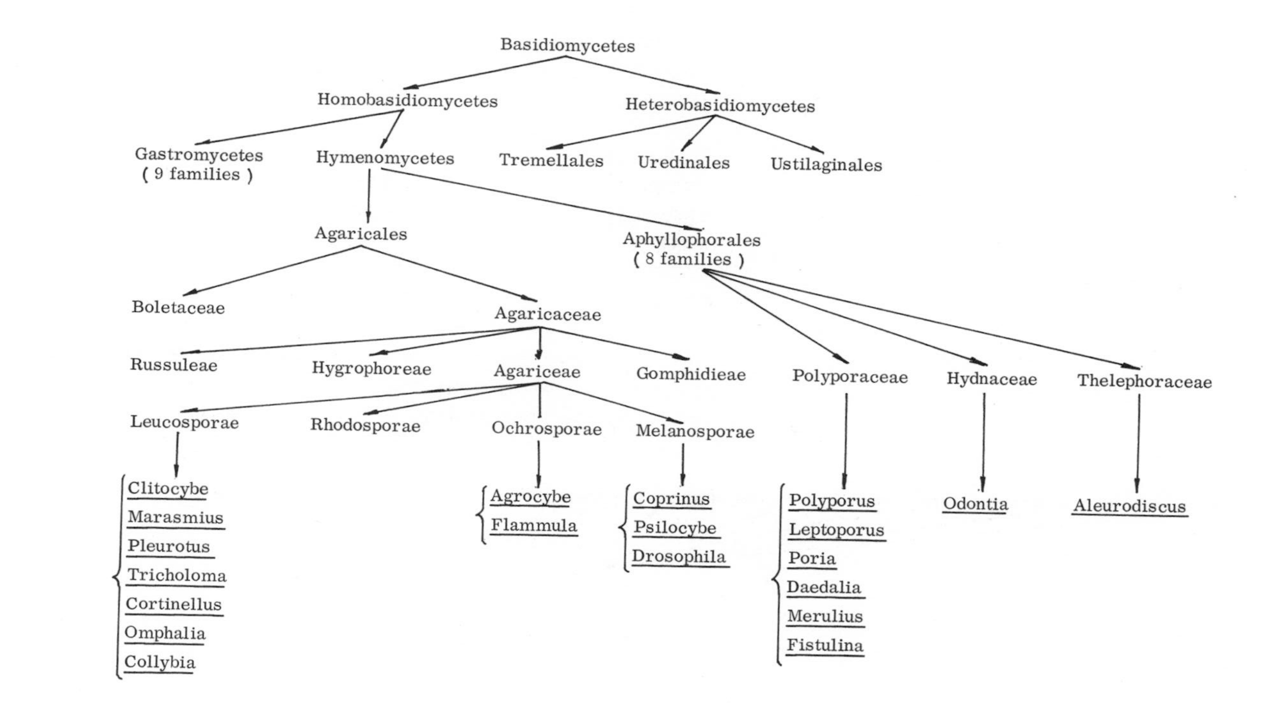

Fig. 8. The occurrence of polyacetylenes in basidiomycete fungi according to the classification of J. A. Nannfeldt.

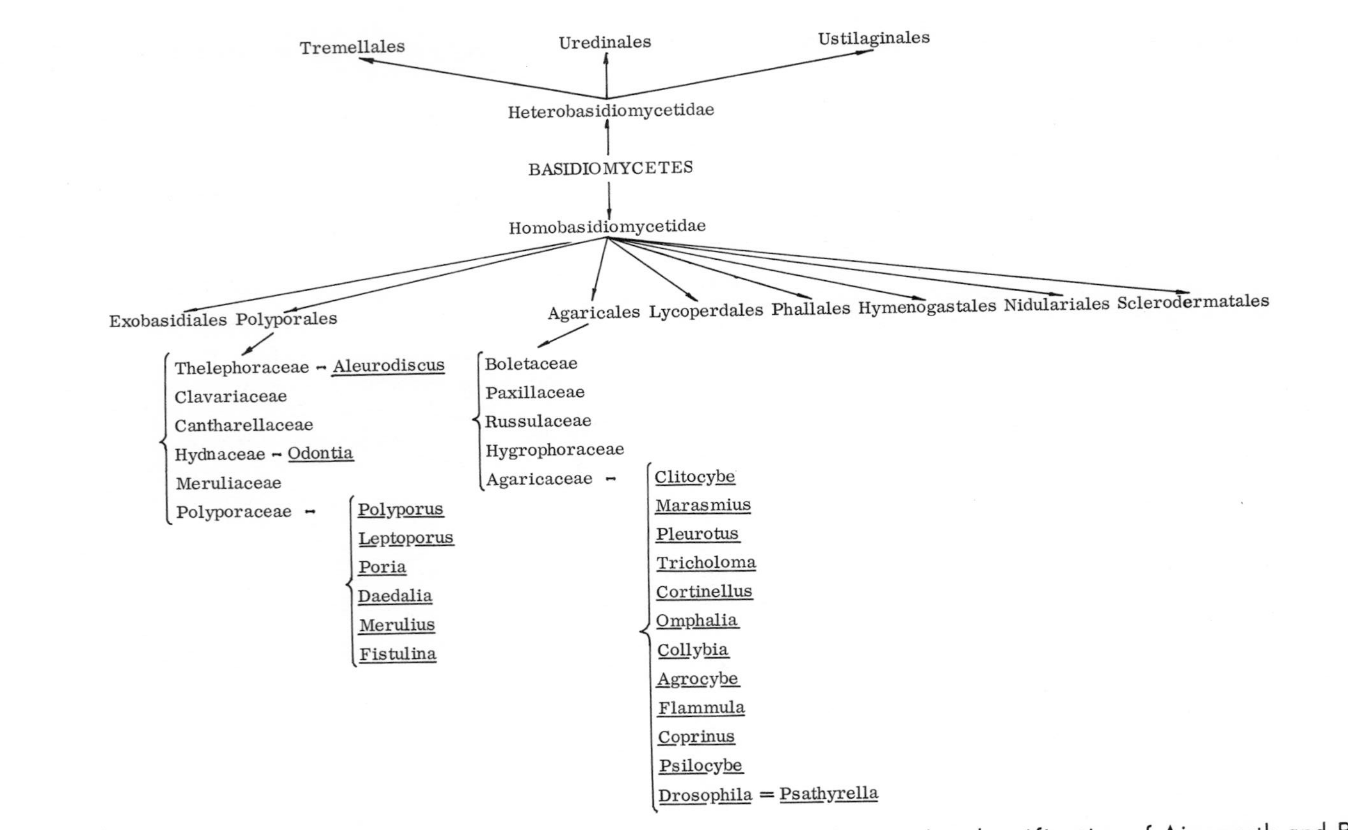

Fig. 9. The occurrence of polyacetylenes in basidiomycete fungi according to the classification of Ainsworth and Bisby.

studies to most of the basidiomycetes available in the culture collections of the world.

It is seldom that we get much information about the absence of compounds; hence, absence in all its aspects is one of the most questionable points of chemotaxonomy. Most of the acetylenes of the basidiomycetes, shown in Figs. 3 to 7, may be detected spectroscopically down to one part in 10^5 and we may state with some degree of certainty that acetylenic compounds of these types occur only in a limited number of the divisions and subdivisions of this class of fungi.

Figure 8 represents the taxonomic scheme of the class Basidiomycetes of J. A. Nannfeldt, Stockholm. The Nannfeldt system has more categories than is usual in botanical classifications. The taxonomic treatment of Ainsworth and Bisby conforms more closely to the usual number of taxonomic categories. By this scheme of classification, as shown in Figure 9, the occurrence of acetylenes is restricted still more. The important point for the present discussion is that dozens of species and strains belonging to the negative subclasses, orders, and families have been tested both at the New York Botanical Garden and at the Dyson Perrins Laboratory in Oxford.

The situation presented in Figures 8 and 9 has changed only a little in the past five years, and so there is good reason to state that there is a taxonomically important connection between the occurrence of acetylenes and the systematic treatment developed by the botanists for these fungi.

One further observation should be made. Although most of these chemical products are excreted into the culture medium, experiments with labeled compounds have revealed that the substances are sometimes metabolized rapidly. The possibility still exists, of course, that the acetylenes are intermediates in the synthesis of other products and that only in the family Aragicaceae and three families of the Polyporales are some of these intermediates lost into the culture medium, whereas in some of the supposedly negative families the internal yield is 100 percent. This is a fascinating point, and it does not make acetylenes less interesting from a chemotaxonomic point of view.

						Tribe	
H_3C	≡	≡	≡	=	$COOCH_3$	III	VII
H_3C	≡	≡	=(furanone ring)=O				VII
H_3C	=	≡	≡	=	$COOCH_3$	III	VII
H_3C	=	≡	=(furanone ring)=O			III	VII
H_3C	=	≡	≡	=	CH_2OR	III	
H_3C	=	≡	≡	$(CH_2)_2$	$COOCH_3$	III	VII
H_3C	=	≡	≡	$(CH_2)_2$	CH_2OR		VII
H_3C	$(CH_2)_2$	≡	≡	=	$COOCH_3$	III	
H_3C	$(CH_2)_2$	≡	=(furanone ring)=O			III	
H_3C	$(CH_2)_2$	≡	≡	=	CH_2OR	III	
H_3C	=	=	≡	=	CH_2OR	III	
H_3C	=	≡	=	=	CH_2OR	III	
H_3C	CH_2CH (with $OCOC(H_3C){=}CHCH_3$ on CH)	≡	≡	=	$COOCH_3$	III	
H_3C	≡	≡	≡	=	$CONHCH_2CH(CH_3)_2$		VII
H_3C	≡	≡	= (with $H_3C{-}S$)	=	$COOCH_3$		VII
H_3C	≡	=	= (with $H_3C{-}S$)	=	$COOCH_3$		VII
H_3C	≡	= (with $H_3C{-}S$)	≡	=	$COOCH_3$		VII
H_3C		≡	= (with $H_3C{-}S$)	=(furanone ring)=O			VII
H_3C	= (with $H_3C{-}S$)	≡	≡	=	$COOCH_3$		VII

Fig. 10. Aliphatic C_{10}-polyacetylenes from the Compositae.

IV. ACETYLENES (OTHER THAN ACETYLENIC FATTY ACIDS) FROM HIGHER PLANTS

A. Family Compositae

The polyacetylenes isolated from the family Compositae by far outnumber those compounds of this group isolated from the rest of the plant kingdom. A number of excellent reviews of this topic have appeared (Bohlmann et al., 1963; Jones, 1959; Bu'Lock, 1964). A concise compilation is given in Figs. 10 to 21, together with the tribes in which individual compounds have been detected.

In Figs. 10–13, which illustrate the aliphatic acetylenes from Compositae, there occur a few overlaps with the acetylenes from fungi. In addition to these identical compounds there occur some other acetylenes where the chemical difference is due only to configuration. For example, the *cis* configuration has been found only occasionally among fungal acetylenes, whereas, the acetylenes of the Compositae often occur either with *cis* configuration, or in both *cis* and *trans* pairs.

As noted recently (Jones, 1966), the polyacetylenes from the Compositae have statistically longer chains C_9–C_{17} (C_{18} in isobutylamides) than those from microorganisms, with an abundance of C_{13} compounds (Fig. 22).

The most remarkable property of the acetylenes of the Compositae lies in the tendency of the aliphatic poly-ynes of the Compositae to cyclize into ring compounds of various types such as benzene rings, mono-, and dithiophenes. In terthienyl, all acetylene bonds have disappeared, but this remarkable natural product (bottom of Fig. 16), discovered by Zechmeister and Sease (1947), undoubtedly arises from an aliphatic acetylene. Furanoid compounds are among the oldest acetylenes known (carlina oxide, Fig. 19). More recently, a number of other oxygen-containing heterocyclics have been isolated as five-membered ring lactones, pyranes, and a number of spiranes with two oxygen-containing five-membered rings or one five- and one six-membered ring (Fig. 20). Bohlmann has isolated many combinations of heterocyclics, partly with a consumption of all acetylenic bonds in the formation of the rings as with terthienyl. Such compounds should be classified as further metabolic derivatives of acetylenes. As is clear from the figures, the transition from the tridecenepenta-yne is so gradual that it is to some extent tempting to place the compound with no triple bonds left as the last compound in the series.

							Tribe					
							IV	V	VII	VIII	IX	XI
H_3C	$\equiv$	$\equiv$	$\equiv$	$\equiv$	$\equiv$	$=H$	IV	V		VIII	IX	
H_3C	$\equiv$	$\equiv$	$\equiv$	$\equiv$	$\equiv$	$\underset{OR}{CH}-\underset{OR}{CH_2}$		V				
H_3C	$\equiv$	$\equiv$	$\equiv$	$\equiv$	$=$	$=H$		V				
H_3C	$\equiv$	$\equiv$	$\equiv$	$\equiv$	$=$	$\underset{\diagdown O \diagup}{CH-CH_2}$						XI
H_3C	$\equiv$	$\equiv$	$\equiv$	$\equiv$	$=$	$CHCl-CH_2OR$						XI
H_3C	$=$	$\equiv$	$\equiv$	$\equiv$	$\equiv$	$=H$		V		VIII	IX	
ROH_2C	$=$	$\equiv$	$\equiv$	$\equiv$	$\equiv$	$=H$		V				
OHC	$=$	$\equiv$	$\equiv$	$\equiv$	$\equiv$	$=H$		V				
ROH_2C	$=$	$\equiv$	$\equiv$	$\equiv$	$\equiv$	$\underset{OR}{CH}-\underset{OR}{CH_2}$		V				
H_3C	$\equiv$	$\equiv$	$\equiv$	$=$	$=$	$=H$			VII			
H_3C	$\equiv$	$\equiv$	$\equiv$	$=$	$=$	CH_2-CH_2OR			VII			
H_3C	$=$	$\equiv$	$\equiv$	$\equiv$	$=$	$=H$		V				
ROH_2C	$=$	$\equiv$	$\equiv$	$\equiv$	$=$	$=H$		V				XI
OHC	$=$	$\equiv$	$\equiv$	$\equiv$	$=$	$=H$		V				
ROH_2C	$=$	$\equiv$	$\equiv$	$\equiv$	$=$	$\underset{\vert}{CH}-\underset{\vert}{CH_2}$		V				

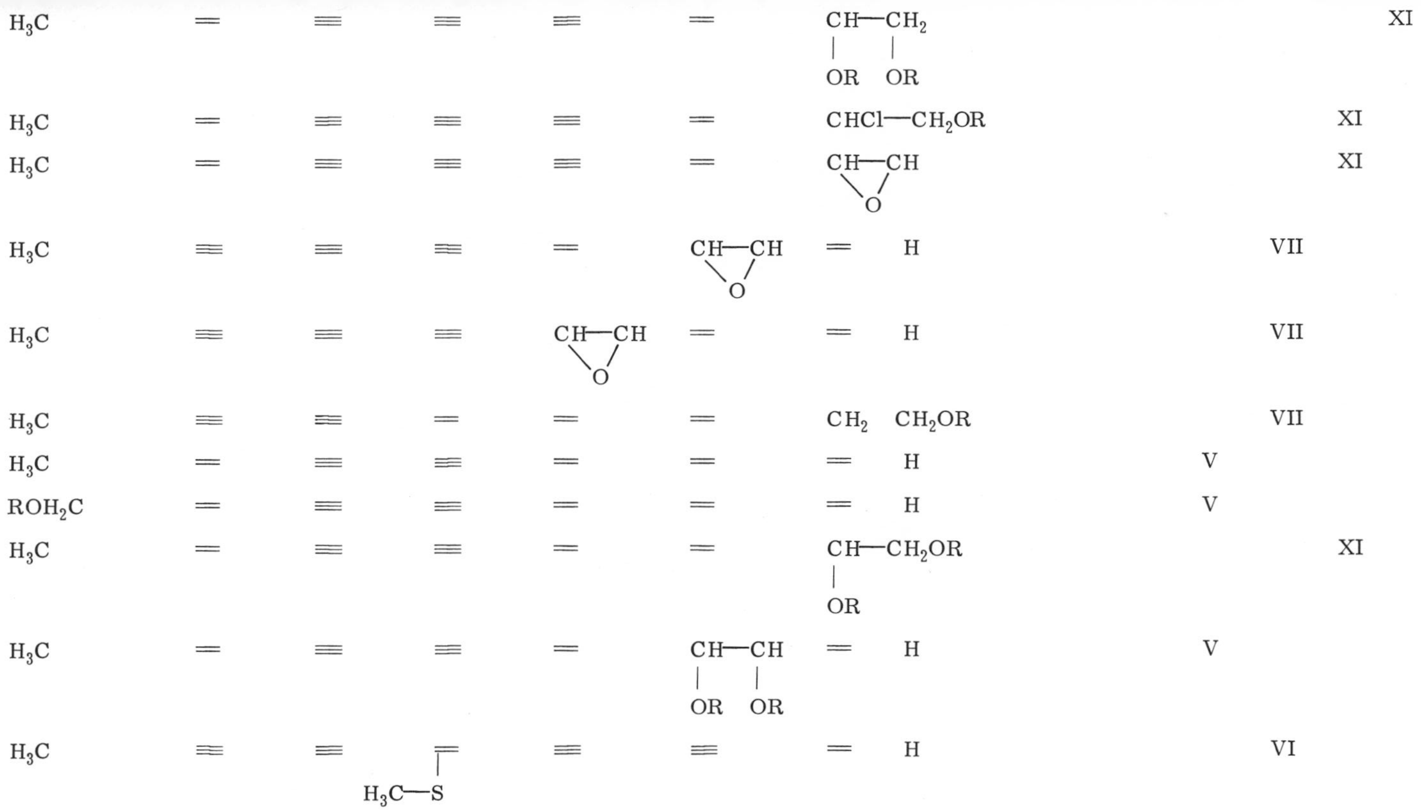

H_3C = ≡ ≡ ≡ = $CH(OR)—CH_2(OR)$ XI

H_3C = ≡ ≡ ≡ = $CHCl—CH_2OR$ XI

H_3C = ≡ ≡ ≡ = $CH—CH$ (epoxide, O) XI

H_3C ≡ ≡ ≡ = $CH—CH$ (epoxide, O) = H VII

H_3C ≡ ≡ ≡ $CH—CH$ (epoxide, O) = = H VII

H_3C ≡ ≡ = = = CH_2 CH_2OR VII

H_3C = ≡ ≡ = = = H V

ROH_2C = ≡ ≡ = = = H V

H_3C = ≡ ≡ = = $CH(OR)—CH_2OR$ XI

H_3C = ≡ ≡ = $CH(OR)—CH(OR)$ = H V

H_3C ≡ ≡ =($S—CH_3$) ≡ ≡ = H VI

Fig. 11. Aliphatic C_{13}-polyacetylenes from the Compositae.

C_{14}

Structure	Tribe		
$H_3C{-}C{\equiv}C{-}C{\equiv}C{-}C{\equiv}C{-}CH{=}CH{-}CH{=}CH{-}(CH_2)_2CH_2OR$		VII	
$H_3C{-}C{\equiv}C{-}C{\equiv}C{-}C{\equiv}C{-}CH{=}CH{-}CH{=}CH{-}CH(OR){-}CH_2CH_2OR$	V	VII	
$H_3C{-}C{\equiv}C{-}C{\equiv}C{-}C{\equiv}C{-}CH{=}CH{-}CH_2CH_2COCH_2CH_3$		VII	
$H_3C{-}CH{=}CH{-}C{\equiv}C{-}C{\equiv}C{-}CH{=}CH{-}CH{=}CH{-}(CH_2)_2CH_2OR$	V	VII	
$H_3C{-}CH{=}CH{-}C{\equiv}C{-}C{\equiv}C{-}CH{=}CH{-}CH{=}CH{-}CH(OH){-}CH_2{-}CH_2(OH)$	V		
$H_3C{-}CH_2{-}CH{=}CH{-}CH{=}CH{-}CH{=}CH{-}CH{=}CH{-}(CH_2)_3CH_2OR$			XI
$H_3C{-}CH{=}CH{-}C{\equiv}C{-}C{\equiv}C{-}CH_2{-}CH_2{-}CH{=}CH{-}CH{=}CH{-}CONHCH_2CH(CH_3)_2$		VII	

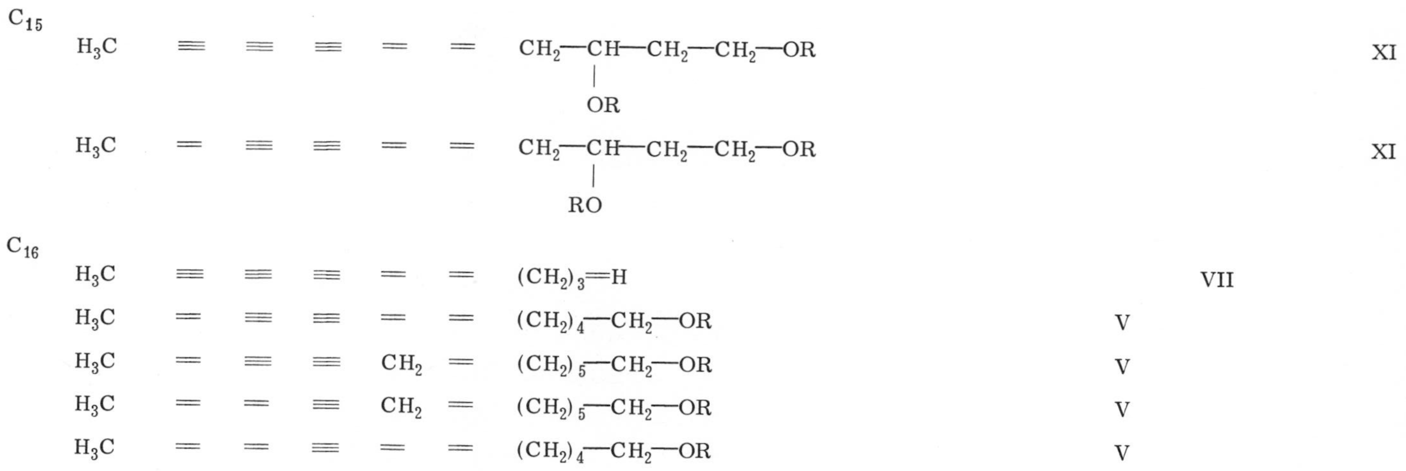

C_{15}

$$H_3C \;\equiv\; \equiv\; \equiv\; =\; =\; CH_2{-}\underset{\displaystyle OR}{\underset{|}{CH}}{-}CH_2{-}CH_2{-}OR \qquad \text{XI}$$

$$H_3C \;=\; \equiv\; \equiv\; =\; =\; CH_2{-}\underset{\displaystyle RO}{\underset{|}{CH}}{-}CH_2{-}CH_2{-}OR \qquad \text{XI}$$

C_{16}

$$H_3C \;\equiv\; \equiv\; \equiv\; =\; =\; (CH_2)_3{=}H \qquad \text{VII}$$

$$H_3C \;=\; \equiv\; \equiv\; =\; =\; (CH_2)_4{-}CH_2{-}OR \qquad \text{V}$$

$$H_3C \;=\; \equiv\; \equiv\; CH_2\; =\; (CH_2)_5{-}CH_2{-}OR \qquad \text{V}$$

$$H_3C \;=\; =\; \equiv\; CH_2\; =\; (CH_2)_5{-}CH_2{-}OR \qquad \text{V}$$

$$H_3C \;=\; =\; \equiv\; =\; =\; (CH_2)_4{-}CH_2{-}OR \qquad \text{V}$$

Fig. 12. Aliphatic C_{14}- to C_{16}-polyacetylenes from the Compositae.

	Formula	Tribe			
C_{17}	$H_3C \equiv \equiv \equiv = = (CH_2)_4 = H$	III	V	VII	XI
	$H_3C \equiv \equiv \equiv = = CH{-}(CH_2)_3 = H$ (with OR on CH)	III	V		
	$H_3C \equiv \equiv \equiv = CH_2(CH_2)_5 = H$		V	VII	
	$H_3C \equiv \equiv \equiv = (CH_2)_2\,CO\,(CH_2)_3 = H$	III			
	$H_3C \equiv \equiv \equiv = (CH_2)_2CHOH(CH_2)_3 = H$	III			
	$H_3C \equiv \equiv \equiv = = = (CH_2)_2\,COCH_3$	III			
	$H_3C = \equiv \equiv = = (CH_2)_4 = H$	III	V	VII	
	$ROH_2C = \equiv \equiv = = (CH_2)_4 = H$		V		
	$OHC = \equiv \equiv = = (CH_2)_4 = H$		V		
	$H_3C = \equiv \equiv = = CH(OH)(CH_2)_3 = H$		V		
	$H_3C = \equiv \equiv CH_2 = (CH_2)_5 = H$			VII	
	$H_3C = \equiv \equiv (CH_2)_4\,CO\,(CH_2)_3 = H$	III			
	$H_3C = \equiv \equiv = CH(OH)(CH_2)_5 = H$	III			XI
	$H = (CH_2)_5 = CH_2 \equiv \equiv CO = H$		V	VII	
	$H_3C{-}CH_2{-}(CH_2)_5 = CH_2 \equiv \equiv CO = H$		V		
C_{18}	$H_3C \equiv \equiv \equiv = = (CH_2)_4{-}CO \cdot CH_2{-}CH_2{-}OR$		V		

						Tribe
C_{11}						
		CH_2	≡	≡	H	VII
		CH_2	≡	=	SCH_3	VII
		CO	≡	≡	H	VII
		CO	≡	=	SCH_3	VII
C_{12}						
		CH_2	≡	≡	CH_3	VII
		CH_2	≡	=	CH_3	VII
		CO	≡	≡	CH_3	VII
		CH | OR	≡	≡	CH_3	VII
C_{13}						
		≡	≡	≡	CH_3	V
		≡	≡	=	CH_3	V
		≡	≡	=	CH_2OR	V

Fig. 14. Nonsubstituted aromatic polyacetylenes from the Compositae.

Some of these heterocyclics are new to chemistry and, for a few, the correct structural presentation is still uncertain. From Adolf von Baeyer's days it is well known that polyacetylenes are very unstable, even occasionally explosive. Some of the heterocyclics have still other unpleasant properties, and recently Bohlmann isolated the first cumulene from a *Conyza*. Mrs.

Number of C atoms	Constitution	Genus	Tribe
12	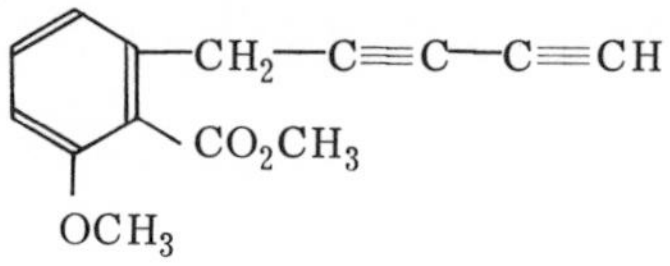 $-CH_2-C\equiv C-C\equiv CH$; CO_2CH_3; OCH_3	*Chrysanthemum*	VII
13	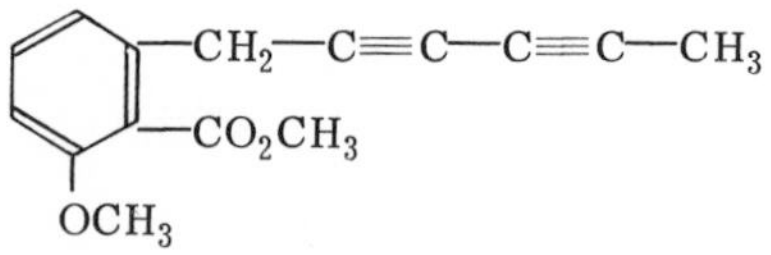 $-CH_2-C\equiv C-C\equiv C-CH_3$; $-CO_2CH_3$; OCH_3	*Chrysanthemum*	VII
12	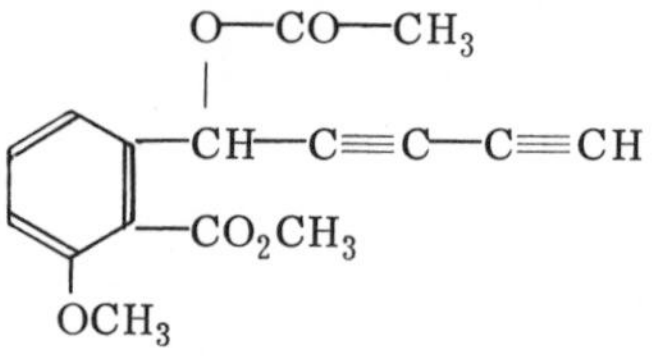 $O-CO-CH_3$; $-CH-C\equiv C-C\equiv CH$; $-CO_2CH_3$; OCH_3	*Chrysanthemum*	VII
13	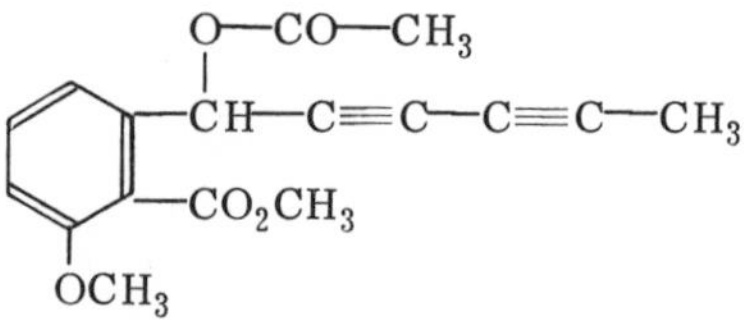 $O-CO-CH_3$; $-CH-C\equiv C-C\equiv C-CH_3$; $-CO_2CH_3$; OCH_3	*Chrysanthemum*	VII
	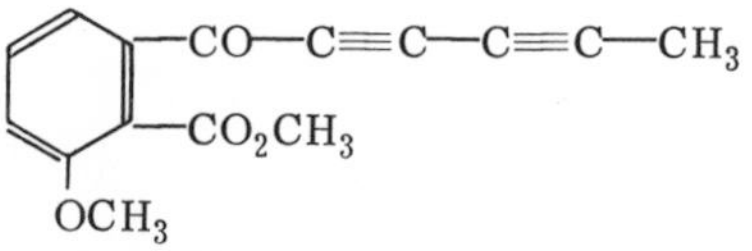 $-CO-C\equiv C-C\equiv C-CH_3$; $-CO_2CH_3$; OCH_3	*Chrysanthemum*	VII
	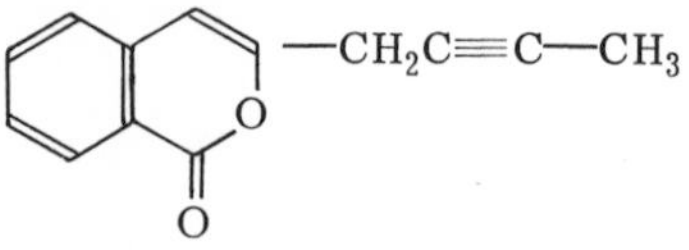 $-CH_2C\equiv C-CH_3$; O; O	*Artemisia* *Chrysanthemum*	VII

Fig. 15. Substituted aromatic acetylenic compounds from the Compositae.

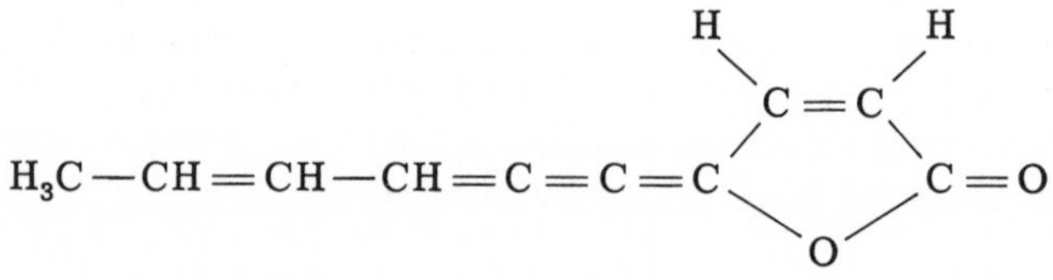

$$H_3C-CH=CH-CH=C=C=C\begin{matrix} H & & H \\ & C=C & \\ & & \\ & O & \end{matrix}C=O$$

	Compound	Tribe V	Tribe VI	Tribe VII	Tribe XI
C_9	H_3C ≡ [thiophene, OR] $CO—CH_3$			VII	
	[thiophene] ≡ = $COOCH_3$			VII	
C_{10}	H_3C [thiophene] ≡ = $COOCH_3$			VII	
	H_3C ≡ [thiophene] = $COOCH_3$			VII	
C_{12}	[thiophene] ≡ = = = H			VII	XI
	H_3C ≡ ≡ [thiophene] ≡ CHO		VI		
	[thiophene]—[thiophene] = = H	V			
	[thiophene]—[thiophene] ≡ = H	V	VI	VII	XI
	[thiophene]—[thiophene] ≡ $CH_2—CH_2OR$		VI		XI
	[thiophene]—[thiophene] ≡ $CH—CH_2$ (RO, OR (Cl))				XI
	[thiophene]—[thiophene]—[thiophene]	V	VI	VII	XI

Fig. 16. C_9-C_{12} thiophene-acetylenic compounds from the Compositae.

Compound	Tribe				
(thiophene, S) ≡ = = $CH_2CH_2CH_2OR$					XI
(thiophene, S) ≡ = = $COCH_2CH_3$				VII	
(thiophene, S) ≡ = $(CH_2)_2\ CO\ CH_2CH_3$				VII	
H_3C ≡ (thiophene, S) ≡ ≡ = H	IV	V	VI		
H_3C ≡ (thiophene, S) ≡ = = H		V			
H_3C ≡ ≡ (thiophene, S) ≡ = H	IV				XI
H_3C ≡ ≡ (thiophene, S) ≡ $CH_2—CH_2—OR$					XI
H_3C ≡ ≡ (thiophene, S) ≡ $CH(RO)—CH_2(OR)$					XI
H_3C ≡ ≡ (thiophene, S) ≡ $CH—CH_2$ (epoxide, O)					XI
H_3C ≡ ≡ (thiophene, S) ≡ $CH(Cl)—CH_2(OR)$					XI

Fig. 17. C_{13}-thiophene-acetylenic compounds from the Compositae.

Sørensen can fully confirm Bohlmann's statements about the properties of this compound. We happened to obtain a small sample of the North Scandinavian *Erigeron unalaschkense* (DC) Vierh. This fleabane is peculiar through its dark-violet colored hairs. If the hairy heads are immersed in spectral hexane one

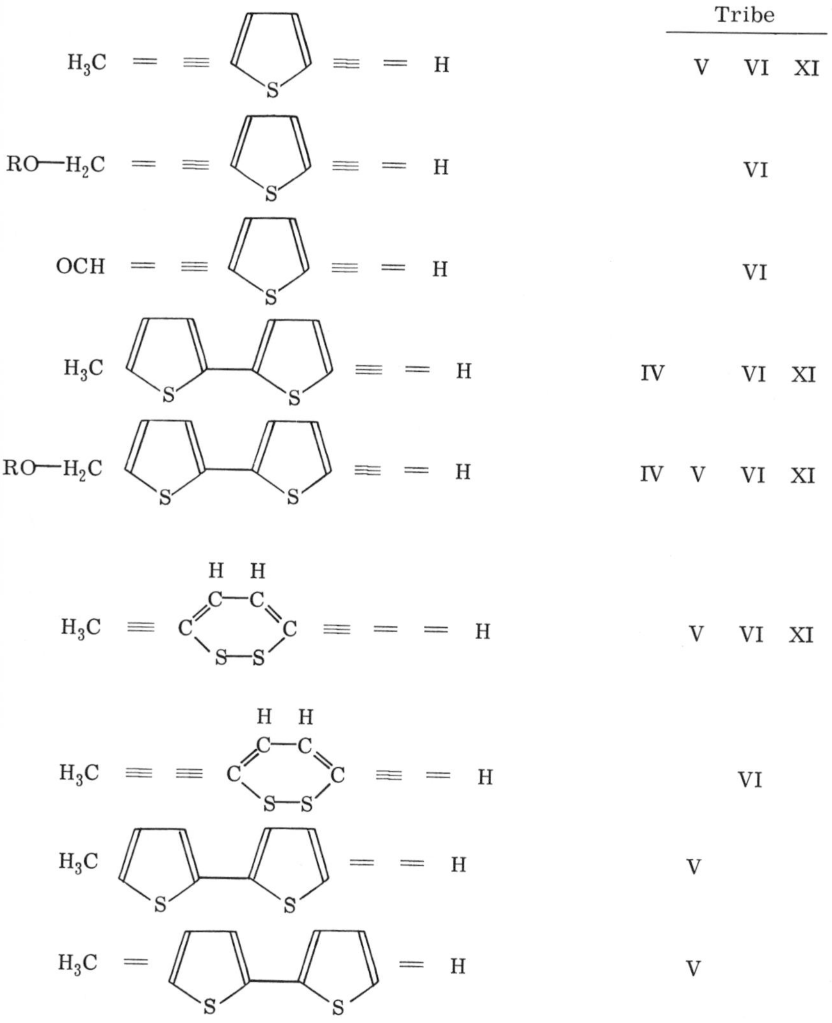

Compound	Tribe			
	IV	V	VI	XI
H_3C = ≡ (thiophene) ≡ = H		V	VI	XI
RO—H_2C = ≡ (thiophene) ≡ = H			VI	
OCH = ≡ (thiophene) ≡ = H			VI	
H_3C (dithienyl) ≡ = H	IV		VI	XI
RO—H_2C (dithienyl) ≡ = H	IV	V	VI	XI
H_3C ≡ (dithiin ring: C, HC—CH, C, S—S) ≡ = = H		V	VI	XI
H_3C ≡ ≡ (dithiin ring: C, HC—CH, C, S—S) ≡ = H			VI	
H_3C (dithienyl) = = H		V		
H_3C = (dithienyl) = H		V		

Fig. 18. C_{13}-thiophene and dithienyl-acetylenic compounds from the Compositae and some related disulfur compounds.

immediately obtains from this solution the spectrum of Bohlmann's cumulene. If, however, one tries to isolate the compound by chromatography, everything is lost.

The number of rather stable acetylenic compounds from the Compositae has been rising steadily especially as a result of the

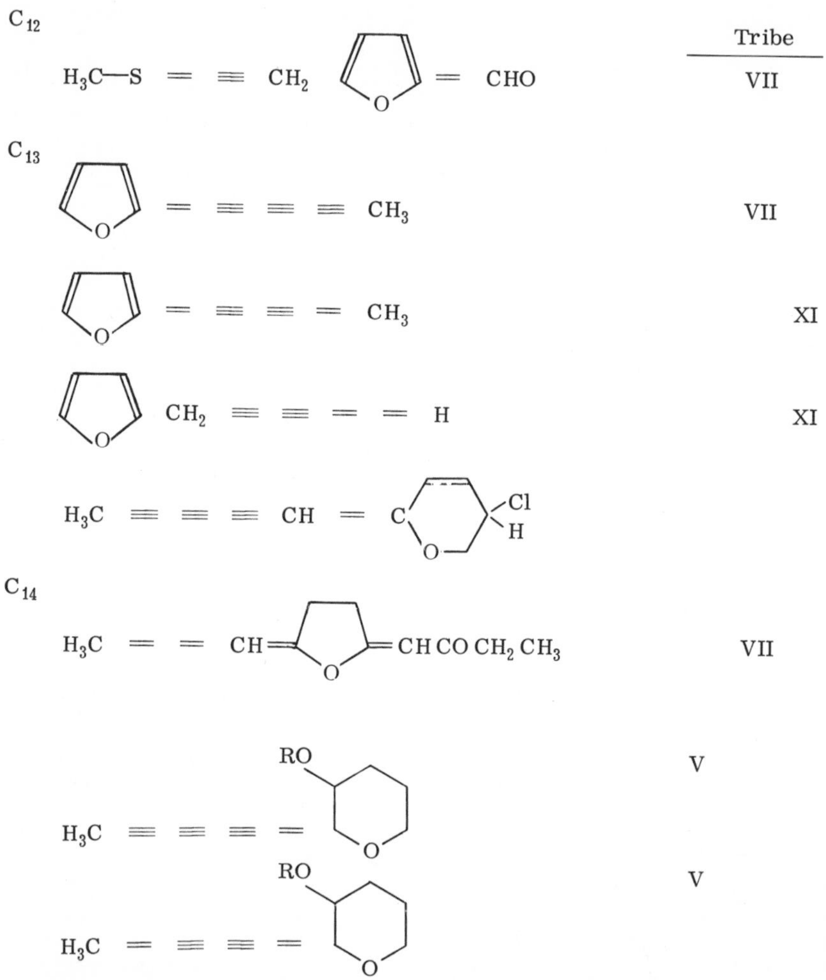

Tribe

C_{12}

H_3C—S = ≡ CH_2 (O) = CHO — VII

C_{13}

(O) = ≡ ≡ ≡ CH_3 — VII

(O) = ≡ ≡ = CH_3 — XI

(O) CH_2 ≡ ≡ = = H — XI

H_3C ≡ ≡ ≡ CH = C (O) Cl H

C_{14}

H_3C = = CH= (O) =CH CO CH_2 CH_3 — VII

RO (O) H_3C ≡ ≡ ≡ = — V

RO (O) H_3C = ≡ ≡ = — V

Fig. 19. Oxygen heterocyclic acetylene compounds from the Compositae.

work of Bohlmann's group, but one should be careful in attempting to elaborate on the importance of these acetylenes to the taxonomy of the Compositae before our chemical knowledge is complete. The intriguing instability of many of these compounds, plus the fact that they are often not detected by ordinary qualitative spectroscopy, gives further warning.

As mentioned earlier, only crepenynic acid has been found in the tribe Cichorieae. This group, which is treated as a subfamily or even a family (*Liguliflorae*) by many botanists, has not been thoroughly investigated. It is unrewarding to a natural product chemist to involve himself in a long program of testing for the absence of a substance. In Trondheim we have carefully tested all those members that the botanists have considered to be possibly related to the other subfamily (the Astereae). In none of them (*Hypochoeris* L., *Scorzonera* L., *Tragopogon* L.) could we detect even traces of acetylenes.

In the Tubuliflorae, acetylenes have been found in all 12 tribes described by O. Hoffmann. In these 12 tribes, wherein nearly 1000 species have been investigated, only a few species have given negative results, *Cosmos bipinnatus*, for example. The type and number of acetylenes vary greatly from one tribe to another. The most widely distributed compound is, curiously, an explosive penta-acetylene:

$$H_3C-C\equiv C-C\equiv C-C\equiv C-C\equiv C-C\equiv C-CH=CH_2$$

In some tribes—I (Vernonieae), II (Eupatorieae), VIII (Senecioneae), IX (Calenduleae), X (Arctotideae), and XII (Mutisieae)—this tridecenepenta-yne is often the dominant acetylene and frequently the only one to be found. In the tribes Astereae (III) and Anthemideae (VII) this penta-yne is stated to be absent. We have only detected it in *Centipeda*, a small Antarctic genus placed in Anthemideae. To a layman it also appears to be a member of that tribe, but the two species investigated [*C. cunninghamii* (D. C.) A. Br. & Aschers. and *C. minima* (L.) A. Br. & Aschers] both contained appreciable amounts of the penta-yne. Since the tribe Anthemideae has been very broadly investigated, this is a case where the chemist might suggest that the botanist reinvestigate the classical description of the genus.

Members of tribes III and VII have very many short-chain compounds in common (C_{10} esters, Fig. 10), but Astereae (III) seems thus far not to develop the chemical variations equivalent to those found in the Anthemideae.

In the tribes Inuleae (IV), Heliantheae (V), Helenieae (VI), and Cynareae (XI) the penta-yne is often found more or less hydrogenated to di-yne-tetra-enes or else is present as a heterocyclic, especially of the thiophene type.

In order to use chemical evidence for taxonomic purposes, it is important to know something about the age of the biochemical processes responsible for the occurrence of certain types of compounds. The Australian flora is famous for its high percen-

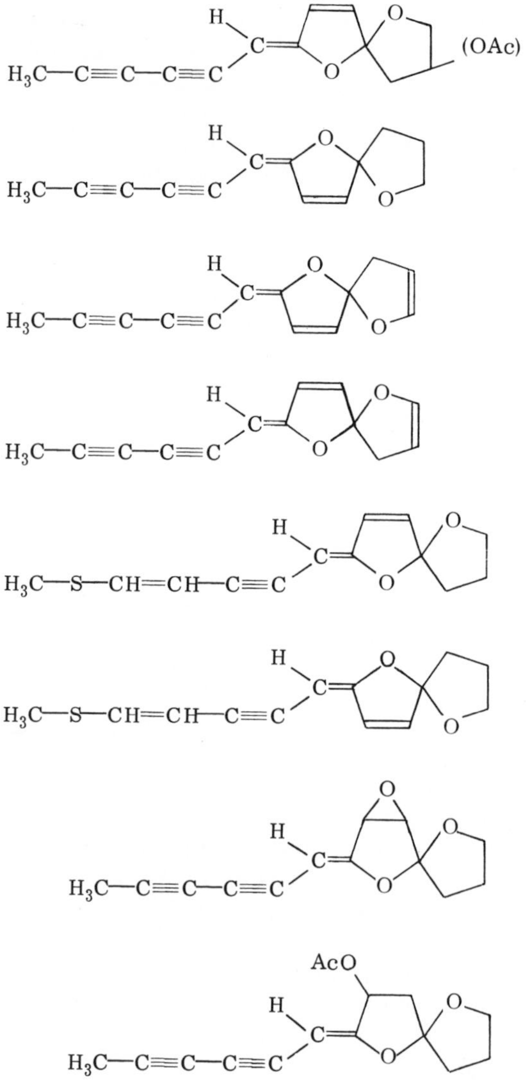

Fig. 20. Spiro-heterocyclic compounds from tribe VII (Anthemideae) of the Compositae.

tage of endemic genera and species and many of these endemics belong to the Compositae. According to geological information, the flora of Australia has been isolated from the rest of the

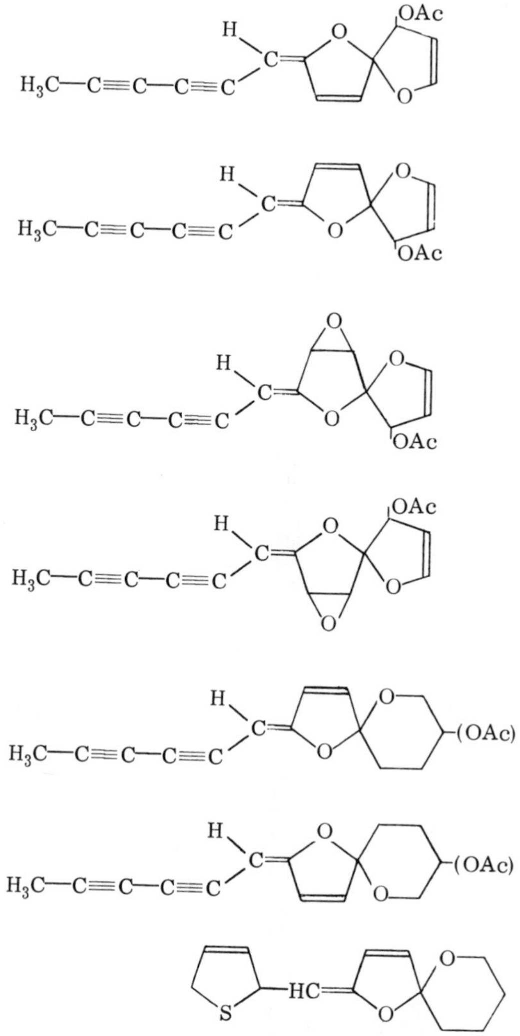

Fig. 20(Cont.)

world since early Cretaceous time. This isolation provides the chemist with an opportunity to investigate members of some tribes of the Compositae that have been isolated geographically over a large period of time. The Australian flora furnishes many examples in the tribes Astereae (III), Anthemideae (VII), and

C_{12} S ≡ = O

C_{13} ROH_2C— S ≡ = O

CH_2 ≡ O

Fig. 21. Acetylenic compounds with two different heterocyclic rings from the Compositae.

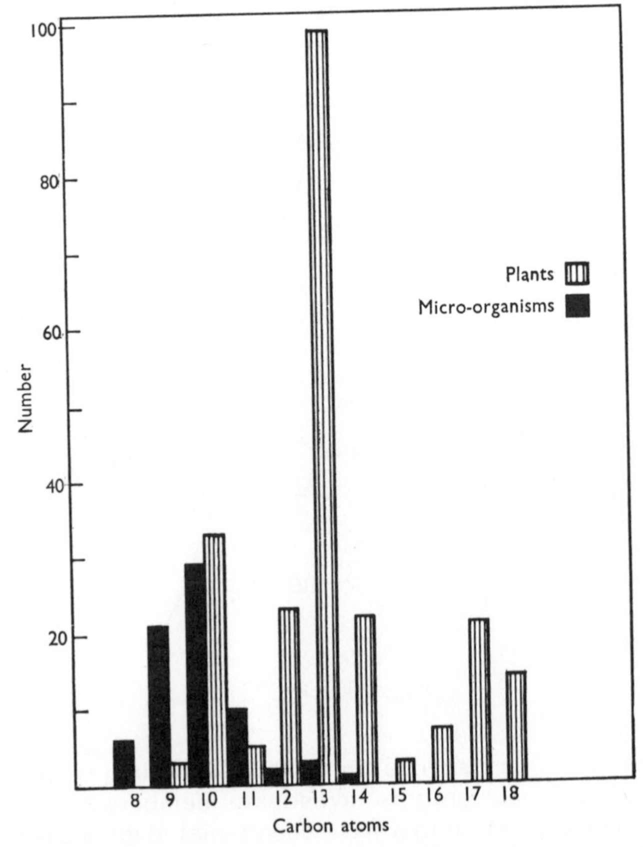

Fig. 22. Chain lengths of natural polyacetylenes according to Jones (1966).

Inuleae (IV). It may be stated safely that the occurrence of acetylenes is a conservative property; a conclusion similar to what may be deduced from our knowledge of the seed fats of the Santalaceae.

The groups shown at the right in Figs. 10–19 demonstrate that there is some agreement between acetylene chemistry and the treatment of the Compositae developed by Bentham and Hooker about 100 years ago. The agreement is not as good if one prefers to use one of the other old taxonomic treatments proposed for the Compositae. Remarkably, no fundamental treatment of the systematics of this family has been published in this century.

If one considers lower taxonomic categories such as subtribes, genera, and sections the situation gets much less clear, primarily because the chemical investigation of the Compositae is still so incomplete. However, there exist some examples that demonstrate the usefulness of acetylene chemistry at these lower taxonomic levels. Several of the Compositae genera are very large, such as the cornflowers (*Centaurea*) with 600 species, the genus *Aster* with 500 species, or *Artemisia* with 280 species. Grossly they are very well-defined genera, and it is quite remarkable that, with these high numbers of species, the individual species generally stands out with remarkable sharpness. Botanists have often divided these large genera into many sections. O. Hoffman divided *Artemisia* into four, *Aster* into 15 and *Centaurea* into no less than 41 sections. A number of such sections earlier had been given the rank of a genus, or had been moved from one genus to a neighboring one. At this level in the taxonomic hierarchy the key characters, which are mostly anatomical, are rather subtle and their validity is sometimes strongly disputed. Here again chemistry may be used to test the various systems followed by botanists.

Bohlmann and his collaborators have investigated a number of cornflowers. The result is that the sections Centaurium and Cassini are distinctive as regards the structure of their acetylenes, whereas the investigated members from other sections are rather homogeneous as to their acetylenic components. The Bohlmann group also has investigated a large number of *Artemisia* species and found many features supporting a separation of the genus on chemical grounds into sections. Since the picture in *Artemisia* is a much more complex one than that in the cornflowers, it would be preferable to have a more comprehensive coverage of this genus before the chemical sections are compared with the sections recognized by the botanists.

TABLE 4

Polyacetylenic Compounds from the Families Umbelliferae and Araliaceae

Number of C Atoms	Constitution	Umbelliferae	Araliaceae
10	$\mathrm{H_3C{-}CH{=}CH{-}C{\equiv}C{-}C{\equiv}C{-}CH{=}CH{-}HC{=}O}$	*Aethusa cynapium* L.	—
13	$\mathrm{H_3C{-}CH_2{-}CH{=}CH{-}CH{=}CH{-}C{\equiv}C{-}C{\equiv}C{-}CH{=}CH{-}HC{=}O}$	*Aethusa cynapium* L.	—
13	$\mathrm{H_3C{-}CH_2{-}CH{=}CH{-}CH{=}CH{-}C{\equiv}C{-}C{\equiv}C{-}CH_2{-}CH_2{-}CH_2OH}$	*Aethusa cynapium* L.	—
13	$\mathrm{H_3C{-}CH{=}CH{-}C{\equiv}C{-}C{\equiv}C{-}CH{=}CH{-}CH{-}CH{-}CH_2{-}CH_3}$ O	*Aethusa cynapium* L.	—
13	$\mathrm{HOCH_2{-}CH{=}CH{-}C{\equiv}C{-}C{\equiv}C{-}CH{=}CH{-}CH{-}CH{-}CH_2{-}CH_3}$ O	*Aethusa cynapium* L.	—
13	$\mathrm{H_3C{-}CH{=}CH{-}C{\equiv}C{-}C{\equiv}C{-}CH{=}CH{-}CH{=}CH{-}CH_2{-}CH_3}$	*Peucedanum verticillare* Koch.	—
13	$\mathrm{H_3C{-}CH_2{-}CH_2{-}CHOH{-}CH{=}CH{-}C{\equiv}C{-}C{\equiv}C{-}CH\quad CH{-}CH_3}$	*Aethusa cynapium* L.	—
13	$\mathrm{H_3C{-}CH_2{-}CHOH{-}CHOH{-}CH{=}CH{-}C{\equiv}C{-}C{\equiv}C{-}CH{=}CH{-}CH_3}$	*Aethusa cynapium* L.	—

13	$H_3C-CH_2-CH=CH-CH=CH-C\equiv C-C\equiv C-CH=CH-CH_2OH$	*Aethusa cynapium* L.	—
17	$H_3C-(CH_2)_5-CH=CH-CH=CH-C\equiv C-C\equiv C-CH=CH-CH_2OH$	*Oenanthe crocata* L.	—
17	$H_3C-(CH_2)_3-CH=CH-CH=CH-CH=CH-C\equiv C-C\equiv C-CH_2-CH_2-CH_2OH$	*Cicuta virosa* L.	—
17	$H_3C-(CH_2)_2-CHOH-CH_2-CH_2-CH=CH-CH=CH-C\equiv C-C\equiv C\quad CH=CH-CH_2OH$	*Oenanthe crocata* L.	—
17	$H_3C-(CH_2)_2-CHOH-CH=CH-CH=CH-CH=CH-C\equiv C-C\equiv C-CH_2-CH_2-CH_2OH$	*Cicuta virosa* L.	—
17	$H_3C-(CH_2)_6-CH=CH-CH_2-C\equiv C-C\equiv C-CO-CH=CH_2$	*Carum carvi* L., *Falcaria vulgaris* *Bernh.*	*Hedera helix* L.
17	$H_3C-(CH_2)_6-CH=CH-CO-C\equiv C-C\equiv C-CO-CH=CH_2$	*Carum carvi* L., *Oenanthe pimpinelloides* L.	—
17	$H_3C-(CH_2)_6-CH=CH-CHOH-C\equiv C-C\equiv C-CO-CH=CH_2$	*Oenanthe pimpinelloides* L.	*Aralia nudicaulis* L.
17	$H_3C-(CH_2)_2-CO-CH_2-CH_2-CH=CH-CH=CH-C\equiv C-C\equiv C-CH=CH-CH_3$	*Oenanthe crocata* L.	—
18	$O=CH-(CH_2)_7-CH=CH-CH_2-C\equiv C-C\equiv C-CO-CH=CH_2$	*Pastinaca sativa* L.	—

Our own research has thrown some light on the relationship between the genera *Erigeron*, *Aster*, and *Conyza* (in tribe III). The chemistry of the main sections is for the most part clear, but as usual there exist some sections restricted to very inaccessible localities and mostly absent from botanical gardens. For this reason I think the chemist has to be rather modest in his statements; instead he should try to get some efficient organization to make such important plant materials more readily available.

B. Families Umbelliferae and Araliaceae

The year 1952–1953 was a remarkable one for naturally occurring acetylenes. During that year not only was the diversity of acetylenes produced by Compositae and fungi evident, but through the important work of Lythgoe and his collaborators the toxic principles of *Oenanthe* and *Cicuta* of the Umbelliferae were shown to be acetylenic and to be accompanied by a number of nontoxic acetylenic compounds. The acetylenes from the Umbelliferae and the closely related family Araliaceae are shown in Table 4. Some of these substances are identical with Compositae acetylenes, and all the others, as may be seen, are closely related to the Compositae acetylenes. I am convinced that acetylenes will be found in many more genera of these two families. Although we have restricted ourselves to the Compositae, tests on material readily available in Trondheim from the Umbelliferae and Araliaceae have indicated that acetylenic compounds occur rather regularly in these families.

These results imply two taxonomic consequences. The families Umbelliferae and Araliaceae have been placed together by most botanists in the same order (Umbelliflorae of Engler, Umbellales of Bentham and Hooker). At times, they have been united in one family, the Apiaceae. Only Hutchinson separated them widely because of his primary division between woody and herbaceous plants, a distinction which has given much discredit to Hutchinson's otherwise brilliant work. The acetylenes common to these two families fully confirm their close relationship.

The next question is more intriguing. Do these results prove a relationship of these two families to the Compositae? One of the main differences between the 12 tribes of the Tubuliflorae and the tribe Cichoreae is that all the Cichoreae contain a milky sap, and all members of the 12 tribes of the Tubuliflorae contain resin channels. The families that most botanists arrange as related to the Compositae lack resin channels; some of them have milky sap, like the Cichoreae (wherein we thought acetylenes were definitely absent until crepenynic acid was detected). Both the Umbellifereae and the Araliaceae have resin channels. However, one may ask, Are resin channels merely a prerequisite for the occurrence of the extremely unsaturated polyacetylenes so easily detected? That is, are they merely a correlated character and not as such an indication of close botanical relationship? I can only state that the acetylenes occur in the resin channels. The disulfur compound

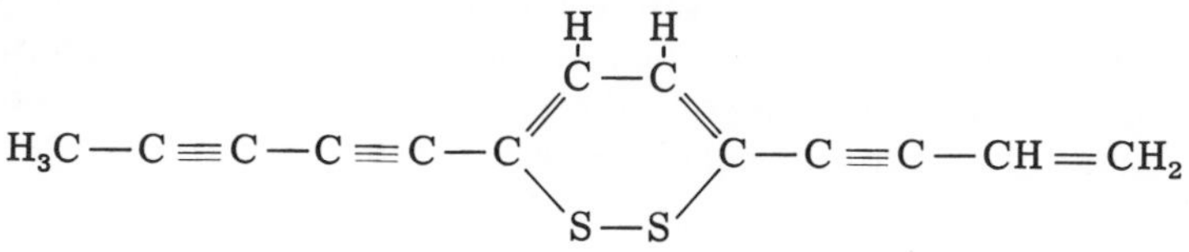

$$\begin{array}{c} \text{H}\quad\text{H} \\ \text{C}-\text{C} \\ H_3C-C\equiv C-C\equiv C-C\diagup\!\!\!\!\!\!\quad\quad\diagdown C-C\equiv C-CH=CH_2 \\ \text{S}-\text{S} \end{array}$$

is about as red as potassium permanganate. If one cuts the fine roots of an *Eriophyllum* species the resin channels may be seen to be colored by this acetylene.

TABLE 5

Some Mono-acetylene Compounds Isolated from the Rest of Plant Kingdom

$H_2NOC-C\equiv C-CONH_2$	*Streptomyces*
$HOOC-(CH_2)_8C\equiv CH$	*Rhodotorula glutinis* (Cryptococcaceae)
$H_3C-CH(OCH_3)-(CH_2)_7C\equiv CH$	*Litsea odorifera* Val. (Lauraceae)
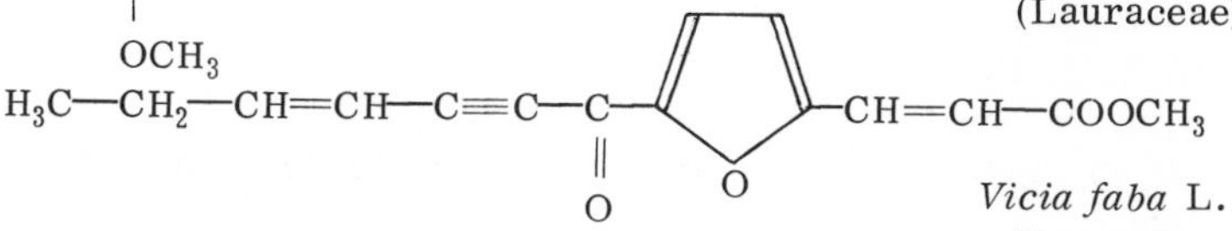 $H_3C-CH_2-CH=CH-C\equiv C-C(=O)-$[furan ring, O]$-CH=CH-COOCH_3$	*Vicia faba* L. (Leguminosae)
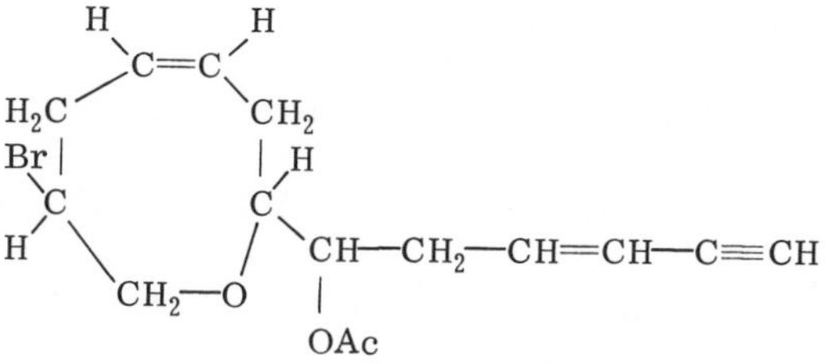 [ring: $H_2C-CH=CH-CH_2-CH(-O-CH_2-CHBr-)$]; $CH(OAc)-CH_2-CH=CH-C\equiv CH$	*Laurentia glandulifera* (Rhodomelaceae)

C. Other Families

In Table 5, I have tabulated the occurrence of acetylenic compounds in other families for those instances where the acetylenes were isolated and their constitutions definitely proven. One may realize immediately that these groups are spread widely throughout the plant kingdom. They all belong to those compound types where present day spectroscopic methods mostly give no qualitative indication at all. How this picture will change when the Raman spectrophotometer becomes a standard instrument in all natural products laboratories, I should very much like to know. As a consequence of the possible changes, I think the acetylene chemist should be very careful, at the present time, in making taxonomic conclusions beyond the few families where a reasonable quantity of definite results are available.

REFERENCES

Anchel, M. 1952. J. Amer. Chem. Soc., 74: 1588.

Bew, R. E., R. C. Cambie, E. R. H. Jones, and G. Lowe. 1966a. J. Chem. Soc.: 135.

_____, J. R. Chapman, E. R. H. Jones, B. E. Lowe, and G. Lowe. 1966b. J. Chem. Soc.: 129.

Birkinshaw, J. H., and P. Chaplen. 1955. Biochem. J., 60: 255.

Bohlmann, P., H. Bornowski, and Chr. Arndt. 1963. Fortschr. Chem. Forsch., 4: 138-272.

Bu'Lock, J. D. 1961. Advances in Applied Microbiology, 3: 293.

_____. 1964. Progress in Organic Chemistry, 85-135, London, Butterworths.

_____, and A. J. Powell. 1965. Experientia, 21: 55.

_____, and G. N. Smith. 1963. Phytochemistry, 2: 289.

Celmer, W. D., and J. A. Solomons. 1952. J. Amer. Chem. Soc., 74: 1870, 2245, 3838; 75: 1372, 3430.

Craig, J. Cymmerman, R. E. Lack, and W. Treibs. 1959. Chem. Industr.: 952.

Earle, F. R., I. A. Wolff, and Q. Jones. 1960. Amer. Oil Chem. Soc., 37: 254.

_____, K. L. Mikolajczak, and I. A. Wolff. 1964. J. Amer. Oil Chem. Soc., 41: 345.

Johnson, E. A., and K. L. Burdon. 1947. J. Bact., 54: 281.

Jones, E. R. H. 1959. Pedler Lecture, February 12; cf. 1960, Proc. Chem. Soc.: 199.

_____. 1966. Chemistry in Britain: 6.

_____, G. Lowe, and P. V. R. Shannon. 1966. J. Chem. Soc.: 139.

McLean, J., and A. H. Clark. 1956. J. Chem. Soc.: 777.

Mikolajczak, K. L., C. R. Smith, Jr., M. O. Bagby, and I. A. Wolff. 1964. J. Org. Chem., 29: 318.

Powell, R. G., and C. R. Smith, Jr. 1966. Biochemistry: 625.

_____, C. R. Smith, Jr., C. A. Glass, and I. A. Wolff. 1966. J. Org. Chem., 31: 528.

Robbins, W. J., F. Kavanagh, and A. Hervey. 1947. Proc. Nat. Acad. Sci. U. S. A., 33: 176.

Schulte, K. E., J. Reisch, and J. Rheinbay. 1965. Phytochemistry, 4: 481.

Smith, G. N., and J. D. Bu'Lock. 1965. Chem. Industr.: 1840.

Treibs, W. 1947. Chem. Ber., 80: 97.

Wailes, P. C., H. H. Hatt, and A. C. K. Triffitt. 1960. Review given August 24th, 1960, IUPAC symposium, Sydney.

Zechmeister, L., and J. W. Sease. 1947. J. Amer. Chem. Soc., 69: 273.

7

PSEUDOGUAIANOLIDES IN COMPOSITAE

WERNER HERZ
Department of Chemistry
The Florida State University, Tallahassee

I. INTRODUCTION

The known naturally occurring sesquiterpene lactones isolated from plants can, at the time of this writing, be classified as follows:

1. The eudesmanolides, of which santonin (I; Simonsen and Barton, 1952; Barton and de Mayo, 1957b; Asher and Sim, 1965)

and alantolactone (II; Simonsen and Barton, 1952; Marshall and Cohen, 1964) are the oldest and best-known examples.

2. The guaianolides (Barton and de Mayo, 1957b; Nozoe and Ito, 1961), among which the structure of geigerin (III) has been fully documented by X-ray crystallography (Hamilton et al., 1962).

3. The germacranolides (Halsall and Theobald, 1962), the first example of which was pyrethrosin (IV; Barton and de Mayo, 1957a; Barton et al., 1960).

4. The lactones derived from drimane, whose structure is illustrated by iresin (V; Djerassi and Burstein, 1958; Rossmann and Lipscomb, 1958).

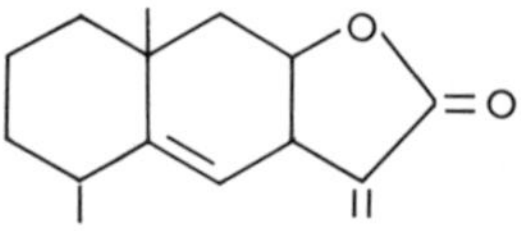

I 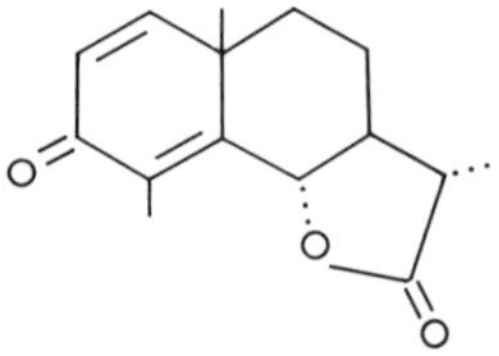 II

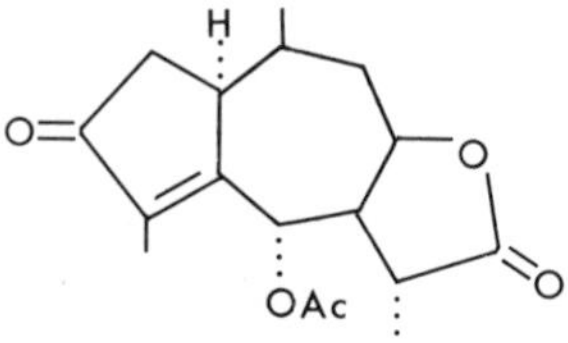

III

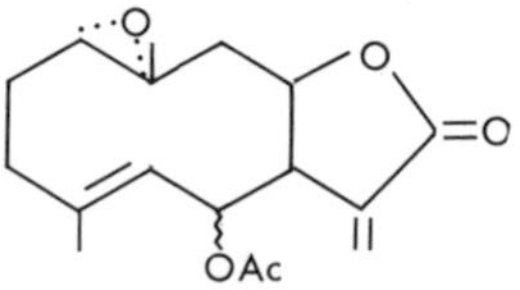

IV 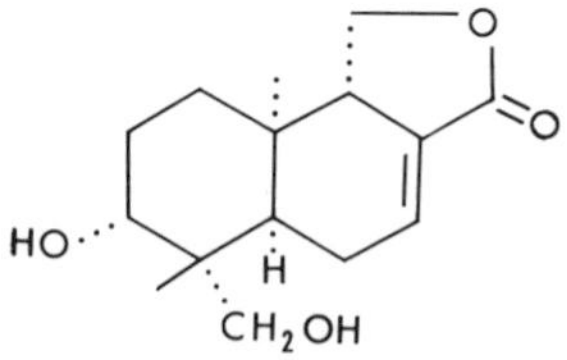 V

The formation of the hydrocarbon precursors of compounds of types I–III may be rationalized by assuming a cyclization process that is initiated by ionization of all-*trans*-farnesyl pyrophosphate (Ruzicka, 1953, 1959; Hendrickson, 1959; Richards and Hendrickson, 1964). Although this scheme has not yet been documented experimentally, the laboratory conversions of some germacranolides to eudesmanolides (Barton and de Mayo, 1957;

Suchý et al., 1959; Rao et al., 1960; Kulkarni et al., 1964) and of a germacranolide to a guaianolide (Govindachari et al., 1965) offer support for steps c and d of Scheme A. The direct conversion of a farnesol derivative to a germacrane (step a), eudesmane (step b), or guaiane (step e) has not yet been realized.

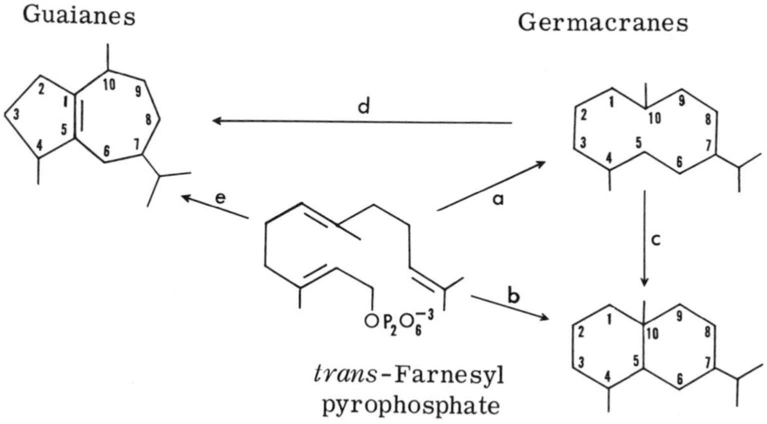

Guaianes

Germacranes

d

e

a

c

b

$OP_2O_6^{-3}$

trans-Farnesyl pyrophosphate

Eudesmanes

Scheme A

On the other hand, the lactones derived from drimane, only a few examples of which are known, may be thought of as arising by the electrophile-catalyzed cyclization of farnesol pyrophosphate typical of di- and triterpenes (Scheme B; Richards and Hendrickson, 1964). A laboratory analogy (not involving lactones) is known (Van Tamelen et al., 1963).

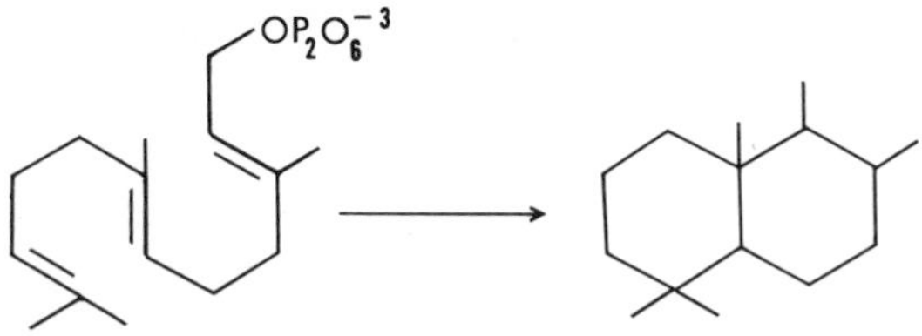

$OP_2O_6^{-3}$

Farnesyl pyrophosphate

Drimane

Scheme B

Further transformation of the classes listed in Scheme A ma give rise to other groups of naturally occurring sesquiterpene lactones. Methyl migration from C_{10} to C_5 of eudesmane derivatives leads to the eremophilanolides, such as eremophilenolide (VI; Novotny et al., 1962, 1963), found so far only in *Petasites* species. Valence bond isomerization of germacran-1,10,4,5-dienolides may result in monocyclic compounds, for instance, the thermal conversion of dihydrocostunolide (VII) to saussurea lactone (VIII) (Rao et al., 1961), the only lactone of this type encountered so far. Naturally occurring cleavage of the C_4–C_5 bon of guaianolides furnishes compounds of the type of xanthumin

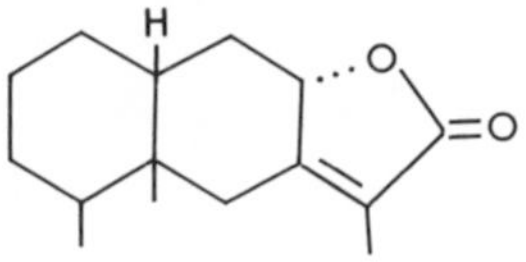

VI 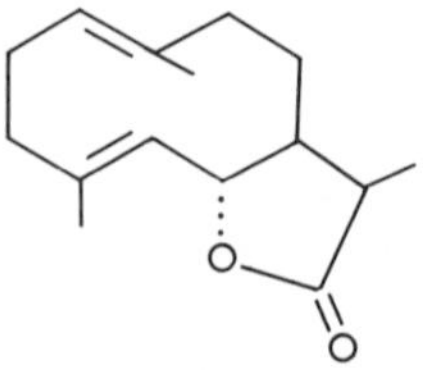 VII 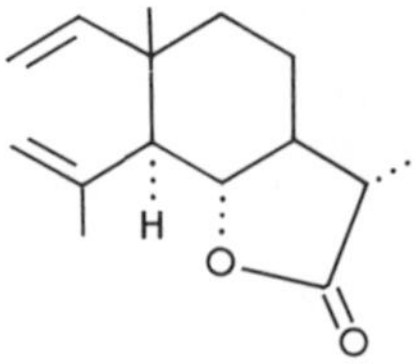 VIII

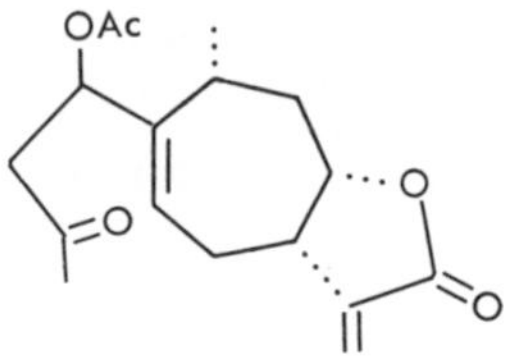

IX 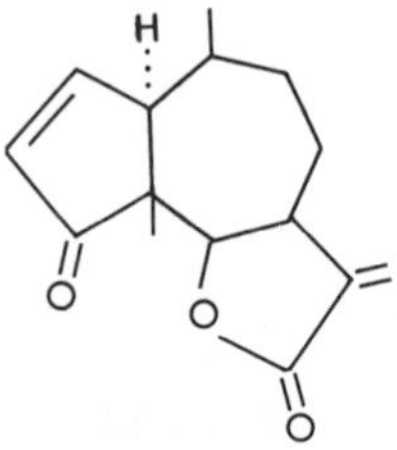 X 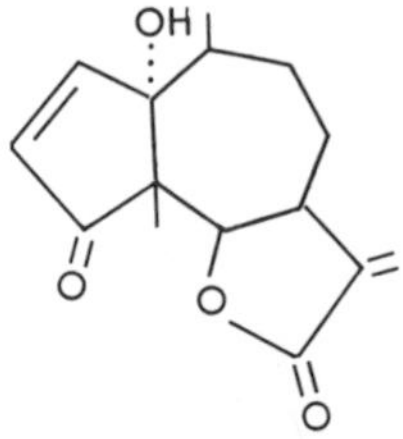 XI

(IX; Minato and Horibe, 1965). Last, methyl migration from C_4 to C_5 of the guaianolide carbon skeleton generates the large clas of pseudoguaianolides, such as ambrosin (X; Herz et al., 1962c), with which this paper is primarily concerned.[1]

[1] We include compounds such as aristolactone (i; Martin-Smith et al., 1964) and linderalactone (ii; Takeda et al., 1964) in the class of germacranolides.

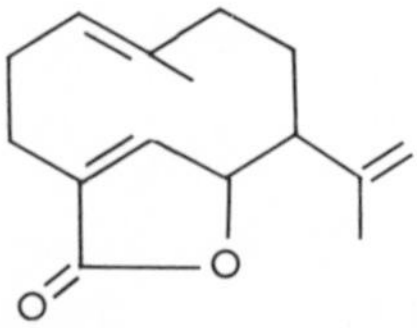

i 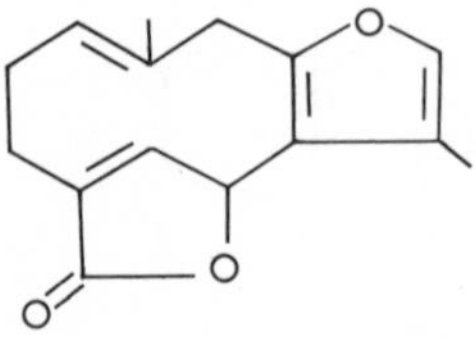 ii

Historically, the existence of the pseudoguaianolide carbon skeleton was first recognized (Herz et al., 1961; Herz et al., 1962c) in ambrosin (X), a constituent of *Ambrosia maritima* L. (Abu-Shady and Soine, 1953, 1954; Bernardi and Büchi, 1957; Sorm et al., 1959; Herz et al., 1959c), and parthenin (XI; Herz et al., 1962c), the main sesquiterpene lactone of *Parthenium hysterophorus* L. Chemical studies coupled with X-ray crystallographic analysis of 3-bromoambrosin (Emerson et al., 1966) have established unequivocally the relative and absolute stereochemistry depicted in formulas X and XI.

Almost simultaneously, it was realized (Herz et al., 1962b); Herz et al., 1963a) that the reactions and properties of the sesquiterpene lactones tenulin from *Helenium amarum* Raf. (Clark, 1939, 1940; Ungnade and Hendley, 1948; Ungnade et al., 1950; Barton and de Mayo, 1956; Braun et al., 1956) and helenalin from both *Helenium autumnale* L. (Lamson, 1913; Clark, 1936; Adams and Herz, 1949; Büchi and Rosenthal, 1956) and *Balduina angustifolia* (Pursh.) Robins (Herz and Mitra, 1958) were best met by the analogous formulas XII and XIII. Subsequent work (Herz et al., 1963b; Rogers and Mazhar-ul-Haque, 1963; Emerson et al., 1964) confirmed these structures and established the relative and absolute stereochemistry depicted here.

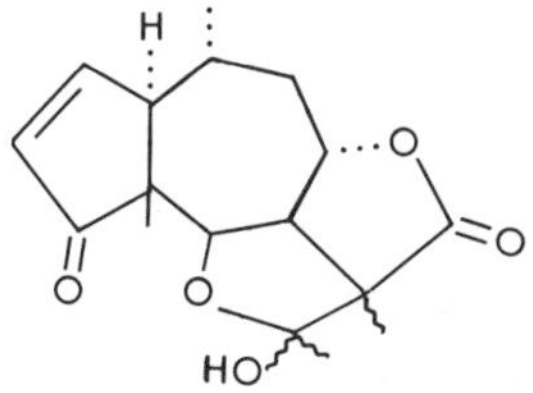

XII 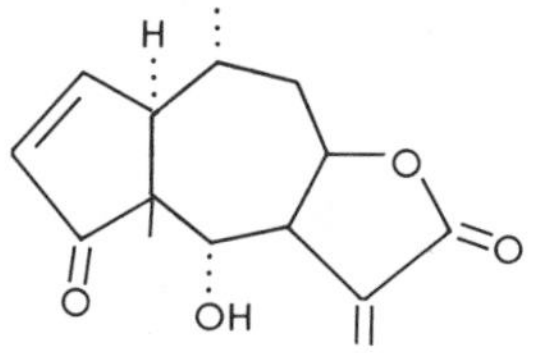 XIII

II. DISTRIBUTION OF PSEUDOGUAIANOLIDES IN HELIANTHEAE

The discovery of compounds with the "abnormal" pseudoguaianolide skeleton in both *Ambrosia* and *Parthenium* species was of interest because of the debate on the position of *Ambrosia* and its relatives in the general taxonomic scheme of the *Compositae*. The matter has been reviewed by Cronquist (1955) and briefly by Solbrig (1963). As a consequence it appeared desirable

TABLE 1

Sesquiterpene Lactones of Ambrosia and Related Species

Species	Compound	Formula	Reference
Heliantheae Cass.			
Ambrosiinae Less.			
Ambrosia artemisifolia L.	Coronopilin	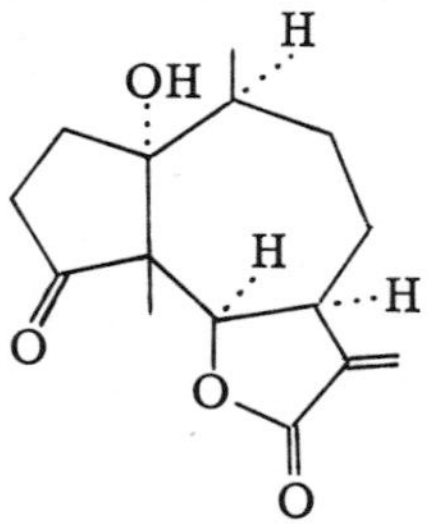	Herz and Högenauer (1961)
Ambrosia bidentata Michx.	No crystalline lactones	—	Herz and Högenauer (1961)
Ambrosia chamissonis (Less.) Greene*[a]	Chamissonin	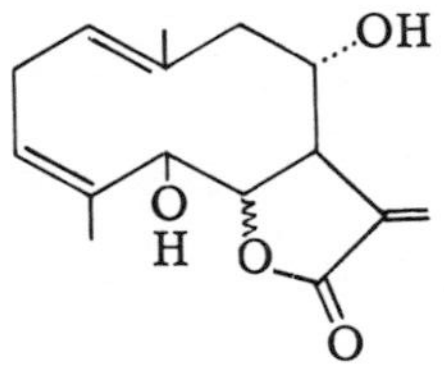	Geissman et al. (1966)

[a]Starred species are members of the genus *Franseria*, whose inclusion in *Ambrosia* is proposed by Payne (1964), a

Ambrosia confertiflora DC.*	Psilostachyin	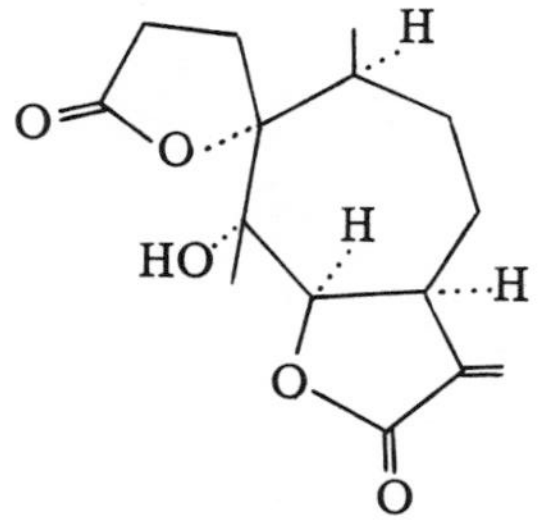	Herz and Raulais (unpublished)
Ambrosia cumanensis Kunth.	Psilostachyin Psilostachyin B	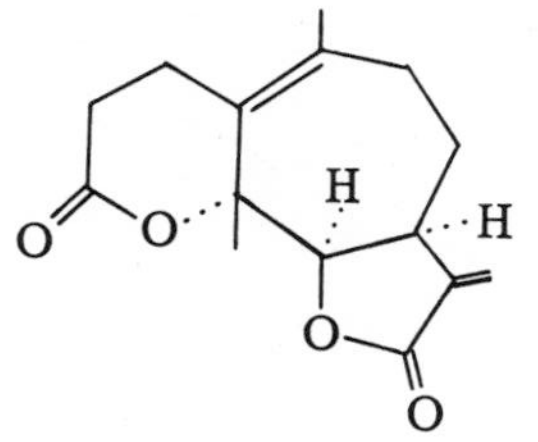	Miller and Mabry, 1967
	Psilostachyin C	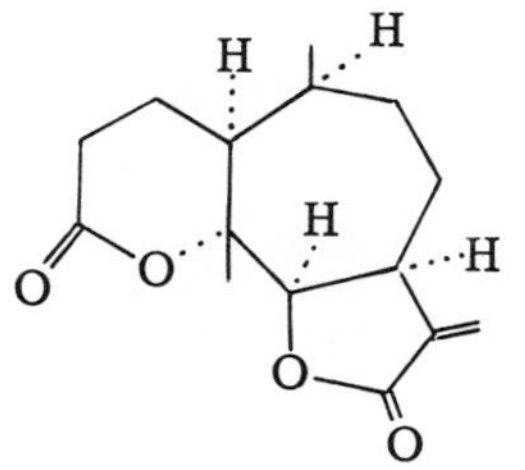	

TABLE 1 (continued)

Species	Compound	Formula	Reference
	Cumanin	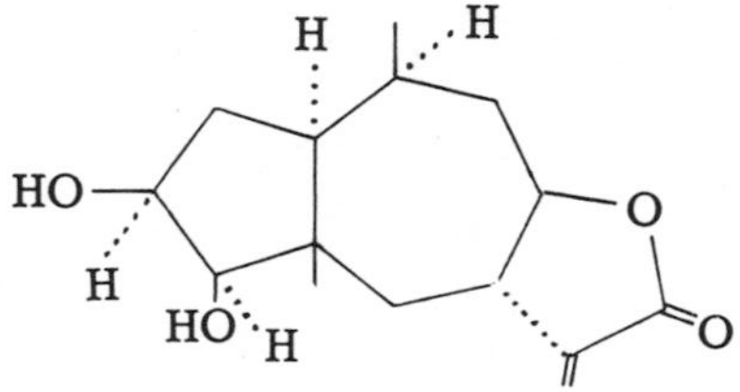	Romo et al. (1966a)
Ambrosia deltoidea (Torr.) Payne*	Damsin	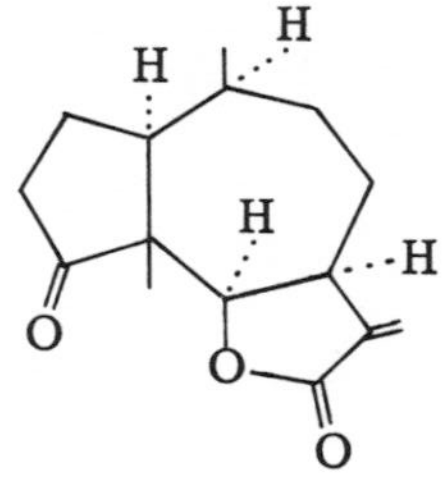	Kagan et al., 1966
	Psilostachyin C	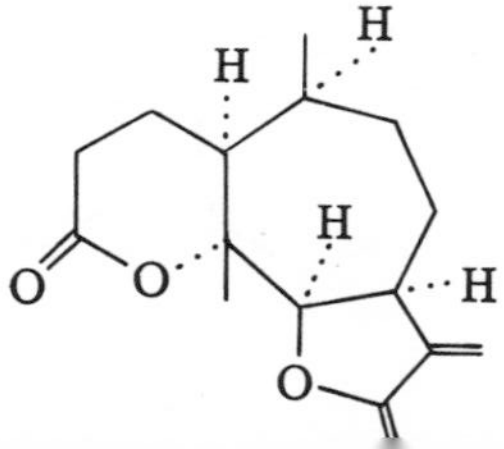	

Ambrosia dumosa (Gray) Payne*	Coronopilin		Geissman and Turley (1964)
Ambrosia hispida Pursh.	Ambrosin		Herz and Sumi (1964)
	Damsin		
Ambrosia ilicifolia (Gray) Payne*	Ilicic acid[b]	H, CO_2H, HO, H	Herz et al. (1966a)
	Costic acid[b]	H, CO_2H, H	Bawdekar and Kelkar (1965)
Ambrosia maritima L.	Ambrosin		Abu-Shady and Soine (1953)
	Damsin		Bernardi and Büchi (1957); Sorm et al. (1959); Herz et al. (1959a); Herz et al. (1961); Herz et al. (1962c); Emerson et al. (1966)
Ambrosia peruviana Willd.	Psilostachyin C		Kagan et al. (1966)

TABLE 1 (continued)

Species	Compound	Formula	Reference
	Tetrahydroambrosin	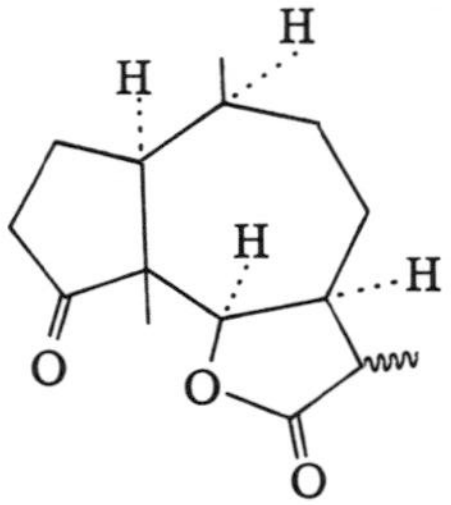	Herz and Anderson (unpublished)
Ambrosia psilostachya DC. (Galveston Is., Texas)	Psilostachyin		Miller et al. (1965)
	Psilostachyin B		Mabry et al. (1966a, b)
	Psilostachyin C		Kagan et al. (1966)
Ambrosia psilostachya DC. (Austin, Texas)	Coronopilin		Mabry et al. (1966c)
	Parthenin	XI	
	Ambrosiol	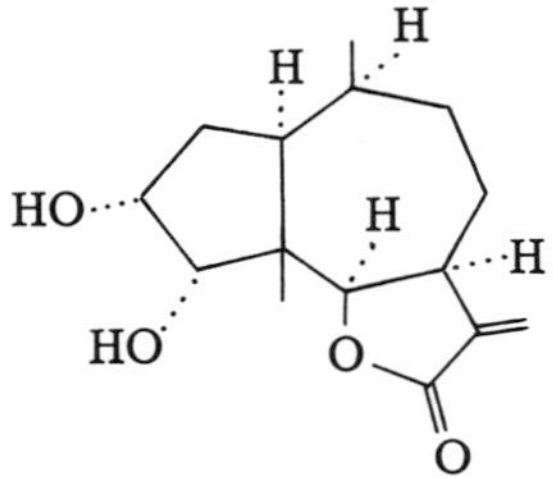	

Ambrosia psilostachya DC. (Calif.)	Coronopilin	Geissman and Turley (1964)
Ambrosia psilostachya DC. var. *coronopifolia* Farw. (Kansas)	Coronopilin	Herz and Högenauer (1961)
Ambrosia trifida L.	——	Herz and Högenauer (1961)
Iva acerosa (Nutt.) Jacison	Coronopilin	Farkas et al. (1966)
Iva angustifolia Nutt.	Ivangustin	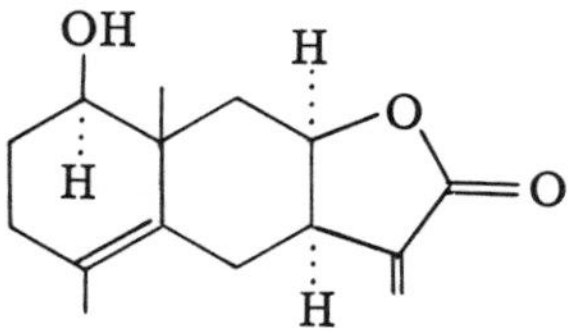 Herz et al. (1967c)
	Ivangulin	
Iva annua var. *caudata* (Small) Jackson	——	Herz (unpublished)
Iva annua C. var. *annua*	——	Herz (unpublished)
Iva asperifolia Less.	Asperilin	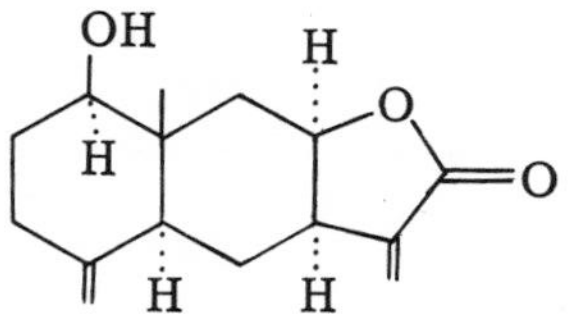 Herz and Viswanathan (1964)

TABLE 1 (continued)

Species	Compound	Formula	Reference
	Ivasperin	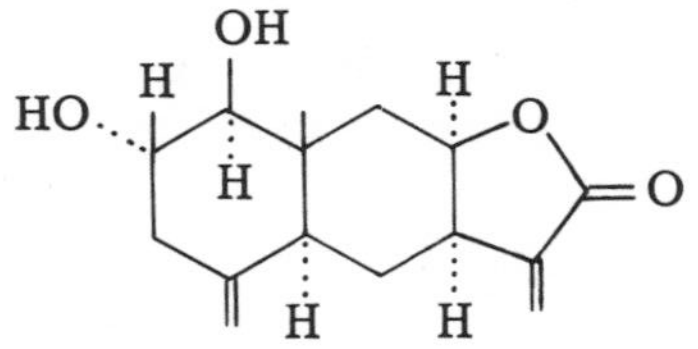	
Iva axillaris Pursh. subsp. *axillaris*	Ivaxillarin	—	Herz and Sudarsanam (unpublished)
Iva axillaris Pursh. subsp. *robustior* (Hook) Bassett	Anhydroivaxillarin	—	
	Axivalin	—	
	Ivaxilin	—	
Iva cheiranthifolia H.B.K.	Ivalin	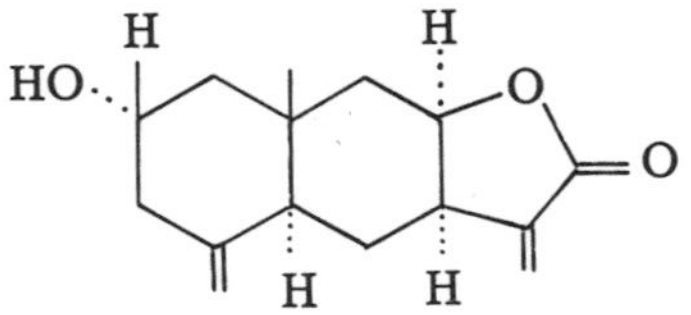	Herz and Sudarsanam (unpublished)
Iva dealbata Gray	Ivalbin	—	Herz et al. (1967)

Iva frutescens subsp. oraria (Bartlett) (Jackson)	——	——	Herz and Sudarsanam (unpublished)
Iva frutescens L. subsp. *frutescens*	——	——	Herz and Sudarsanam (unpublished)
Iva hayesiana Gray	Lactone which polymerizes	——	Herz and Sudarsanam (unpublished)
Iva imbricata Walt.	Ivalin		Herz and Högenauer (1962)
Iva microcephala Nutt. var. A	Ivalin		Herz and Högenauer (1962)
Iva microcephala Nutt. var. B	Microcephalin	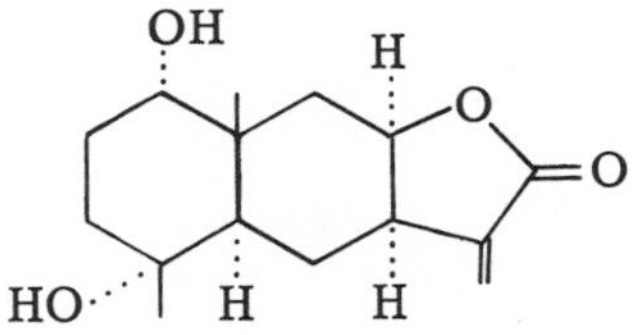	Herz et al. (1964a)
	Pseudoivalin	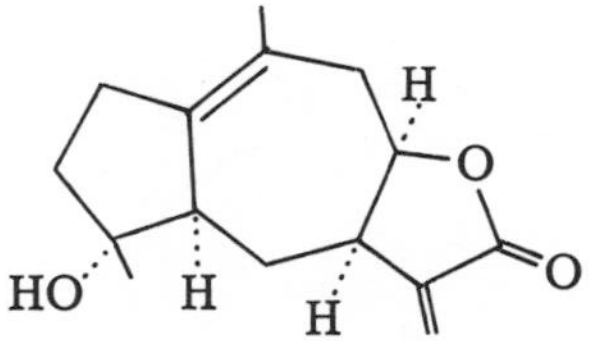	Herz et al. (1965)

TABLE 1 (continued)

Species	Compound	Formula	Reference
	Dihydropseudo-ivalin	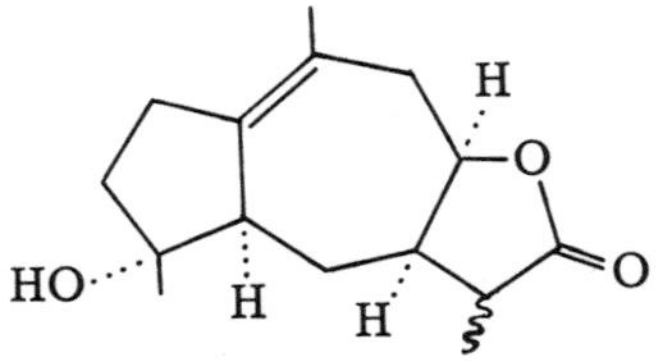	
Iva nevadensis M.E. Jones	Coronopilin Parthenin Nevadivalin	Unknown	Farkas et al. (1966)
Iva texensis Jackson	Asperilin Ivasperin		Herz and Viswanathan (1964)
Iva xanthifolia Nutt.	Coronopilin		Herz (unpublished); Novikov et al. (1964)
	Lactone[c]	—	
Xanthium italicum L.	Xanthinin		Toth et al. (1962)
Xanthium pennsylvanicum Wallr.	Xanthinin	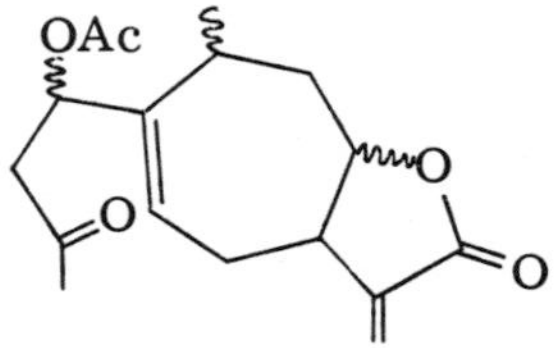	Geissman et al. (1954); Geissman and Deuel (1957); Dolejs et al. (1958); Geissman (1962)

Xanthium strumarium L.	Xanthinin		Plourde and Mockle (1960)
Xanthium strumarium L. (Japan)	Xanthumin (isomer of xanthinin)	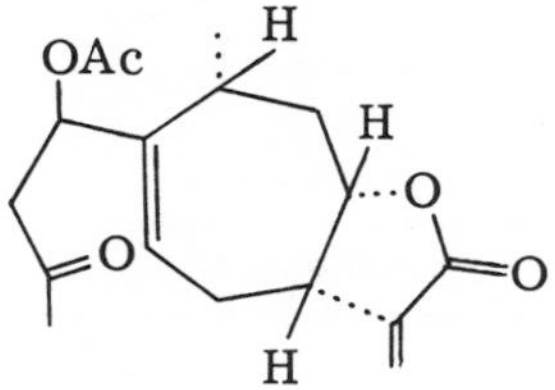	Minato and Horibe (1965)
Melampodiinae Less.			
Parthenium argentatum Gray	Partheniol cinnamate[d]	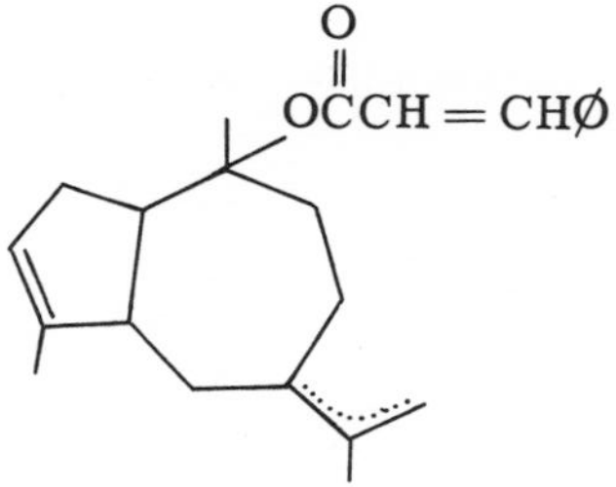	Hendrickson and Rees (1962); Haagen-Smit and Fong (1948)
Parthenium hysterophorus L.	Parthenin		Arny (1890, 1897); Herz et al. (1959); Herz et al. (1961); Herz et al. (1962c)

TABLE 1 (continued)

Species	Compound	Formula	Reference
Parthenium hysterophorus L. (*P. bipinnatifidum* Ortega?) (Valley of Mexico)[e]	Ambrosin Hysterin	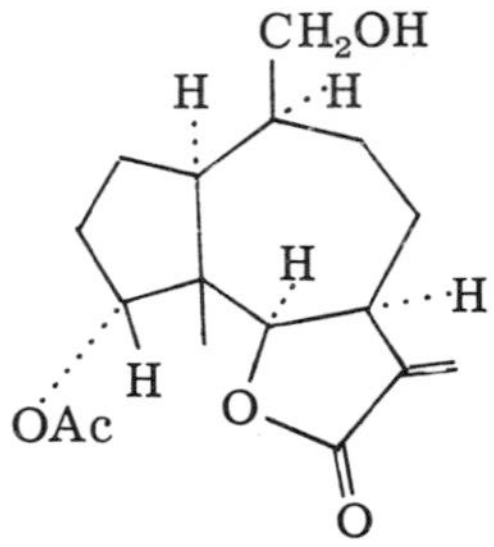	Romo de Vivar et al. (1966a)
Parthenium incanum H.B.K.	Coronopilin		Herz and Högenauer (1961)
	Ambrosin		Herz et al. (1962c)

[d]Not a lactone but listed here to indicate type of sesquiterpene constituent.

[e]The place of origin and chemical composition of this collection suggests that it may actually be *P. bipinnatifidum* Ortega, which according to Rollins (1950) is difficult to distinguish from *P. hysterophorus* L.

Addendum to Table 1

Species	Compound	Formula	Reference
Ambrosia ambrosioides (Cav.) Payne*[a]	——	——	Herz et al. (unpublished)
Ambrosia confertiflora DC.* (Kennedy and Kingsville collection)	Confertiflorin	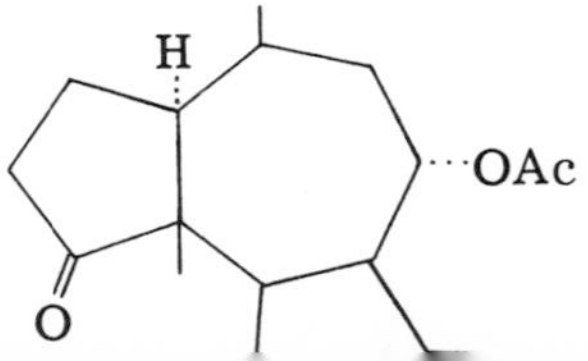	Fischer and Mabry (1967)

	Desacetylconfertiflorin		
Ambrosia cordifolia (Gray) Payne*	Psilostachyin C Stereoisomer of psilostachyin		Herz and Raulais (unpublished)
Ambrosia cumanensis H.B.K. (Colombia)	Ambrosin Psilostachyin Coronopilin		Herz and Fitzhenry (unpublished)
Ambrosia peruviana Willd. (State of Hidalgo, Mex.)	Peruvin	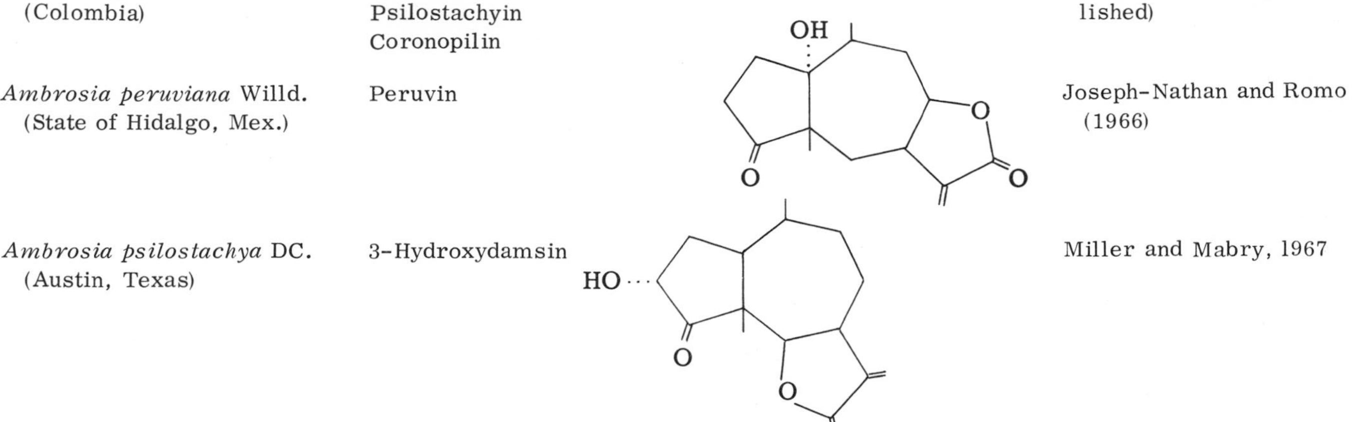	Joseph-Nathan and Romo (1966)
Ambrosia psilostachya DC. (Austin, Texas)	3-Hydroxydamsin		Miller and Mabry, 1967

[a]Starred species are members of the genus *Franseria*, whose inclusion is proposed by Payne (1964), a procedure which is followed here.

Addendum to Table 1 (continued)

Species	Compound	Formula	Reference
Iva axillaris Pursh. sabsp. *robustior* (Hook) Bassett	Ivaxillarin	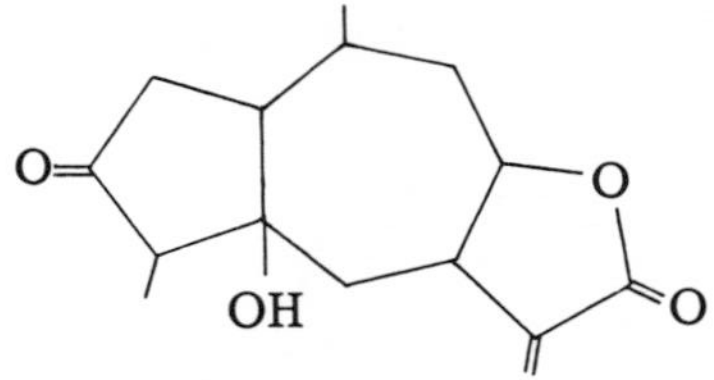	Herz et al. (1966e)
	Anhydroivaxil-larin	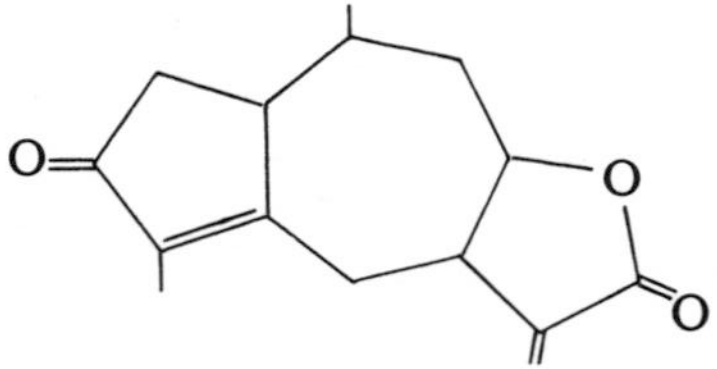	
	Axivalin	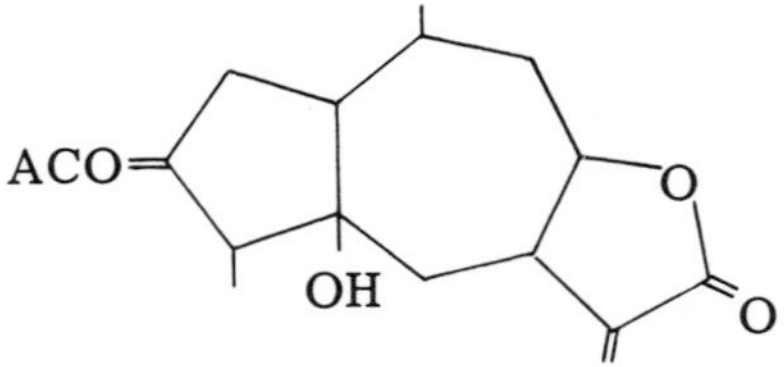	
Iva dealbata Gray	Ivalbin	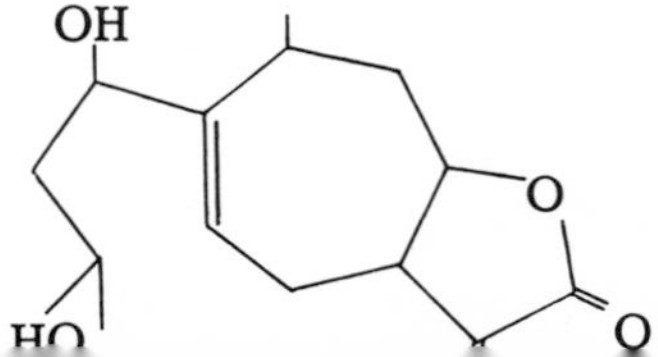	Herz et al. (1967a)

to investigate the sesquiterpene lactone content of *Ambrosia* and related species. The results that have been accumulated so far are listed in Table 1.

Although the experimental material from *Parthenium* is still quite meager, the sesquiterpene lactone chemistry of most *Ambrosia* and *Parthenium* species on which work has been brought to a conclusion is practically indistinguishable. This may perhaps be interpreted as lending support to the school that incorporates *Ambrosia* and related genera in the tribe Heliantheae. The formation of a fairly limited number of closely related pseudoguaianolides appears to be typical. Noteworthy is the group of psilostachyins (Miller et al., 1965; Mabry et al., 1966a, b; Kagan et al., 1966), whose biogenesis must involve pseudoguaianolide precursors. Other sesquiterpene skeletal types have so far been encountered only in the two "franserioid" (Payne, 1964) species, *A. chamissonis* and *A. ilicifolia.* It is hoped that further work now in progress may permit extension of this generalization.

Among *Iva* species, however, only members of the section *Cycloachena*, as defined by Jackson (1960), elaborate pseudoguaianolides.[2] All other species studied so far elaborate eudesmanolides or closely related guaianolides. Whether this criterion is sufficient to distinguish the sections is now under study.

A minor point of some interest is that two subspecies of *I. axillaris* that have recently been differentiated on morphological grounds (Bassett et al., 1962) also differ in sesquiterpene lactone content.

Observations of modified guaianolides of the type of xanthumin (IX) in Heliantheae have so far resulted only from investigations dealing with *Xanthium* species (but see footnote 2). However, the occurrence of carabrone (XIV; Minato et al., 1964) in the fruits of *Carpesium abrotanoides* L. (Compositae, tribe Inuleae) suggests that the distribution of such compounds may be more general.

Noteworthy in the additions to Table 1 is the discovery, in *Ambrosia* species, of lactones closed to C-8 instead of C-6, and the chemical variability in species related to *Ambrosia psilostachya* D.C. Ivalbin from *Iva dealbata* Gray is a new structural variant previously confined to *Xanthium* species.

[2] An exception is *Iva dealbata* Gray, whose main constituent, ivalbin, appears to be chemically similar to the sesquiterpene lactones found in *Xanthium* species.

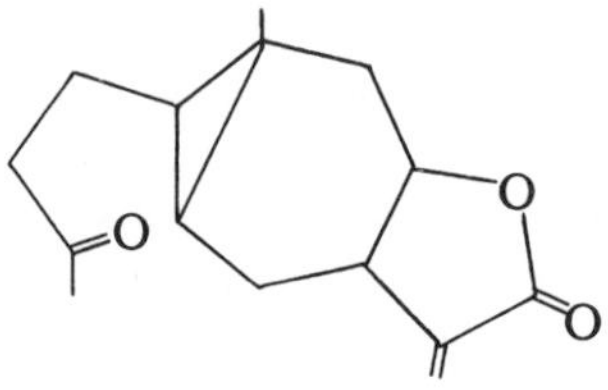

XIV

III. DISTRIBUTION OF PSEUDOGUAIANOLIDES IN HELENIAE

As a consequence of the discovery of tenulin (XII) and helenalin (XIII), the genera *Helenium* (Heleniae, subtribe Heleniinae) and *Gaillardia* (which adjoins *Helenium* in the taxonomic scheme of Compositae) have received careful scrutiny. This attention has resulted in the isolation of a second, considerably more varied, group of pseudoguaianolides, the representatives of which are enumerated in Table 2. Identical or very similar compounds have also been isolated from the ditypic genus *Balduina* (Heliantheae, subtribe Galinsoginae), which is therefore included in Table 2 as well.

According to a private communication from the late H. F. L. Rock, the morphological and cytological characteristics of *Balduina* species suggest that they are closely related to *Helenium* and *Gaillardia*, a conclusion that seems to be borne out by the sesquiterpene lactone chemistry. It should be remarked that there is no unanimity on the separate tribal status of Heleniae, which appears to be a rather artificial assemblage. It has been stated (Solbrig, 1963) that most of the genera of Heleniae belong with Heliantheae and that the others are better placed elsewhere. The findings with respect to *Balduina* may be considered to lend substance to this supposition, and investigations of related genera are now in progress to shed further light on the problem.

Inspection of Tables 1 and 2 reveals two fundamental differences between the pseudoguaianolides of Ambrosiinae and those of Heleniae. In the former the lactone ring is invariably closed to C_6 and the C_{10}-methyl group is β. In the latter the lactone ring is invariably closed to C_8 and the C_{10}-methyl group is α. The significance of this difference in terms of the biogenetic scheme outlined earlier (Scheme A) is not clear, but no simple cyclization and migration process can be written to accommodate the stereochemistry of both classes of compounds.

TABLE 2

Sesquiterpene Lactones of *Helenium* and Related Species

Species	Compound	Formula	Reference
Heleniae Benth. & Hook.			
Heleniinae Less.			
Helenium amarum Raf. (*Helenium tenuifolium* Nutt.)	Tenulin	XII	Clark (1939, 1940); Ungnade and Hendley (1948); Ungnade et al. (1950); Barton and de Mayo (1956); Braun et al. (1956); Herz et al. (1962b); Herz et al. (1963b); Rogers and Mazharhul-Haque (1963)
Helenium amarum Raf.	Tenulin		Lucas et al. (1964a)
	Aromaticin	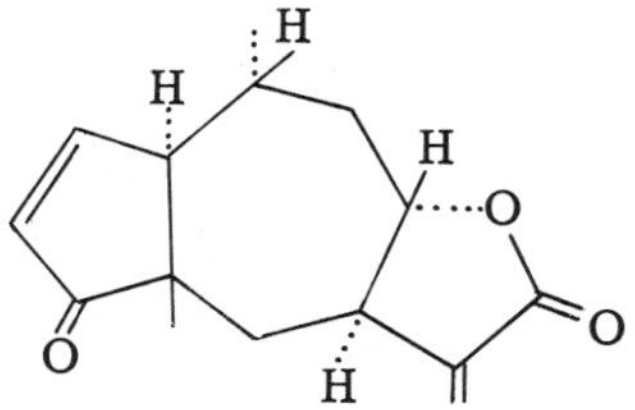	

TABLE 2 (continued)

Species	Compound	Formula	Reference
	Amaralin	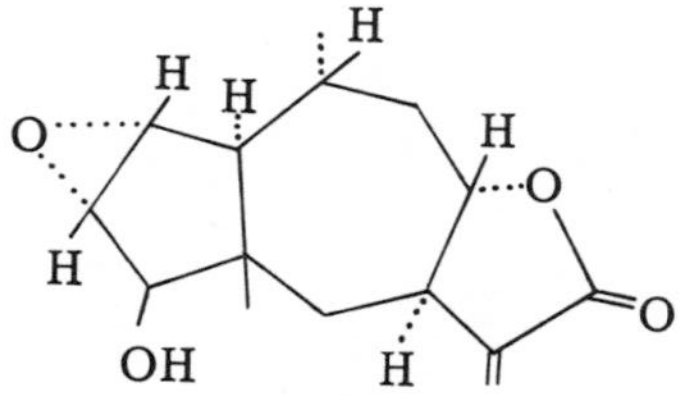	
Helenium arizonicum Blake	Isotenulin	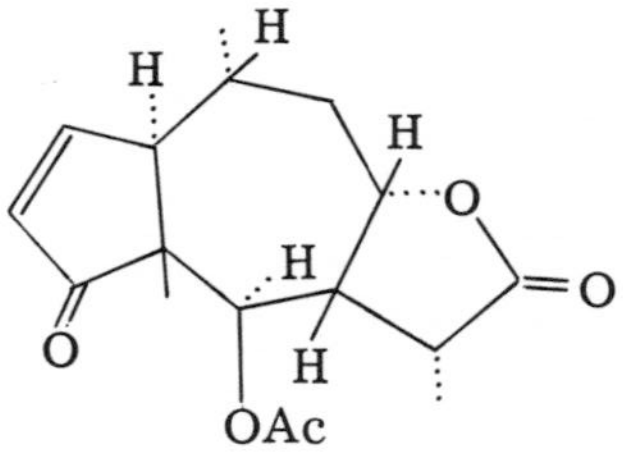	Herz (1962)
Helenium aromaticum (Hook) Bailey	Helenalin	XIII	Romo et al. (1964)
	Mexicanin I	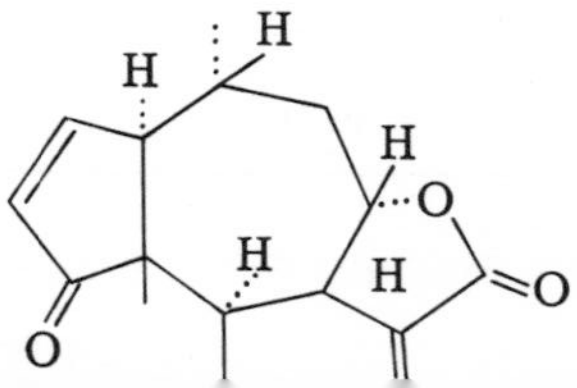	

	Aromatin	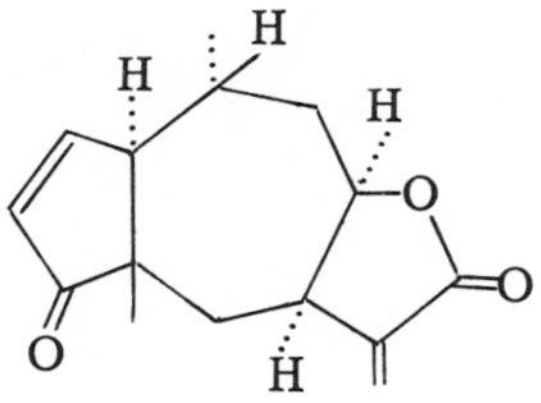	
	Aromaticin		
Helenium autumnale L.	Helenalin		Lamson (1913) ; Clark (1936) ; Adams and Herz (1949) ; Büchi and Rosenthal (1956) ; Herz et al. (1963a)
Helenium autumnale L.	Helenalin		Lucas et al. (1964b)
	Dihydromexi-canin E	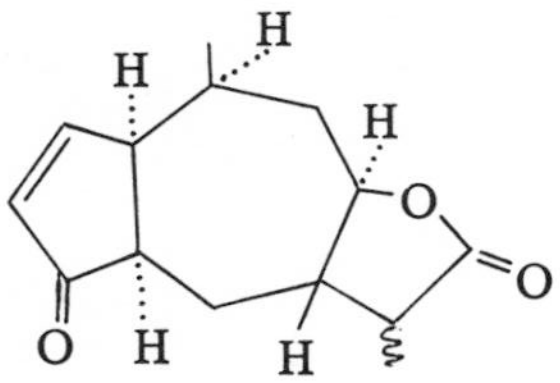	
Helenium badium Greene	Tenulin		Clark (1940)

TABLE 2 (continued)

Species	Compound	Formula	Reference
Helenium Bigelovii Gray	Tenulin Isotenulin Desacetyliso-tenulin Bigelovin	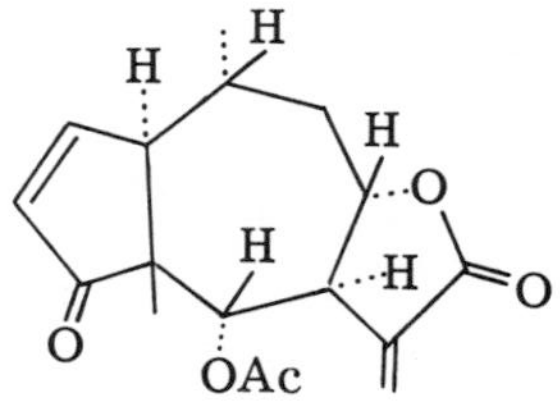	Parker and Geissman (1962); Herz and Lakshmikantham (1965)
Helenium Bloomquistii Rock	Tenulin		Herz (1962)
Helenium brevifolium (Nutt.) A. Wood	Brevilin A,B,C	Unknown	Herz et al. (1959b)
Helenium campestre Small	Helenalin		Herz et al. (1960)
Helenium Drummondii Rock	——		Herz (1962)
Helenium elegans DC.	Tenulin		Clark (1939)
Helenium flexuosum Raf.	Flexuosin A	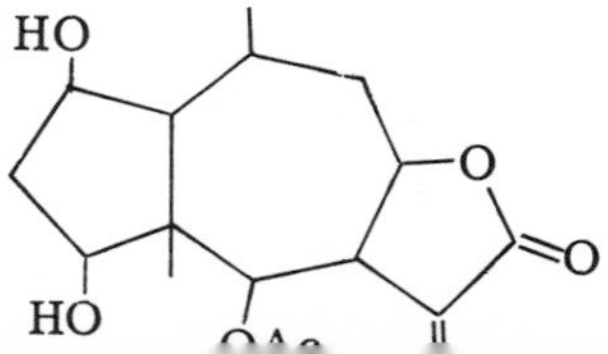	Herz et al. (1960)

	Flexuosin B	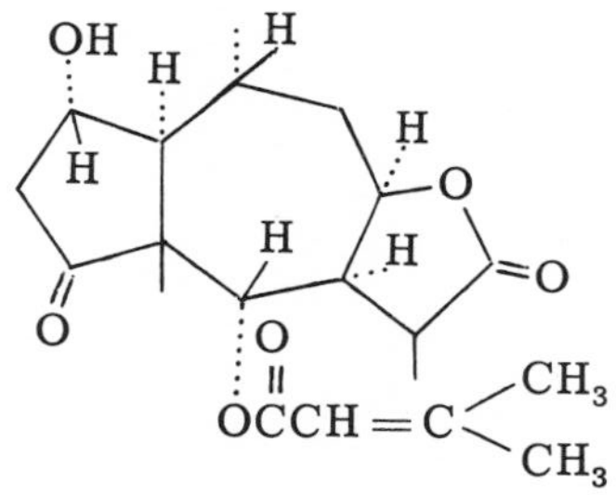 OH, H, H, H, H, H, O, O, O, O, OCCH=C, CH_3, CH_3	Herz et al. (1964b)
Helenium laciniatum Gray	Helenalin		Herz (1962)
Helenium linifolium Rydb.	Linifolin A	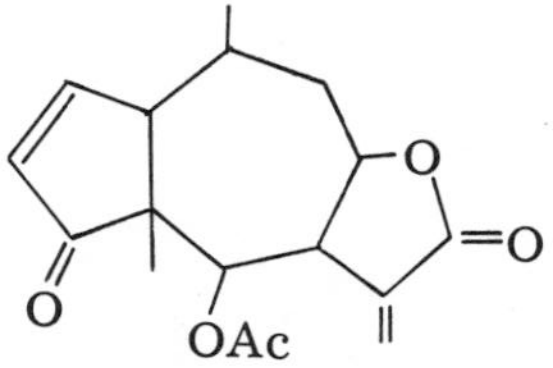 O, O, O, OAc	Herz (1962)
	Linifolin B	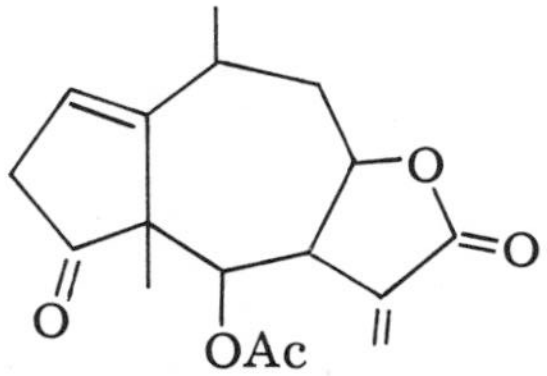 O, O, O, OAc	
Helenium mexicanum H.B.K. (State of Oaxaca)	Helenalin		Romo de Vivar and Romo (1961a)

TABLE 2 (continued)

Species	Compound	Formula	Reference
Helenium mexicanum H.B.K. (State of Mexico)	Helenalin		Romo de Vivar and Romo (1959, 1961b)
	Mexicanin A	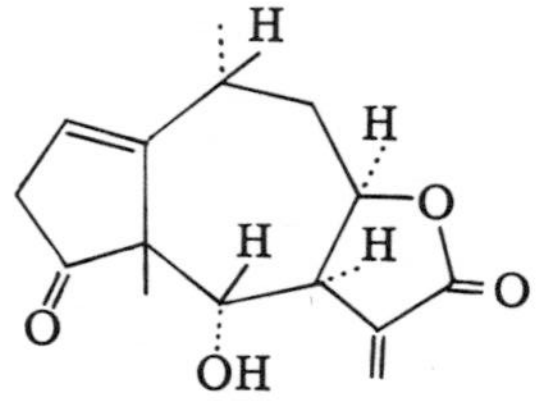	Romo de Vivar and Romo (1959, 1961b); Herz et al. (1963a, b, 1966b)
	Mexicanin B		Romo de Vivar and Romo (1959, 1961b)
	Mexicanin C	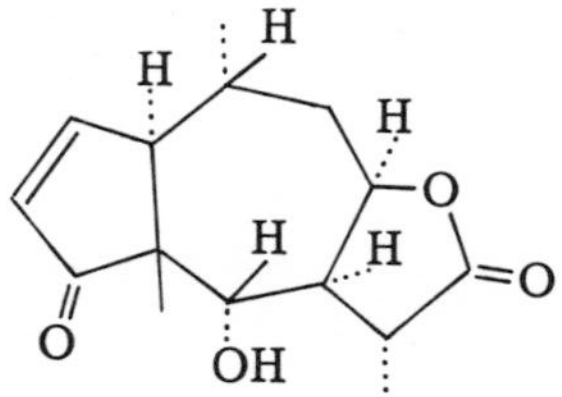	Romo de Vivar and Romo (1959, 1961b); Herz et al. (1963)
	Mexicanin E	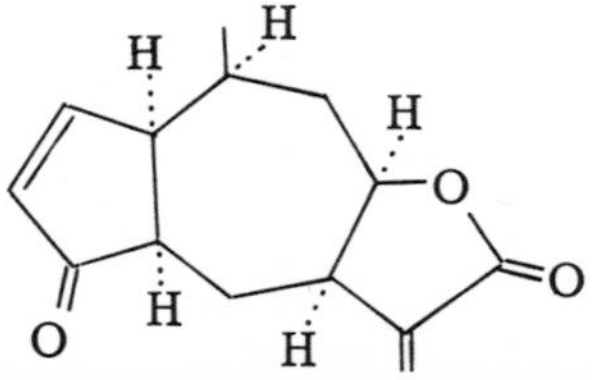	Romo de Vivar and Romo (1961a, b); Romo et al. (1963); Caughlan et al. (1966)

	Mexicanin F	Unknown	Romo de Vivar and Romo (1961b)
	Mexicanin G	Unknown	Romo de Vivar and Romo (1961b)
	Mexicanin I	Unknown	Dominguez and Romo (1963)
	Neohelenalin (mexicanin D)	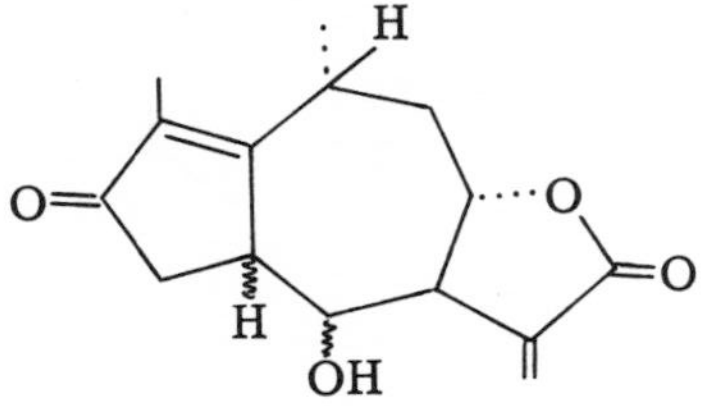	Romo de Vivar and Romo (1961b); Herz et al. (1963)
	Mexicanin H	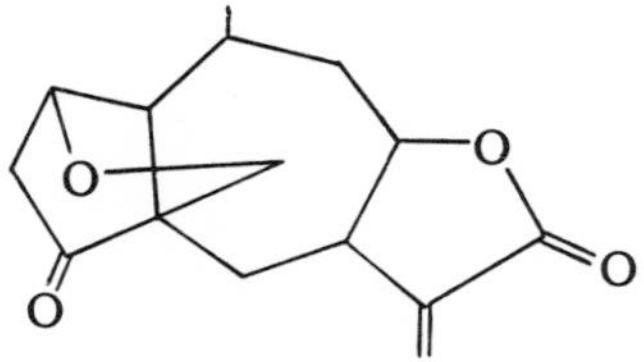	Romo de Vivar and Romo (1961b); Romo et al. (1966b)
Helenium microcephalum M. A. Curt. ex Gray	Helenalin		Adams and Herz (1949)
Helenium montanum Nutt.	Tenulin		Clark (1940)
Helenium ooclinium Gray	Neohelenalin Mexicanin E		Herz (1962)

TABLE 2 (continued)

Species	Compound	Formula	Reference
Helenium pinnatifidum (Nutt.) Rydb.	Pinnatifidin	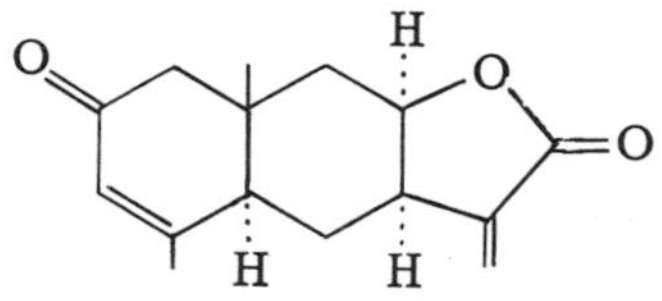	Herz et al. (1959b, 1962a)
Helenium quadridentatum Labill.	Helenalin		Clark (1940); Giral and Ladabaum (1961)
Helenium scorzoneraefolium DC. (Gray)	Helenalin isomer		Herz (1962)
Helenium Thurberi Gray	Tenulin		Herz and Lakshmikantham (1965); Romo de Vivar et al. (1966b)
	Thurberilin	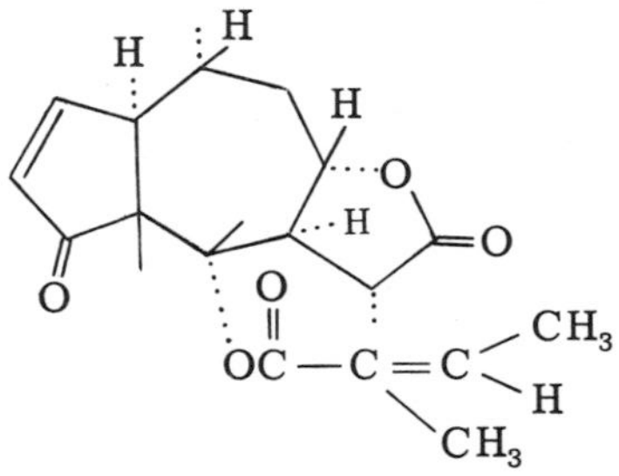	
Helenium vernale Walt.	Helenalin		Herz et al. (1959b)

Helenium virginicum Blake	Virginolide	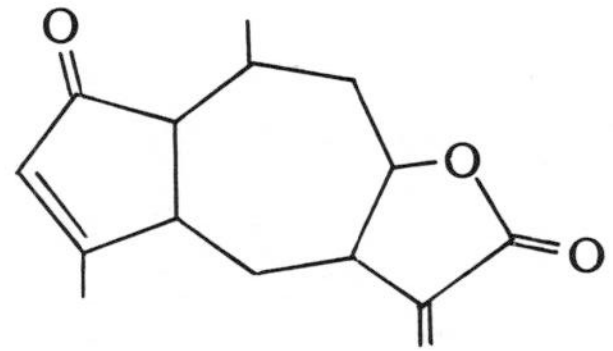	Herz and Santhanam (1967)
Gaillardia aristata Pursh.	Spathulin	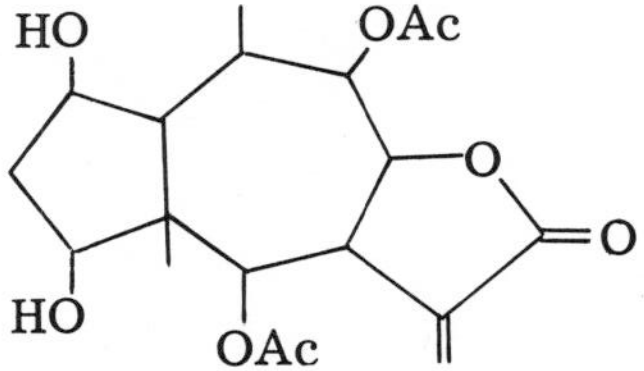	Herz et al. (1967b)
	Aristalin	Unknown	
Gaillardia arizonica Gray	Gaillardilin	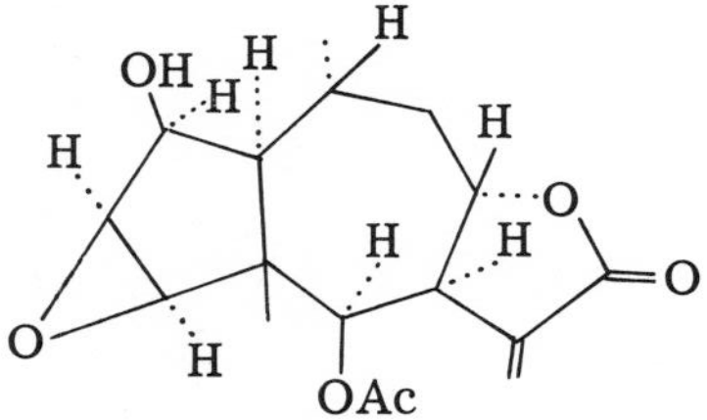 Probable stereochemistry	Herz et al. (1966c)

TABLE 2 (continued)

Species	Compound	Formula	Reference
Gaillardia fastigiata Greene	Fastigilin A	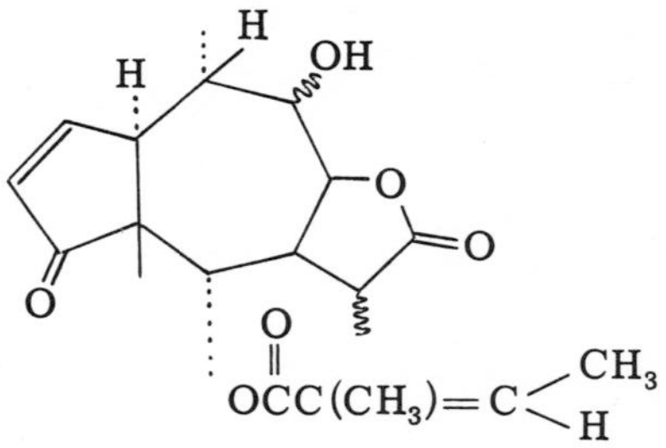 Probable stereochemistry	Herz et al. (1966d)
	Fastigilin B	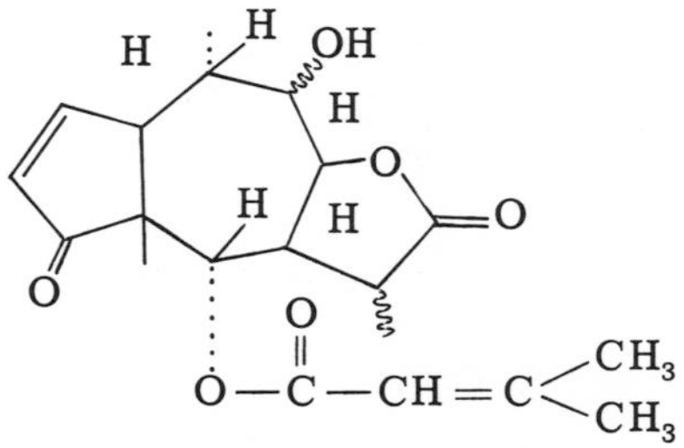 Probable stereochemistry	

	Fastigilin C	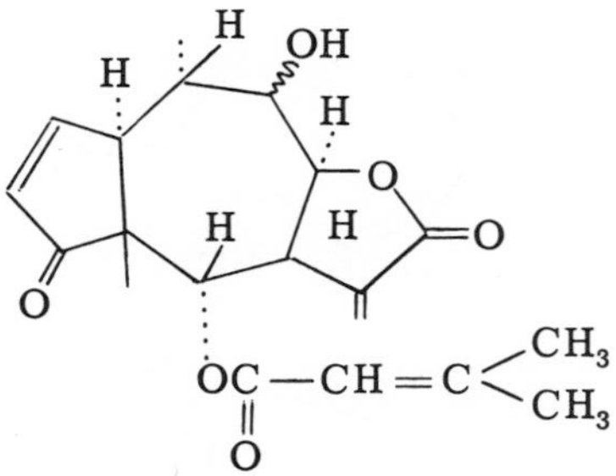 Probable stereochemistry	
Gaillardia grandiflora van Houtte (hybrid of G. *aristata* and *G. pulchella*)	Spathulin	——	Herz et al. (1967b)
	Aristalin	Unknown	
Gaillardia megapotamica (Spreng.) Baker	Helenalin		Herz and Inayama (1964)
Gaillardia mexicana Gray	Spathulin		Herz et al. (1967b)
	Mexilin	——	
Gaillardia multiceps Greene	Helenalin	——	Herz and Inayama (1964)
Gaillardia parryi Greene	Flexuosin A	——	Herz et al. (1967b)
Gaillardia pinnatifida Torr.	Gaillardilin		Herz et al. (1966c, d)

TABLE 2 (continued)

Species	Compound	Formula	Reference
	Gaillardipinnatin	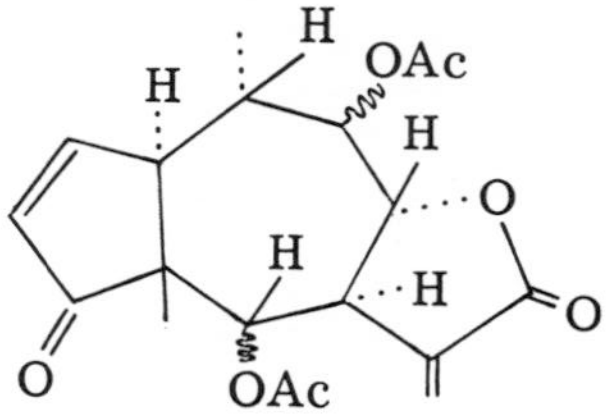 Probable stereochemistry	
Gaillardia pulchella Foug. coast race	Helenalin Pulchellin	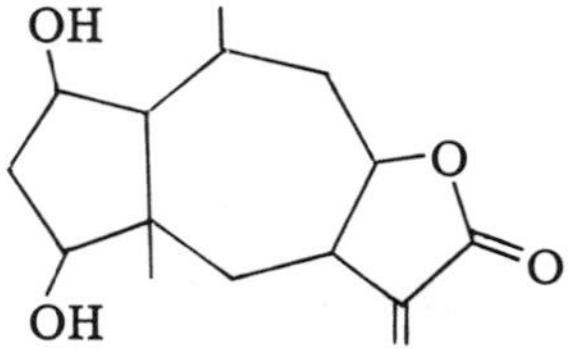	Herz et al. (1963c)
Gaillardia pulchella Foug. plains race	Pulchellin B	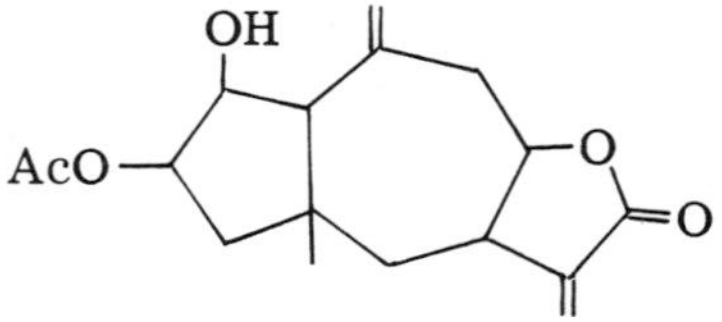	Herz and Inayama (1964); Herz and Roy (unpublished)

Gaillardia pulchella Foug. plains race (cont.)	Pulchellin C	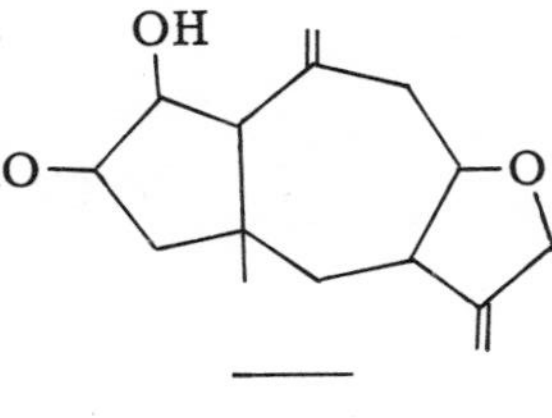	
	Pulchellin D	—	
	Pulchellin E	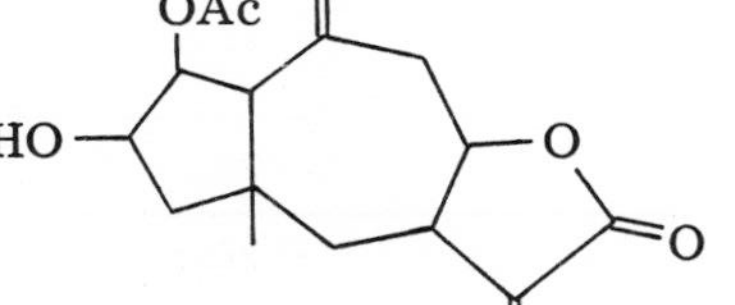	
	Pulchellin F	$CH_3C(H)=C(CH_3)-C(=O)O-$ (attached to the structure)	
Gaillardia spathulata Gray	Spathulin		Herz et al. (1967b)
Gaillardia suavis (Gray and Engelm.) Britt.	Unknown lactone	—	Herz and Raūlais (unpublished)

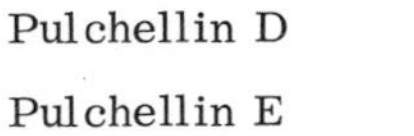

TABLE 2 (continued)

Species	Compound	Formula	Reference
Heliantheae Cass.			
Galinsoginae Benth. & Hooker			
Balduina angustifolia (Pursh.) Robins.	Helenalin		Herz and Mitra (1958)
Balduina uniflora Nutt.	Balduilin	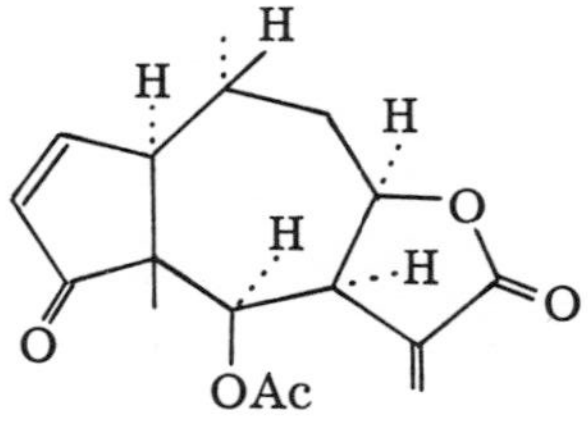	Herz et al. (1959a, 1963a,b)

Addendum to Table 2

Species	Compound	Formula	Reference
Helenium alternifolium (Spreng.) Cabr.	linifolin A tenulin brevilin A alternilin		Herz and Gast, unpubl.
Gaillardia pulchella Foug. (Live Oak County, Texas)	gaillardin	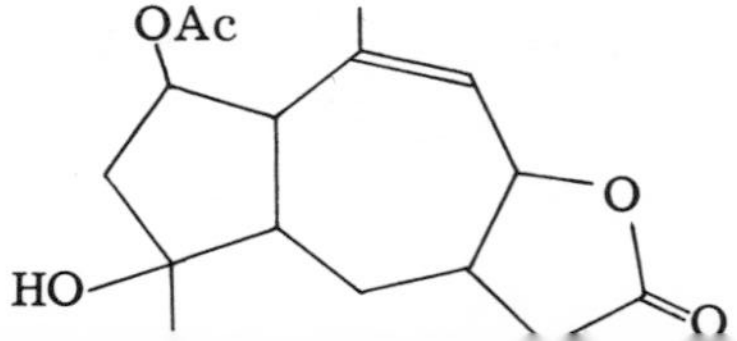	Kupchan et al. (1965); Kupchan et al. (1966)

Only two of the 35 *Helenium* and *Gaillardia* species studied elaborate norpseudoguaianolides. First, *H. pinnatifidum*, unlike the other members of the section *Leptopoda* (*H. brevifolium*, *H. campestre*, *H. flexuosum*, *H. Drummondii*, and *H. vernale*), yields the eudesmanolide pinnatifidin. It is striking that *H. vernale*, with which this taxon is easily confused (Rock, 1957) and frequently intermingled, should differ so markedly in sesquiterpene lactone content.

A second discrepancy exists with respect to *H. virginicum* Blake, a species (Blake, 1936) that according to Rock (private communication) exhibits no sound morphological features distinctive from *H. autumnale* and is extremely restricted in geographical distribution. Nevertheless, it is the only species among those listed that has consistently furnished a guaianolide.

A third point of interest is the occurrence, in several *Helenium* species or varieties thereof, of the norsesquiterpene lactone mexicanin E or its derivatives, the first such substance to be isolated and recognized (Romo de Vivar and Romo, 1961a,b; Romo et al., 1963). Mexicanin E and its congeners may derive from pseudoguaianolides such as XV or XVI, formed by biological oxidation, which can undergo retroaldol or decarboxylation reactions. Support for this is found in the structure assigned to mexicanin H (Romo et al., 1966b). The stereochemistry, elucidated by X-ray crystallography (Caughlan et al., unpublished), possesses unusual features in comparison with the stereochemistry of other constituents of *Helenium* species.

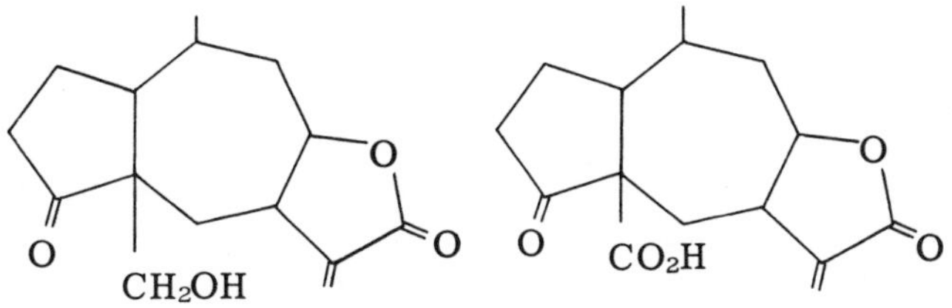

O CH_2OH O O CO_2H O

XV XVI

Note. The sesquiterpene lactone geigerinin, which accompanies the guaianolide geigerin (III) in *Geigeria aspera* Harv. and *Geiperia africana* Griessel (Inuleae Cass.; de Villiers, 1959), has been assigned a pseudoguaianolide skeleton (de Villiers and Pachler, 1963). This is the first report of a pseudoguianolide in a tribe other than Heliantheae or Heleniae.

In the additions to Table 2, attention should be drawn to the isolation of the normally constituted guaianolide gaillardin from another collection of *Gaillardia pulchella* Foug., which illustrates the great variability of this taxon.

IV. CONCLUSION

The relatively limited distribution of pseudoguaianolides in Compositae suggests that their presence has potential utility in taxonomy when used in conjunction with evidence based on morphology and cytology. Although some broad features are apparently beginning to emerge from a systematic investigation of a few *genera* in Heliantheae and Heleniae, generalizations, particularly at the infrageneric level, must await the outcome of more extended phytochemical investigations.

ACKNOWLEDGMENTS

The work reviewed on these pages was supported in part by grants from the United States Public Health Service (RG-05814), the National Science Foundation (GP-1962), and the Research Council of the Florida State University. Thanks are due to the following persons for collections and valuable correspondence: R. J. Barr (Tucson, Arizona), Mitchell Beauchamp (San Diego State College, California), W. Payne (University of Illinois, Ubana, Illinois), B. H. Braun (Takoma Park, Maryland), R. B. Channell (Vanderbilt University, Nashville, Tennessee), A. Clewell and R. K. Godfrey (Florida State University), Tallahassee, Florida), S. C. Harvey (University of Utah, Salt Lake City, Utah), N. Henderson (University of Missouri, Kansas City, Missouri), Ira La Rivers (University of Nevada, Reno, Nevada), E. Lehto (Arizona State University, Temple, Arizona), June McCaskill (University of California, Davis, California), T. J. Mabry (University of Texas, Austin, Texas), Elizabeth C. Norland (San Diego, California), Paul Redfearn (Southwest Missouri State College, Springfield, Missouri), the late H. F. L. Rock (Vanderbilt University, Nashville, Tennessee), W. P. Stoutamire, (University of Akron, Akron, Ohio), Paul Tueller (University of Nevada, Reno, Nevada), B. L. Turner (University of Texas, Austin, Texas), C. S. Wallis (Connors State College, Warner, Oklahoma), and B. H. Warnock (Sul Ross State College, Alpine, Texas)

REFERENCES

Abu-Shady, H., and T. D. Soine. 1953. J. Amer. Pharm. Ass., 42: 387.

_____, and T. D. Soine. 1954. J. Amer. Pharm. Ass., 43: 365.

Adams, R., and W. Herz. 1949. J. Amer. Chem. Soc., 71: 2546, 2551, 2554.

Arny, H. V. 1890. J. Pharm.: 121.

_____. 1897. J. Pharm.: 169.

Asher, J. D. M., and G. A. Sim. 1965. J. Chem. Soc.: 6041.

Barton, D. H. R., and P. de Mayo. 1956. J. Chem. Soc.: 143.

_____, and P. de Mayo. 1957a. J. Chem. Soc.: 150.

_____, and P. de Mayo. 1957b. Quart. Rev. Biol., 11: 189.

_____, O. C. Böckman, and P. de Mayo. 1960. J. Chem. Soc.: 2263.

Bassett, I. J., G. A. Mulligan, and C. Frankton. 1962. Canad. J. Bot., 40: 1243.

Bawdekar, A. S., and G. R. Kelkar. 1965. Tetrahedron, 21: 1521.

Bernardi, L., and G. Büchi. 1957. Experientia, 13: 466.

Blake, S. F. 1936. Claytonia, 3: 5.

Büchi, G., and D. Rosenthal. 1956. J. Amer. Chem. Soc., 78: 3860.

Braun, B. H., W. Herz, and K. Rabindran. 1956. J. Amer. Chem. Soc., 78: 4423.

Caughlan, C. N., Mazhur-ul-Haque, and M. T. Emerson. 1966. Chem. Commun: 151.

Clark, E. P. 1936. J. Amer. Chem. Soc., 58: 1982.

_____. 1939. J. Amer. Chem. Soc., 61: 1836.

_____. 1940. J. Amer. Chem. Soc., 62: 597.

Cronquist, A. 1955. Am. Midl. Nat., 53: 478.

Djerassi, C., and S. Burstein. 1958. J. Amer. Chem. Soc., 80: 2593.

Dolejs, L., V. Herout, and F. Sorm. 1958. Collection Czech. Chem. Commun., 23: 504.

Dominguez, E., and J. Romo. 1963. Tetrahedron, 19: 1415.

Emerson, M. T., C. N. Caughlan, and W. Herz. 1964. Tetrahedron Lett.: 621.

_____, C. N. Caughlan, and W. Herz. 1966. Tetrahedron Lett.: 3151.

Farkas, L., M. Nogradi, V. Sudarasanam, and W. Herz. 1966. J. Org. Chem., 31: 3228.

Fischer, N. H., and T. J. Mabry. 1967. Tetrahedron, 23: 2529.

Geissman, T. A. 1962. J. Org. Chem., 27: 2692.

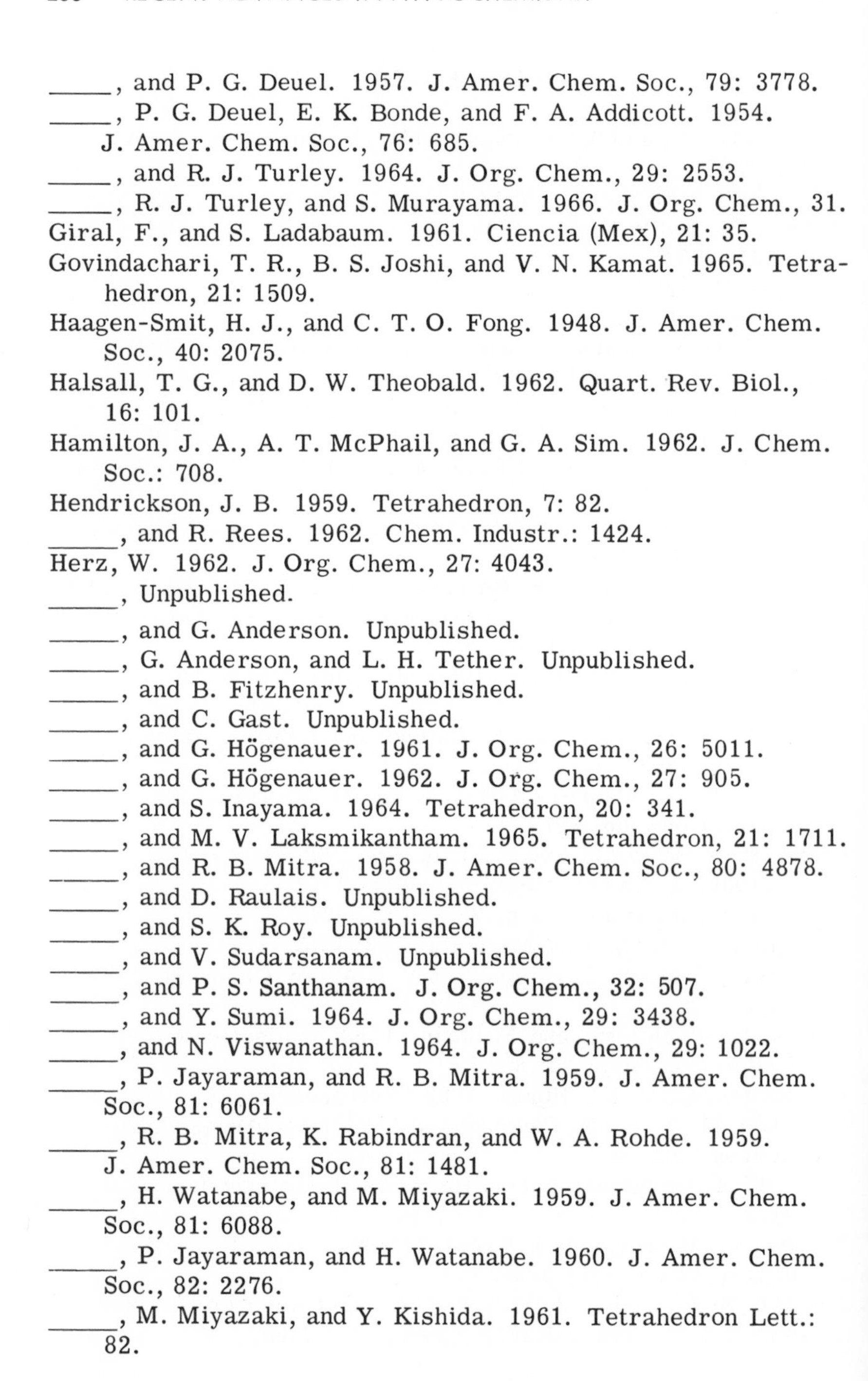

_____, and P. G. Deuel. 1957. J. Amer. Chem. Soc., 79: 3778.
_____, P. G. Deuel, E. K. Bonde, and F. A. Addicott. 1954. J. Amer. Chem. Soc., 76: 685.
_____, and R. J. Turley. 1964. J. Org. Chem., 29: 2553.
_____, R. J. Turley, and S. Murayama. 1966. J. Org. Chem., 31.
Giral, F., and S. Ladabaum. 1961. Ciencia (Mex), 21: 35.
Govindachari, T. R., B. S. Joshi, and V. N. Kamat. 1965. Tetrahedron, 21: 1509.
Haagen-Smit, H. J., and C. T. O. Fong. 1948. J. Amer. Chem. Soc., 40: 2075.
Halsall, T. G., and D. W. Theobald. 1962. Quart. Rev. Biol., 16: 101.
Hamilton, J. A., A. T. McPhail, and G. A. Sim. 1962. J. Chem. Soc.: 708.
Hendrickson, J. B. 1959. Tetrahedron, 7: 82.
_____, and R. Rees. 1962. Chem. Industr.: 1424.
Herz, W. 1962. J. Org. Chem., 27: 4043.
_____, Unpublished.
_____, and G. Anderson. Unpublished.
_____, G. Anderson, and L. H. Tether. Unpublished.
_____, and B. Fitzhenry. Unpublished.
_____, and C. Gast. Unpublished.
_____, and G. Högenauer. 1961. J. Org. Chem., 26: 5011.
_____, and G. Högenauer. 1962. J. Org. Chem., 27: 905.
_____, and S. Inayama. 1964. Tetrahedron, 20: 341.
_____, and M. V. Laksmikantham. 1965. Tetrahedron, 21: 1711.
_____, and R. B. Mitra. 1958. J. Amer. Chem. Soc., 80: 4878.
_____, and D. Raulais. Unpublished.
_____, and S. K. Roy. Unpublished.
_____, and V. Sudarsanam. Unpublished.
_____, and P. S. Santhanam. J. Org. Chem., 32: 507.
_____, and Y. Sumi. 1964. J. Org. Chem., 29: 3438.
_____, and N. Viswanathan. 1964. J. Org. Chem., 29: 1022.
_____, P. Jayaraman, and R. B. Mitra. 1959. J. Amer. Chem. Soc., 81: 6061.
_____, R. B. Mitra, K. Rabindran, and W. A. Rohde. 1959. J. Amer. Chem. Soc., 81: 1481.
_____, H. Watanabe, and M. Miyazaki. 1959. J. Amer. Chem. Soc., 81: 6088.
_____, P. Jayaraman, and H. Watanabe. 1960. J. Amer. Chem. Soc., 82: 2276.
_____, M. Miyazaki, and Y. Kishida. 1961. Tetrahedron Lett.: 82.

_____, R. B. Mitra, K. Rabindran, and N. Viswanathan. 1962a. J. Org. Chem., 27: 4041.
_____, W. A. Rohde, K. Rabindran, P. Jayaraman, and N. Viswanathan. 1962b. J. Amer. Chem. Soc., 84: 3857.
_____, H. Watanabe, M. Miyazaki, and Y. Kishida. 1962c. J. Amer. Chem. Soc., 84: 2601.
_____, A. Romo de Vivar, J. Romo, and N. Viswanathan. 1963a. J. Amer. Chem. Soc., 85: 19.
_____, A. Romo de Vivar, J. Romo, and N. Viswanathan. 1963b. Tetrahedron, 19: 1359.
_____, G. Högenauer, and A. Romo de Vivar. 1964a. J. Org. Chem., 29: 1700.
_____, Y. Kishida, and M. V. Lakshmikantham. 1964b. Tetrahedron, 20: 979.
_____, A. Romo de Vivar, and M. V. Lakshmikantham. 1965. J. Org. Chem., 30: 118.
_____, H. Chikamatsu, and L. R. Tether. 1966a. J. Org. Chem., 31: 1632.
_____, M. V. Lakshmikantham, and R. N. Mirrington. 1966b. Tetrahedron, 22: 1709.
_____, S. Rajappa, M. V. Lakshmikantham, and J. J. Schmid. 1966c. Tetrahedron, 22: 693.
_____, S. Rajappa, S. K. Roy, J. J. Schmid, and R. J. Mirrington. 1966d. Tetrahedron, 22: 1907.
_____, V. Sudarsanam, and J. J. Schmid. 1966e. J. Org. Chem., 31: 3232.
_____, H. Chikamatsu, N. Viswanathan, and V. Sudarsanam. 1967a. J. Org. Chem., 32: 682.
_____, S. Rajappa, M. V. Lakshmikantham, D. Raulais, and J. J. Schmid. 1967b. J. Org. Chem., 32: 1042.
_____, Y. Sumi, V. Sudarsanam, and D. Raulais. 1967c. J. Org. Chem. 32: 3658.
_____, G. Anderson, and L. R. Tether. Unpublished.
Jackson, R. C. 1960. J. Kans. Sci. Bull., 41: 793.
Joseph-Nathan, P., and J. Romo. 1966. Tetrahedron, 22: 1723.
Kagan, H. B., H. E. Miller, W. Renold, M. V. Lakshmikantham, L. R. Tether, W. Herz, and T. J. Mabry. 1966. J. Org. Chem., 31: 1629.
Kulkarni, G. H., G. R. Kelkar, and S. C. Bhattacharyya. 1964. Tetrahedron, 20: 2639.
Kupchan, S. M., J. M. Cassady, J. Bailey, and J. R. Knox. 1965. J. Pharm. Sci., 54: 1703.
_____, J. M. Cassady, S. E. Kelsey, H. K. Schnoes, D. H. Smith, and A. L. Burlingame. 1966. J. Amer. Chem. Soc., 88: 5292.

Lamson. 1913. J. Pharmacol. Exp. Ther., 4: 471.

Lucas, R. A., S. Rovinski, R. J. Kiesel, L. Dorfman, and H. B. MacPhillamy. 1964a. J. Org. Chem., 29: 1549.

_____, R. G. Smith, and L. Dorfman. 1964b. J. Org. Chem., 29: 2101.

Mabry, T. J., H. B. Kagan, and H. E. Miller. 1966a. Tetrahedron 22: 1943.

_____, H. E. Miller, H. B. Kagan, and W. Renold. 1966b. Tetrahedron, 22: 1139.

_____, W. Renold, H. E. Miller, and H. B. Kagan. 1966c. J. Org. Chem., 31: 681.

Marshall, J. A., and N. Cohen. 1964. J. Org. Chem., 29: 3727.

Martin-Smith, M., P. de Mayo, S. J. Smith, J. B. Stenlake, and W. D. Williams. 1964. Tetrahedron Lett.: 2391.

Miller, H. E., and T. J. Mabry. 1967. J. Org. Chem., 32: 2929.

_____, H. B. Kagan, W. Renold, and T. J. Mabry. 1965. Tetrahedron Lett.: 3397.

Minato, H., and I. Horibe. 1965. J. Chem. Soc.: 7009.

_____, S. Nosaka, and I. Horibe. 1964. J. Chem. Soc.: 5503.

Novikov, V. I., K. S. Rybalko, and K. E. Koreschchuk. 1964. J. Obshch. Khim., 34 (12): 4120.

Novotny, L., J. Jizba, V. Herout, and F. Sorm. 1962. Collection Czech. Chem. Commun., 27: 1393.

_____, J. Jizba, V. Herout, F. Sorm, L. H. Zalkow, S. Hu, and C. Djerassi. 1963. Tetrahedron, 19: 1101.

Nozoe, T., and S. Ito. 1961. Fortschr. Chem. Org. Naturst., 19.

Parker, B. A., and T. A. Geissman. 1962. J. Org. Chem., 27: 4127.

Payne, W. W. 1964. J. Arnold Arboretum, 45: 401.

Plourde, J. R., and J. A. Mockle. 1960. Canad. Pharm. J., Sci. Sec., 93 (10): 53; (11): 43.

Rao, A. S., G. R. Kelkar, and S. C. Bhattacharyya. 1960. Tetrahedron, 9: 274.

_____, A. Paul, Sadgopal, and S. C. Bhattacharyya. 1961. Tetrahedron, 13: 318.

Richards, J. H., and J. B. Hendrickson. 1964. The Biosynthesis of Steroids, Terpenes and Acetogenins, New York, Benjamin

Rock, H. F. L. 1957. Rhodora, 59: 101, 128, 168, 179.

Rogers, D., and Mazhar-ul-Haque. 1963. Proc. Chem. Soc.: 92.

Rollins, R. C. 1950. Contrib. Gray Herbarium, No. 172.

Romo, J., P. Joseph-Nathan, and A. F. Diaz. 1964. Tetrahedron, 20: 79.

_____, A. Romo de Vivar, and W. Herz. 1963. Tetrahedron, 19: 2317.

_____, P. Joseph-Nathan, and G. Siade. 1966a. Tetrahedron, 22: 1499.
_____, A. Romo de Vivar, and P. Joseph-Nathan. 1966b. Tetrahedron Lett.: 1019.
Romo de Vivar, A., and J. Romo. 1959. Chem. Industr.: 882.
_____, and J. Romo. 1961a. Ciencia (Mex.), 21 (1): 33.
_____, and J. Romo. 1961b. J. Amer. Chem. Soc., 83: 2326.
_____, E. A. Bratoeff, and T. Rios. 1966a. J. Org. Chem., 31: 673.
_____, L. Rodriguez, J. Romo, M. V. Lakshmikantham, R. N. Mirrington, J. Kagan, and W. Herz. 1966b. Tetrahedron, 22.
Rossmann, M. G., and W. N. Lipscomb. 1958. J. Amer. Chem. Soc., 80: 2592.
Ruzicka, L. 1953. Experientia, 9: 357.
_____, 1959. Proc. Chem. Soc.: 341.
Simonsen, J., and D. H. R. Barton. 1952. The Terpenes, vol. III, London, Cambridge Univ. Press.
Solbrig, O. 1963. J. Arnold Arboretum, 44: 436.
Sorm, F., M. Suchy, and V. Herout. 1959. Collection Czech. Chem. Commun., 24: 1548.
Suchý, M., V. Herout, and F. Sorm. 1959. Collection Czech. Commun., 24: 1547.
Takeda, K., H. Minato, and M. Ishikawa. 1964. J. Chem. Soc.: 4578.
Toth, J., S. Holly, L. Ferenczy, and O. Kovacs. 1962. Rev. Chim. Acad. Rep. Populaire Rumaine, 7: 1339.
Ungnade, H. E., and E. C. Hendley. 1948. J. Amer. Chem. Soc., 70: 3921.
_____, E. C. Hendley, and W. Dunkel. 1950. J. Amer. Chem. Soc., 72: 3818.
Van Tamelen, E. E., A. Storni, F. J. Hessler, and M. Schwartz. 1963. J. Amer. Chem. Soc., 85: 3295.
de Villiers, J. P. 1959. J. Chem. Soc.: 2412.
_____, and K. Pachler. 1963. J. Chem. Soc.: 4989.

8

SYSTEMATIC ASPECTS OF THE DISTRIBUTION OF DI- AND TRITERPENES

G. PONSINET, G. OURISSON, and A. C. OEHLSCHLAGER
Institut de Chimie
Strasbourg, France

I. INTRODUCTION

Our present knowledge of natural triterpenes is impressive. In dicotyledons alone these compounds have been found in more than 100 families. To date, about 400 triterpenes have been characterized and the structures of 300 elucidated, allowing one to draw some conclusions concerning their biogenesis.

The use of modern methods of chromatography for analysis and of spectroscopy (NMR, mass spectrometry) for structure elucidation has greatly facilitated the systematic study of these natural compounds in recent years, and increasingly effective instrumentation may make these investigations far more efficie in the future.

II. TAXONOMIC SIGNIFICANCE OF TRITERPENES

The common structural characteristics of these compounds speaks for a common biogenesis, which, since the work of the Ruzicka school (Eschenmoser et al., 1955), has been recognized as proceeding from squalene (I). However experimental proof that squalene is a natural progenitor of triterpenes has been ob tained in only a few cases.

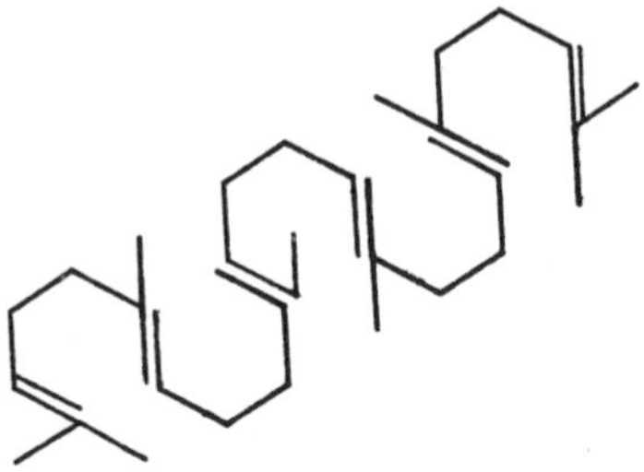

I

The genetic origin of the triterpenes may be explained by th action of specific enzymes that impose on the acyclic precursor (I) a particular conformation that, upon cyclization, yields the observed basic carbon skeletons. Most triterpenes arise from further reactions of these basic skeletons such as rearrangements, degradations, oxidations, and reductions. The complete list of known structural types is too long to be given here. For further details the reader is referred to a number of recently published reviews (Boiteau et al., 1964; Connolly and Overton, 1964; Halsall and Aplin, 1964; Ourisson et al., 1964; Richards and Hendrickson, 1964). The first of these (Boiteau et al., 1964 is an exhaustive report on the occurrence and physiological im plications of triterpenes.

Considerable insight into the taxonomic importance of trite penes can be derived from an examination of the manner by which various plants degrade the basic carbon skeletons. The results concerning the degradations of tetracyclic triterpenes are numerous, each formula in Table 1 being an example of a general process of degradation: alkaloids of Apocynaceae (e.g., conessin), fungal acids (e.g., helvolic acid), alkaloids of Buxaceae (e.g., cyclobuxin), pseudoditerpenes of Simarubaceae (e.g., chaparrin), and bitter principles of Rutaceae and Meliaceae (e.g., gedunin). Much work has been concerned with the formation of sterols (e.g., β-sitosterol), which are distinguished from many of the other triterpene degradation products because of their biological importance. On the other hand, our present knowledge does not allow us to attribute (or to deny) a generic character or a biological importance to the degradation of pentacyclic triterpenes (Table 2), as each known product provides a unique example of degradation:

1. degradation of ring A—2–3 cleavage (acid of *Bursera*), 3–4 cleavage (nyctanthic acid), contraction (ceanothic acid)
2. opening of rings D and E (ebelin)
3. loss of methyl at C-17 (albigenin), C-20 (platanic acid), and C-5 (celastrol).

III. LIMITATIONS OF TRITERPENE DATA FOR SYSTEMATIC PURPOSES

At present, it is not easy to draw phylogenetic conclusions from the known triterpene distribution patterns because they are not representative of the entire plant kingdom. Indeed, most of the known results are not derived from extensive surveys but from isolated studies, often justified by an economic (Rutaceae, Dipterocarpaceae) or pharmaceutical (Euphorbiaceae) interest in a plant. Sometimes even the local abundance (Leguminosae) of a plant is sufficient reason for an investigation. Seldom have plants been investigated solely because of their botanical interest.

The discovery of a chemically interesting product in one species has rarely brought on a more definitive study of the genus or family. Only such extensive investigations have much chemotaxonomic value for categories above the species level. Thus, in 1957, comprehensive reviews by Djerassi for the cactus triterpenes and Rehm et al. for the bitter principles of cucurbits demonstrated that the interests of the chemist, looking for new compounds, and the botanist, concerned with classification and a better understanding of biological evolution, could be combined.

TABLE 1

Degradations of Tetracyclic Skeletons[a]

Squalene Conformation	Basic Skeleton	Degradation Products	
	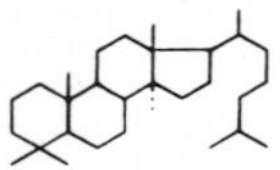 Lanostane	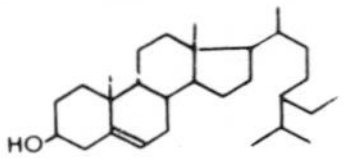 β-Sitosterol	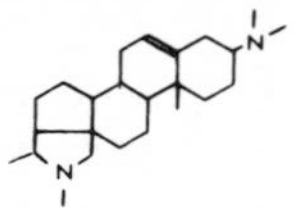 Conessine
Chair-boat-chair-boat-unfolded			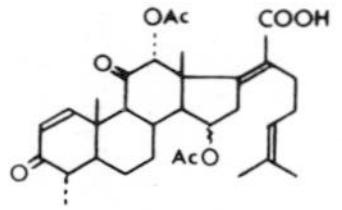 Helvolic acid
	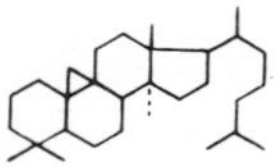 Cycloartane	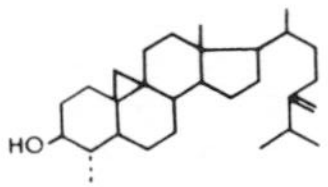 Cycloeucalenol	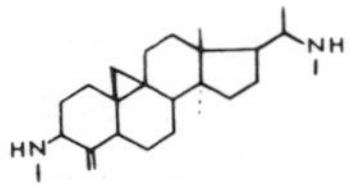 Cyclobuxin
	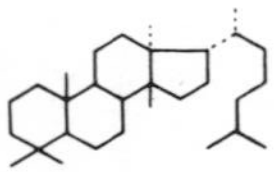 Euphane	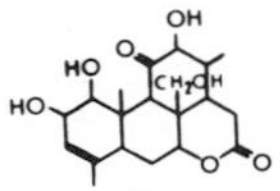 Chaparrin	
Chair-chair-chair-boat-unfolded			
	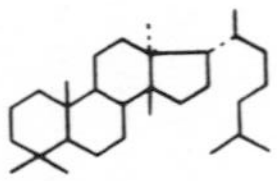 Elemane	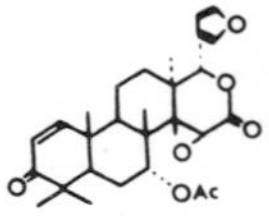 Gedunin	

TABLE 3

Biogenetic Relationships among Cucurbitacins

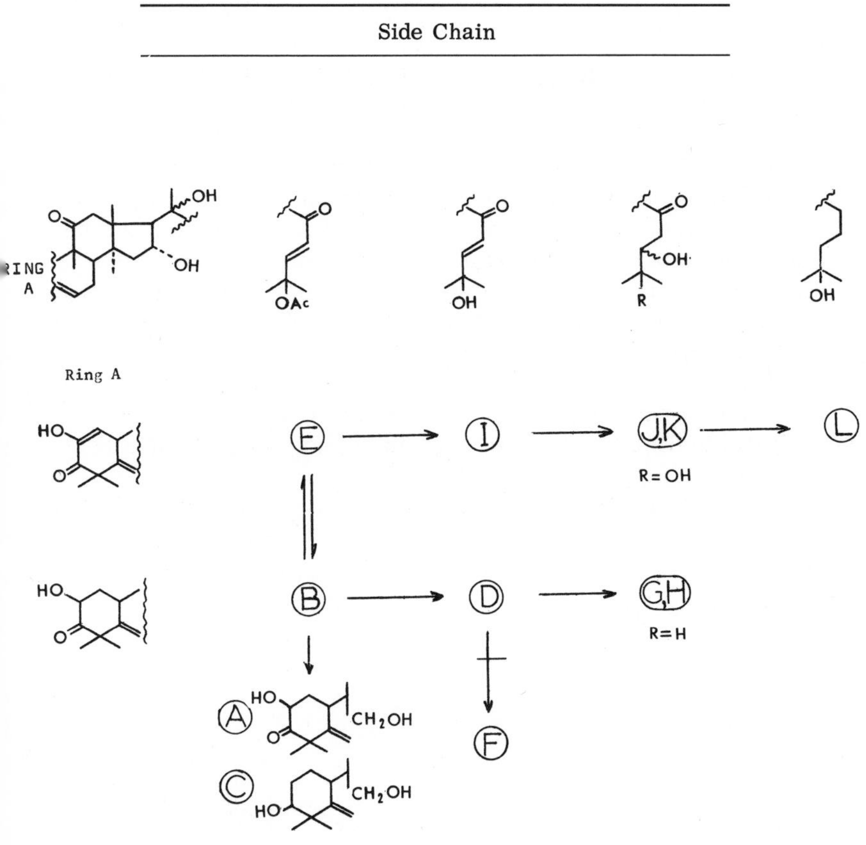

Side Chain

RING A

OH OH O O O OAc O OH O OH R OH

Ring A

HO O E → I → J,K → L
R = OH

HO O B → D → G,H
R = H

A HO O CH_2OH F

C HO CH_2OH

In this paper, we shall give some examples to demonstrate that it is possible to attribute, at each level of classification above the species, common chemical characteristics that have different taxonomic groupings. These characteristics pertain both to the carbon skeletons and to the oxidation patterns.

While such studies may give useful taxonomic information, they must await more definitive studies of the enzyme systems involved before conclusions concerning the hereditary significance of the distribution patterns of these secondary products may be formulated.

IV. APPLICATIONS OF TRITERPENE CHEMISTRY TO TAXONOMY

At the species level, we have observed in the genus *Euphorbia* a rather interesting example of convergence of gross morphological features with respect to chemical constituents. We have studied the latex of more than 70 species of *Euphorbia* (Ponsinet and Ourisson, in press). These latices contain squalene-oxydocyclases, which produce various tetracyclic triterpenes. Specifically we have found that

1. herbaceous euphorbias all contain cycloartenol (IV);
2. cactuslike euphorbias contain euphol (V) and euphorbol (VI);
3. corallike euphorbias contain euphol and tirucallol (VII).

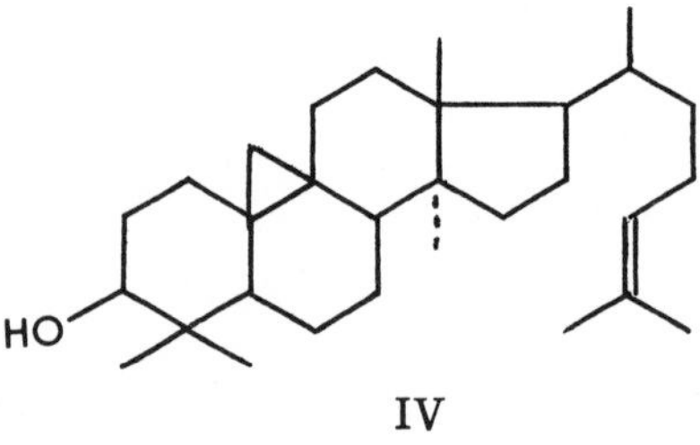

IV

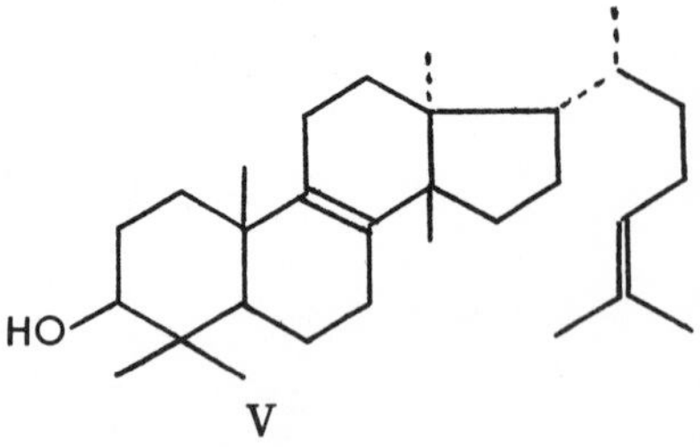

V

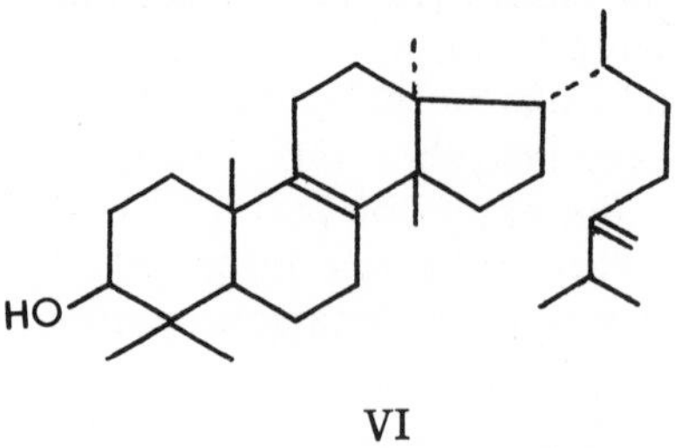

VI

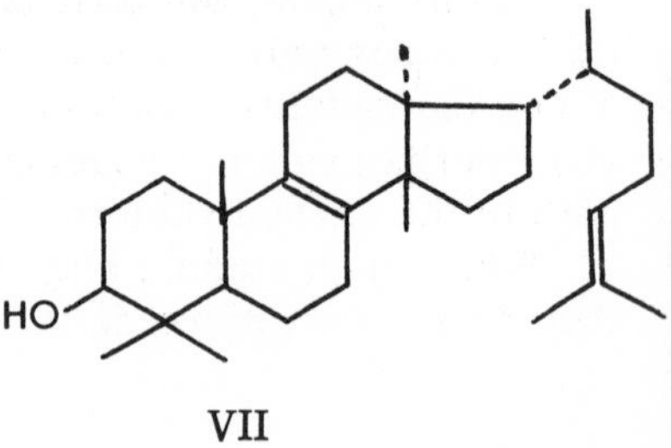

VII

TABLE 3

Biogenetic Relationships among Cucurbitacins

Side Chain

In this paper, we shall give some examples to demonstrate that it is possible to attribute, at each level of classification above the species, common chemical characteristics that have different taxonomic groupings. These characteristics pertain both to the carbon skeletons and to the oxidation patterns.

While such studies may give useful taxonomic information, they must await more definitive studies of the enzyme systems involved before conclusions concerning the hereditary significance of the distribution patterns of these secondary products may be formulated.

IV. APPLICATIONS OF TRITERPENE CHEMISTRY TO TAXONOMY

At the species level, we have observed in the genus *Euphorbia* a rather interesting example of convergence of gross morphological features with respect to chemical constituents. We have studied the latex of more than 70 species of *Euphorbia* (Ponsinet and Ourisson, in press). These latices contain squalene-oxydocyclases, which produce various tetracyclic triterpenes. Specifically we have found that

1. herbaceous euphorbias all contain cycloartenol (IV);
2. cactuslike euphorbias contain euphol (V) and euphorbol (VI);
3. corallike euphorbias contain euphol and tirucallol (VII).

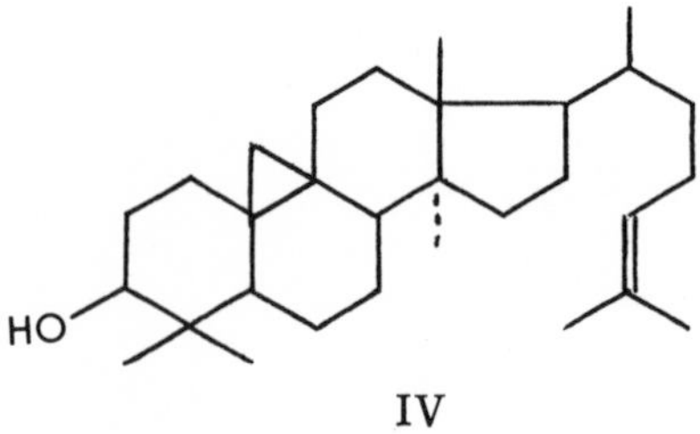

HO

IV

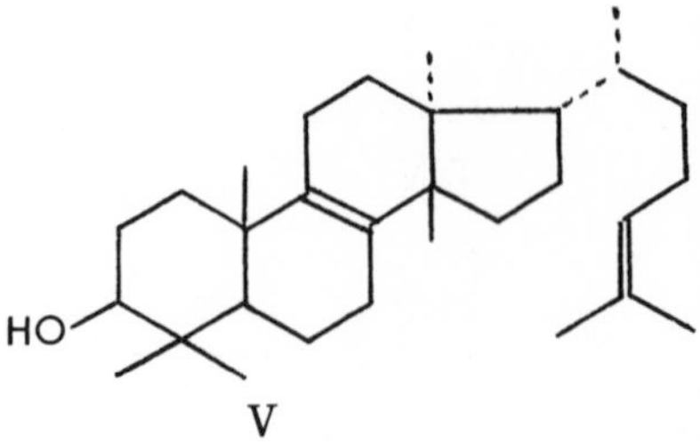

HO

V

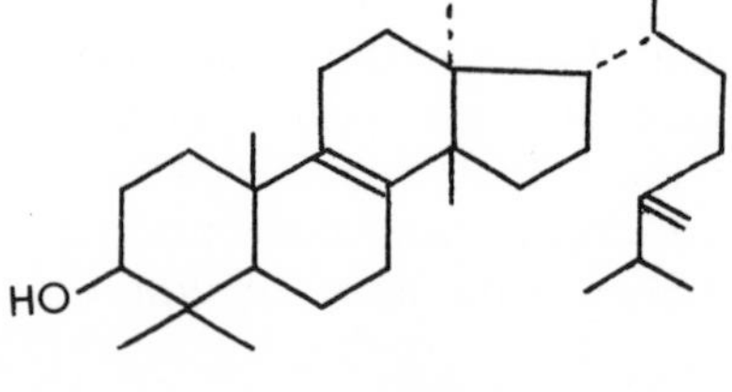

HO

VI

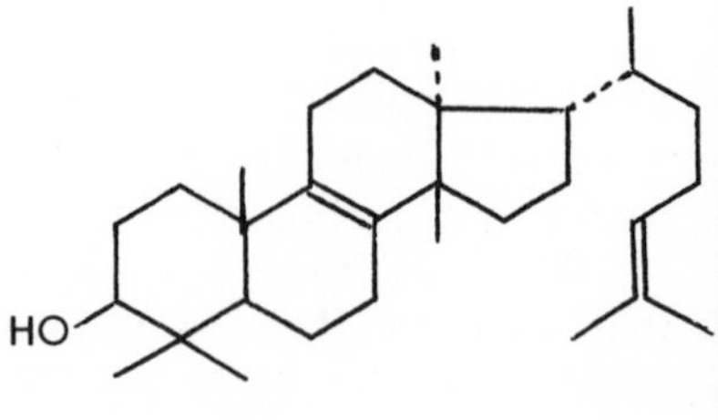

HO

VII

Although the various morphological types mentioned above are certainly produced by an adaptation to climatic conditions, we cannot, at this point, be sure that the variations in chemical composition arise also from this adaptation. In any event, it is an interesting correlation and, as far as we know, unique.

Many isolated studies mention the extraction of a single product from a given species. In considering such studies, it should be pointed out that, as long as we do not know seasonal and geographical influences and, indeed, even the homogeneity of the species, it is difficult to attribute much taxonomic significance to the isolation of a single compound.

However, another example may be mentioned: many genera of Cucurbitaceae contain bitter principles derived only from cucurbitacin B or only from cucurbitacin E. In the genus *Cucurbita*, both series are present as indicated in Table 4 (Rehm, 1960). Thus, it is thought that more primitive species contain only cucurbitacin B, whereas the more highly evolved species contain B and E or only E.

One idealized goal of chemotaxonomists is, of course, to be able to characterize a taxonomic grouping, a genus or a family, by a unique constituent. Such a feature would have great immediate usefulness in families where both definition and recognition of morphological characters are difficult. In these instances, any additional criteria—including chemical ones—should be welcome.

Thus in the fungi (Boiteau et al., 1964), a remarkable chemical difference appears between *Basidiomycetes* and *Ascomycetes*. The former has yielded eleven lanostanic prod-

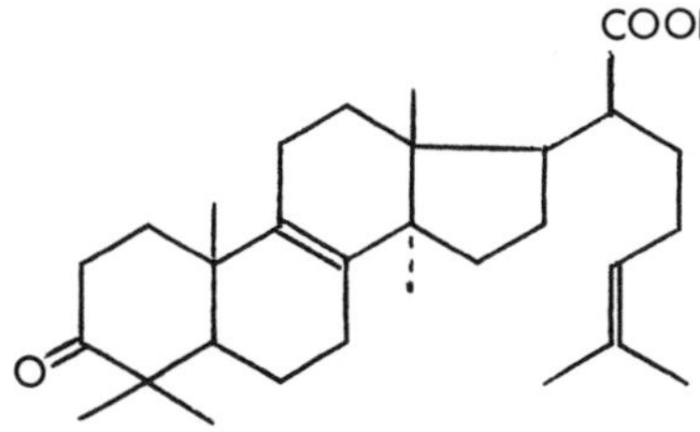

VIII 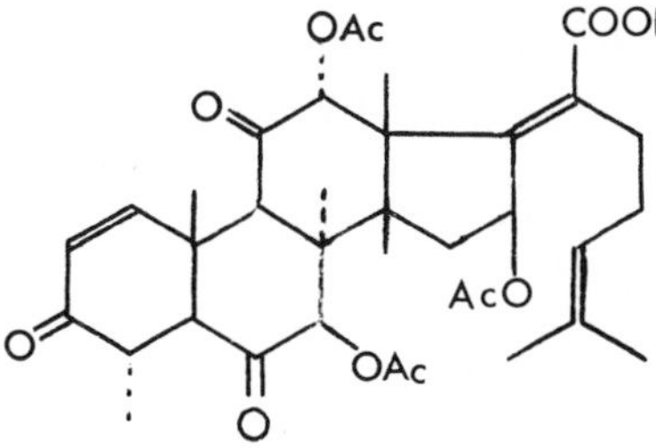 IX

ucts, all with a C-21 carbonyl function (e.g., pinicolic A acid, VIII), while in the latter three analogous acids occur with the following modifications of structure: a 17–20 double bond, additional oxygen functions, and partial demethylation (e.g., helvolic acid, IX).

TABLE 4

Cucurbitacins in Roots of Wild Cucurbita

Species	B, %	E, %
C. andreana	0.06	—
C. sororia	0.03	—
C. lundelliana	0.01	0.001
C. texana	0.02	0.03
C. cylindrica	0.01	0.009
C. foetidissima	0.008	0.04
C. balmata	—	0.02
C. okeechobeensis	—	0.014

The various genera of the subtribe Cereanae of the family Cactaceae also show chemical differences. Thus, Djerassi (1957) pointed out that each of three genera contains its own characteristic compound (Table 5). Another genus (*Lophocereus*) seems to contain no triterpenes but is rich in alkaloids. We cannot give here more details on the distribution of triterpenes in the various species; however, it is noteworthy that the existing chemical data confirm divisions of the subtribe Cereanae accepted by the botanists and provide assistance, in some situations, in the placement of species of uncertain taxonomic position.

As mentioned above, the triterpene constituents of the family Cucurbitaceae are homogeneous with respect to the cucurbitane skeleton and the oxidation patterns of this skeleton. Two other examples may be mentioned: the constituents of the Anacardiaceae are characterized by their oxidation patterns and those of Dipterocarpaceae, by their skeleton.

Several independent investigators have extracted tetracyclic triterpenic acids (Table 6) from secretions of various genera of Anacardiaceae. These acids show the same patterns of oxidation in the side chain on two different skeletons, cycloartenol (IV) and tirucallol (VII). This is the only family in which products arising from oxidation of the terminal end of the side chain to yield a conjugated acid have been found.

Another example of the chemical characterization of a family is given by the derivatives of dammarane (X) in the Dipterocarpaceae. We have undertaken the systematic survey of sesqui- and triterpenes of the resins obtained from members of this

TABLE 5

Chemical Characterization of Genera of the Cactaceae, Subtribe Cereanae

Genus	Specific Constituent
Lemaireocereus	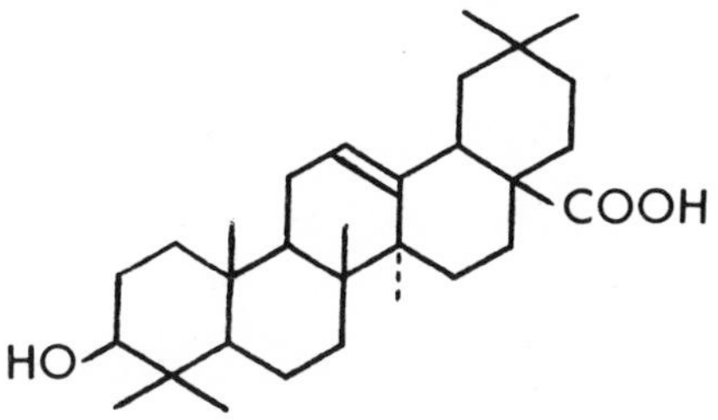 Oleanolic acid
Myrtillocactus	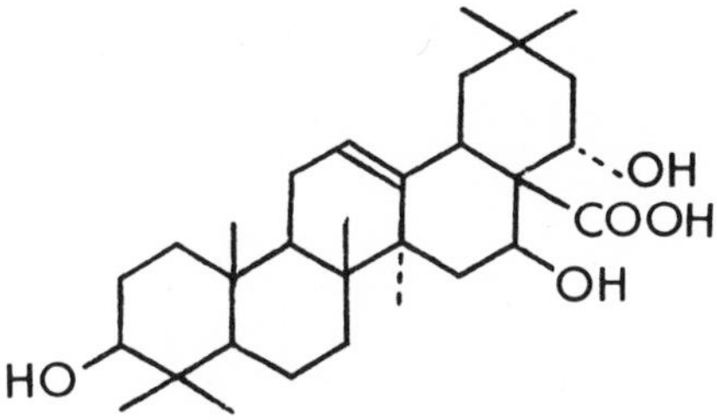 Chichipegenin
Machaerocereus	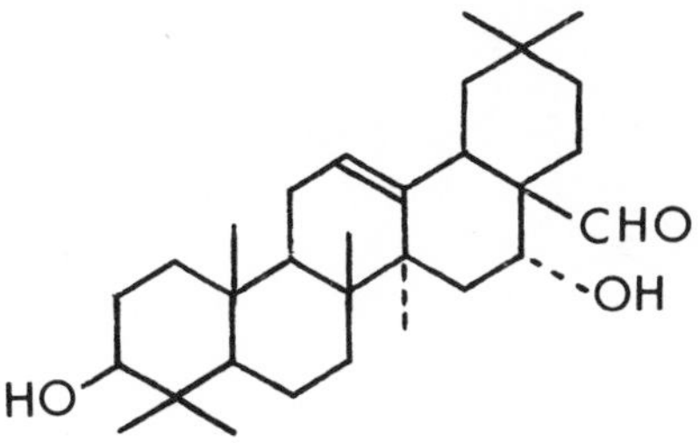 Gummosogenin
Lophocereus	Alkaloids

TABLE 6

Triterpenic Acids of Anacardiaceae

Compound	Genus	Reference
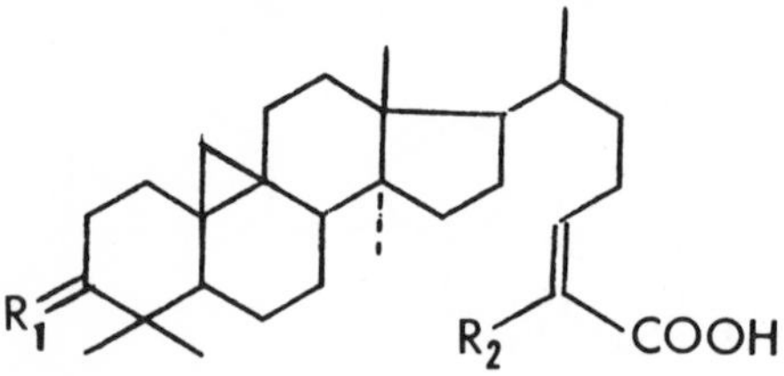 $R_1 = O$, $R_2 = CH_3$; $R_1 = \langle^{H}_{OH}$, $R_2 = CH_3$; $R_1 = O$, $R_2 = CH_2OH$	*Mangifera*	Corsano and Mincione (1965)
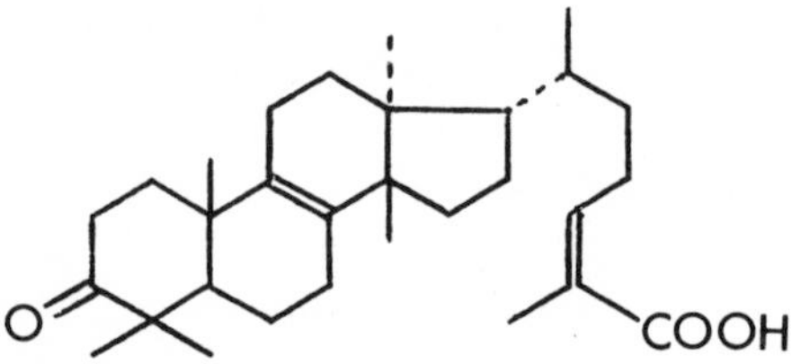 (3-oxo acid, COOH)	*Pistacia*	Barton and Seoane (1956)
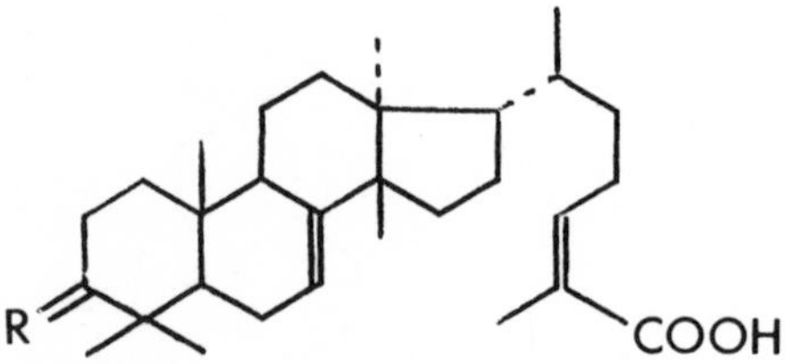 $R = O$; $R = \langle^{H}_{OH}$	*Pistacia* *schinus*	Rao and Rose (1958)

family. From the investigation of 80 species in three genera, it appears that each genus contains one or several derivatives of dammarane (Table 7). In particular, dipterocarpol (**XI**) is the major constituent of all the resins. It has not been isolated as such from any other family.

TABLE 7

Occurrence of Dammarane Derivatives in Dipterocarpacae

Genus	Compound	Reference
Dipterocarpus	XI, XII, XIII, XIV	Bisset et al. (1966)
Vatica	XI	Diaz and Ourisson (unpublished)
Doona	XI, XII	Diaz et al. (1966)

The dammarane skeleton (X) is also present in some other families (Table 8), but these families all belong to orders lacking common phylogenetic characteristics.

On the other hand, derivatives of the related euphane and ele-

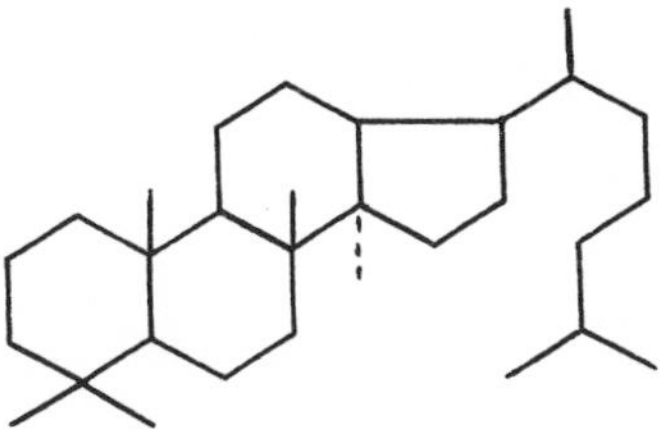

X

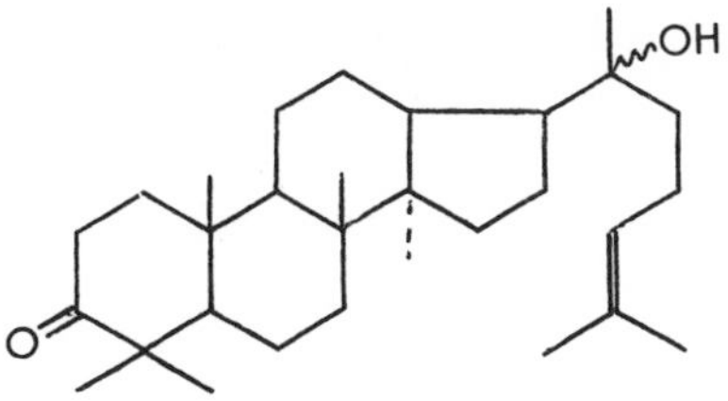

XI

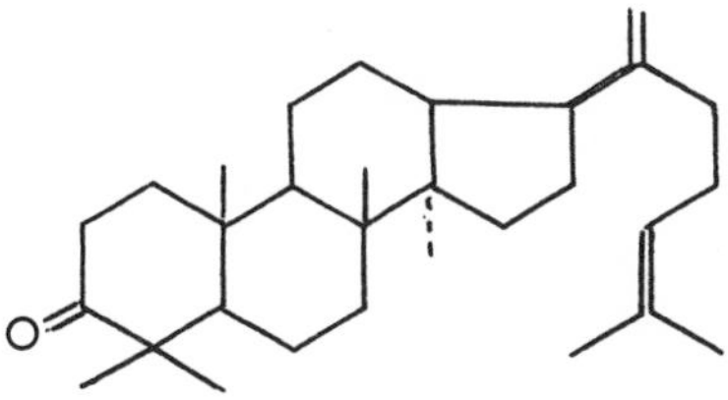

XII

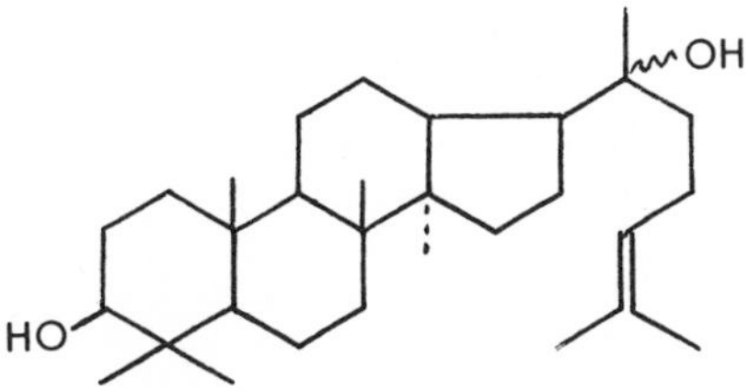

XIII

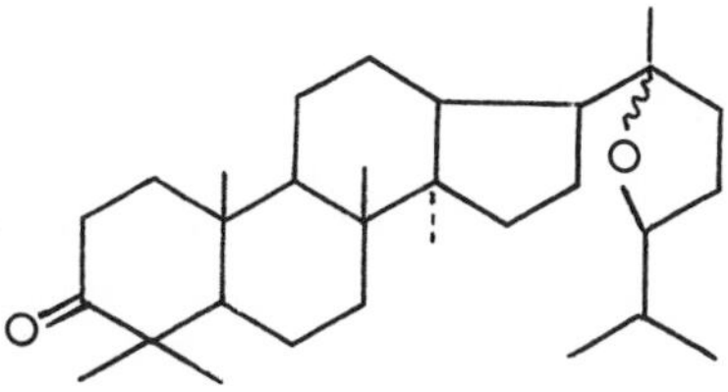

XIV

TABLE 8

Occurrence of Dammarane Derivatives

Compound	Family (and Order)	Reference
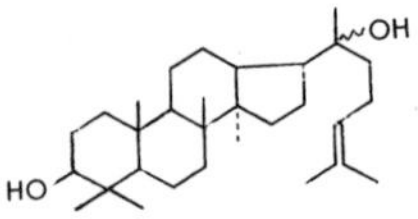 Dammarenediol	Compositae (Asterales)	Asselineau and Asselineau (1957)
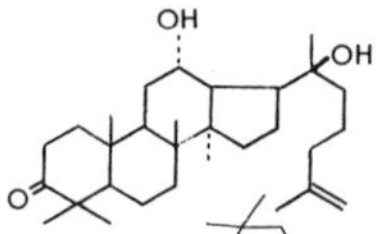	Betulaceae (Fagales)	Fischer and Seiler (1961)
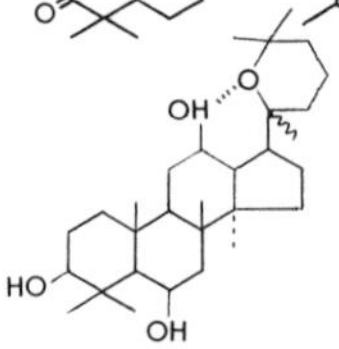 Panaxatriol	Araliaceae (Araliales)	Shibata et al (1965)
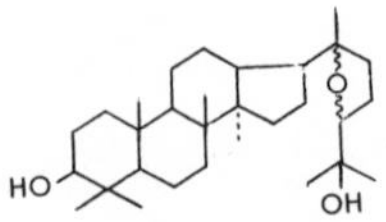 Ocotillol	Fouquieriaceae (Solanales)	Warnhoff and Halls (1965)

TABLE 9

Occurrence of Derivatives of Euphane and Elemane

Typical Constituent	Family	Order
Euphol, Tirucallol, Butyrospermol	Euphorbiaceae	Euphorbiales
Butyrospermol	Sapotaceae	Ebenales
Elemolic acid	Burseraceae	
Pseudo diterpenes	Simarubaceae	Rutales
Degraded triterpenes	Meliaceae, Rutaceae	
Tirucallol	Anacardiaceae	Sapindales

mane skeletons (see Table 1) are of great taxonomic interest since they are found in certain families considered to be closely related from a morphological point of view (Table 9) (see classifications of Hutchinson, 1960, and Engler, 1964). Moreover, among these substances there are a large number of bitter principles that possess great homogeneity with respect to degradation and oxidation patterns (Table 10). At such a distance from squalene along the biosynthetic pathway, it is highly probable that this unity in the type of constituents has a definite evolutionary significance, and that it expresses a more general genetic unity. In addition to offering to the chemist new and interesting structural studies, such coincidence may also be useful to the systematist. It would be rewarding to know the constituents of the other families that have been considered to be related to the ones mentioned above (for instance, those of Averrhoaceae). Such information could perhaps give some aid in the classification of the family Euphorbiaceae, which is presently the subject of much taxonomic controversy. At higher taxonomic levels it is quite impossible to draw conclusions of much systematic usefulness. Cycloartenol appears generally only among phanerogams but this type of generalization is extremely dependent on the thoroughness of the inventory of known substances.*

*An extension of this inventory now enables one to propose cycloartenol appears generally among all living organisms, except animals, fungi, and bacteria. Red and blue algae and Bryophytes are not included in this inventory.

TABLE 10

Structural Relationships among the Bitter Principles of Rutales

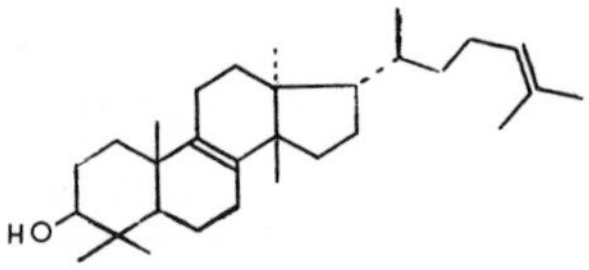

Tirucallol

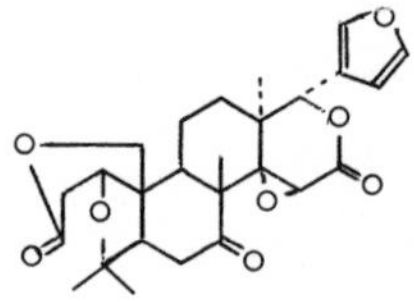

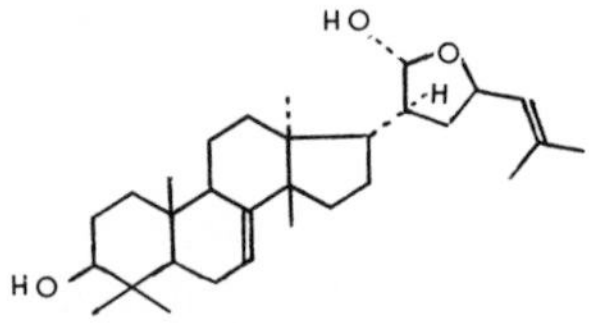

Flindissol
(Birch et al., 1963)

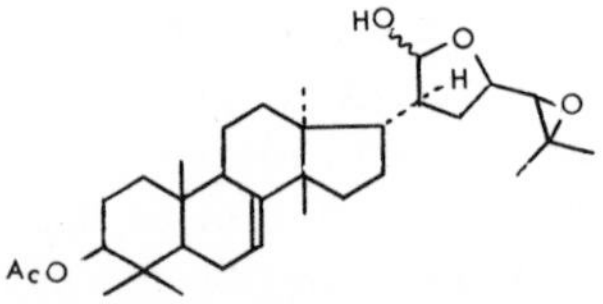

Turraeanthin
(Bevan et al., 1965)

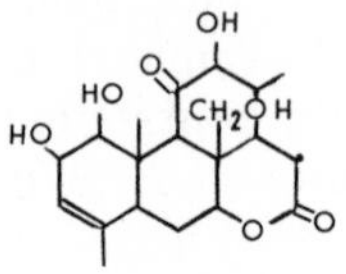

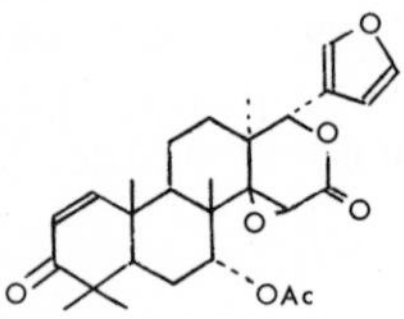

Gedunin
(Akisanya et al., 1960)

Limonin
(Barton, 1961)

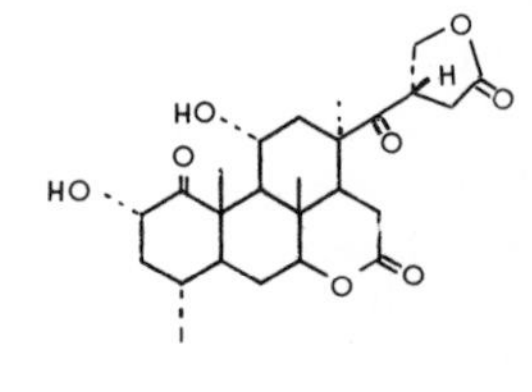

Simarolide
(Polonski, 1964)

Chaparrin
(Geissman and Ellestad, 1962)

The comments above can be illustrated by Table 11, which shows some triterpenes from cryptogams. It appeared until recently that all triterpenes of cryptogams would be 3-desoxy compounds, derived from hopane. Yet this generalization has been shattered by the discovery of pyxinic acid in a lichen in 1966. The first explanation of the lack of oxygen functions at C-3 was the assertion that cryptogams could cyclize squalene with H^+ and not with OH^+ as has been generally proposed for other triterpenes. Another proposal explained this peculiarity by postulating a further reduction at C-3 which was specific in cryptogams. Whatever the explanation, the exception of pyxinic acid remains.

Among the higher categories of the plant kingdom no generalizations may be supported on the basis of current knowledge of the distributions of triterpenes. As far as we know, all living organisms except bacteria have the ability to synthesize and to cyclize squalene.

The recent isolations of tetrahymanol (XV) from a protozoan, *Tetrahymena pyriformis* (Mallory et al., 1963), and of dustanin (XVI) from an *Aspergillus* (Tsuda and Isobe, 1965) provide the

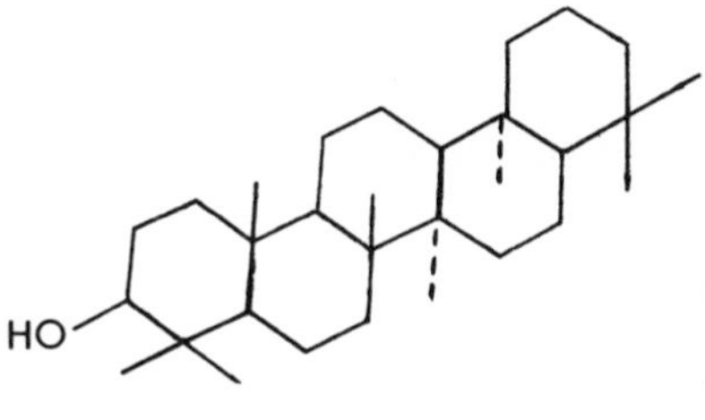

HO

XV

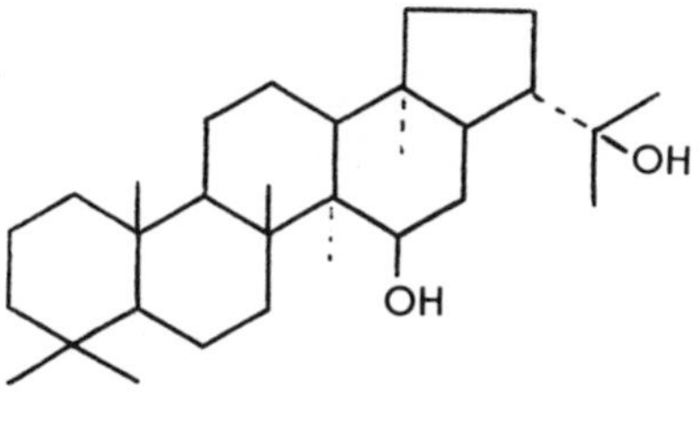

OH

OH

XVI

first examples of the occurrence of pentacyclic triterpenes outside chlorophyll-producing organisms. Until now these substances have been considered characteristic of the chlorophyllians.

V. COMMON DISTRIBUTION OF DI- AND TRITERPENES

As a transition between tri- and diterpenes it is appropriate to mention their cooccurrence in coffee oil. This oil contains one unique triterpene, coffesterol (XVII) (Kaufmann and Sen Gupta, 1964a), and two diterpenes, cafestol (XVIII) (Djerassi et al., 1959) and kahweol (XIX) (Kaufmann and Sen Gupta, 1964b), all of which possess a ring A modified by the transformation of

TABLE 11

Some Triterpenes of Cryptogams

Compound	Reference
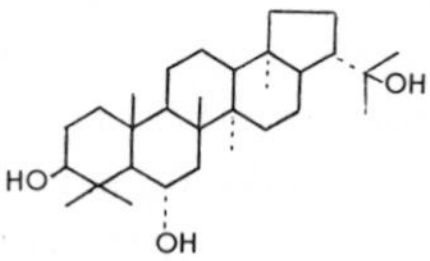 Leucotylin (Lichen)	Yosioka and Nakanishi (1963)
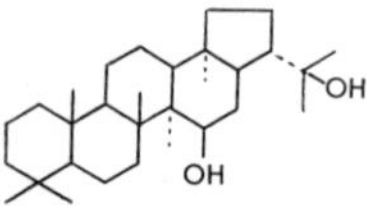 Dustanin (Fungus)	Isuda and Isobe (1965)
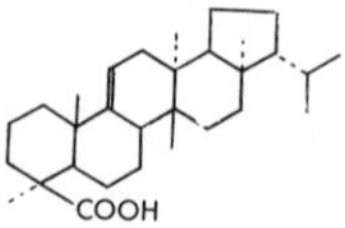 Davallic acid (Fern)	Lin et al. (1965)
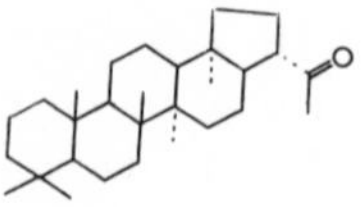 Adiantone (Fern)	Berti et al. (1963)
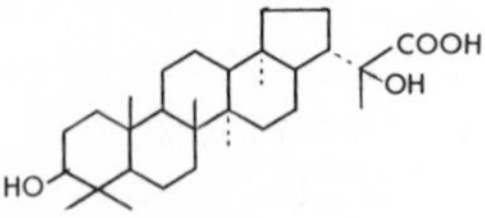 Pyxinic acid (Lichen)	Yosioka et al. (196

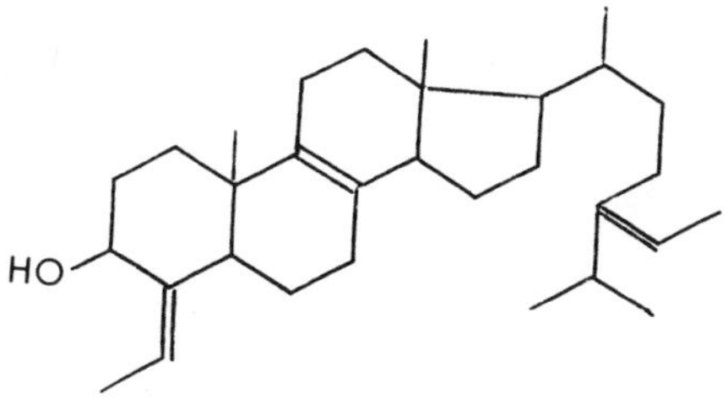

the C-4 methyl to a C-4 ethyl group. This chemical association is quite exceptional for several reasons. First, di- and triterpenes are seldom found in the same species and then only rarely in the same tissues (Nicholas, 1964; Graebe et al., 1965; Goodwin, 1965). The reason for this is apparently that chemical and enzymatic differences in the requirements for the synthesis and the cyclizations of their precursors produce a competition in the synthesis of these two terpene groups. Furthermore, in the absence of information concerning the catabolism of these two groups of substances, it is not possible to explain the differences observed in their accumulation. Second, such a modification of the common C-4 geminal dimethyl group is unique among the di- and triterpenes, and it is quite remarkable to find it in the same plant and on two different types of substances. Some di- and triterpenes are known to contain the same structural elements, as indicated in Table 12, but the plants containing these similar products have no common phylogeny.

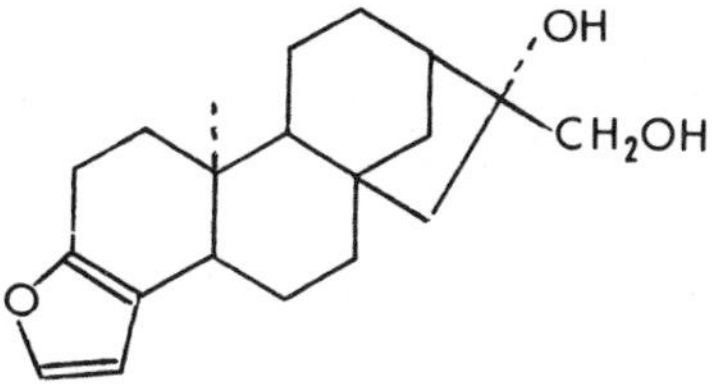

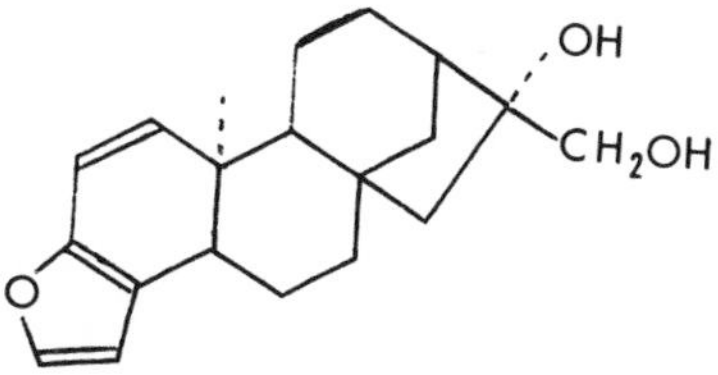

The same remarks made in the introductory statement of this paper for triterpenes may also be applied to diterpenes. In fact, although diterpenes contain only 20 carbon atoms, and although the plants that have been examined for diterpenes are less numerous, a greater variety of the possible cyclization and oxidation patterns has been observed in the diterpenes. Thus the list of

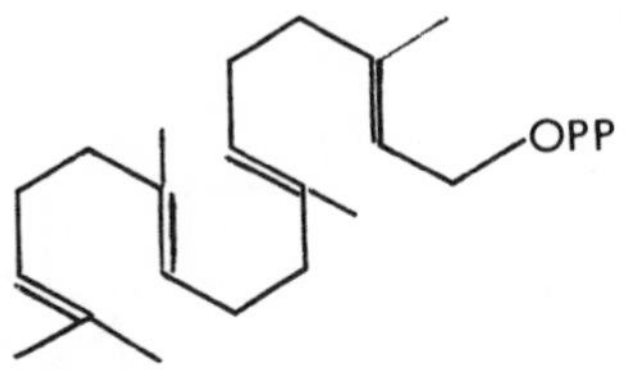

XX

known structures is as large for diterpenes as for triterpenes.

The cyclization of geranylgeranyl pyrophosphate (XX) produces specific configurations at some of the asymmetric atoms, particularly at positions 5, 8, 9, 10, and 13. The diterpenes, like mono- and sesquiterpenes, have often been found in two antipodal forms, whereas triterpenes apparently exist only in the normal (10β-methyl) series. Our present state of knowledge does not allow us to draw any conclusions from the distributions of these configurations. To support this comment we consider some products extracted from the single species *Agathis australis* (Erdtmann, 1963; Enzell and Thomas, 1965). This conifer (family Araucariaceae) contains (–)-kaurene (XXI), agathic acid (XXII), araucarenolone (XXIII), abietic acid (XXIV), isopimaradiol (XXV), and the acid XXVI. These compounds are not chemically homogeneous with respect to the number of rings, the stereochemistry of the A/B ring junction, the configuration at C-13, the stereochemistry of the oxidation at C-4, or the position and degree of other oxidations.

The conifers have been the subject of several systematic studies, which are summarized in such interesting reviews as that of Erdtmann in 1963. The reader is referred to this work for an exhaustive treatment of this classic chapter of chemical taxonomy.

VI. DITERPENES OF LEGUMINOSAE

The investigations of diterpenes of the Leguminosae do not proceed from a systematic effort but from isolated studies. For example, the structures of the alkaloids of the genus *Erythrophleum* and the constituents of resins of the tribe Amherstieae have

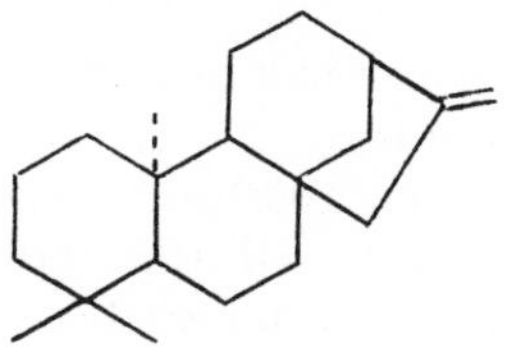

COOH

COOH

XXI

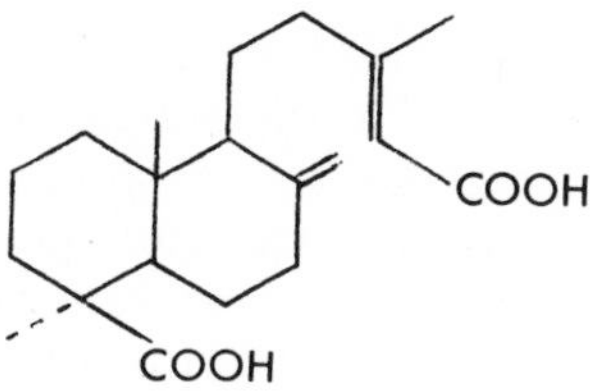

XXII

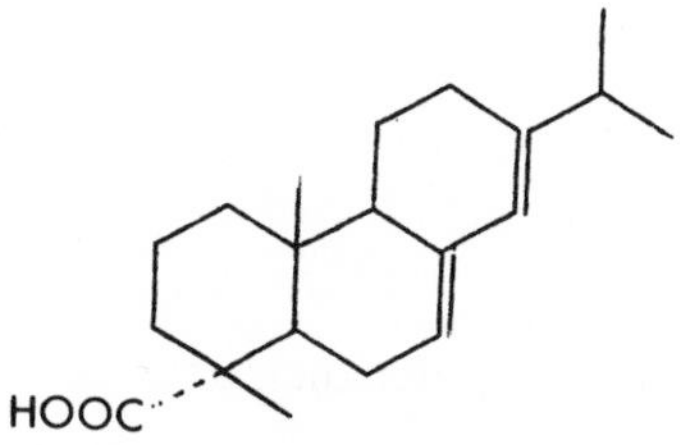

HOOC

HO

CH_2

O

O

XXIV

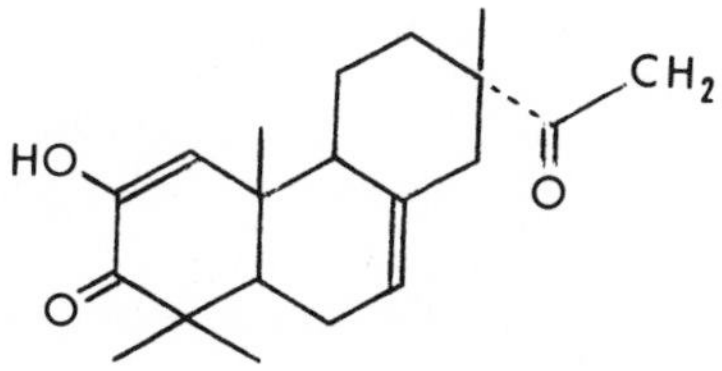

XXIII

HO

COOH

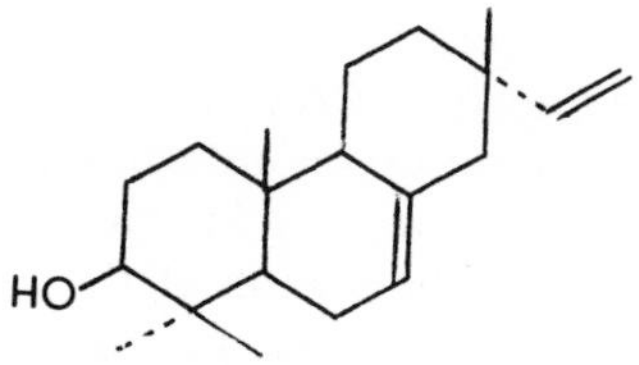

XXV

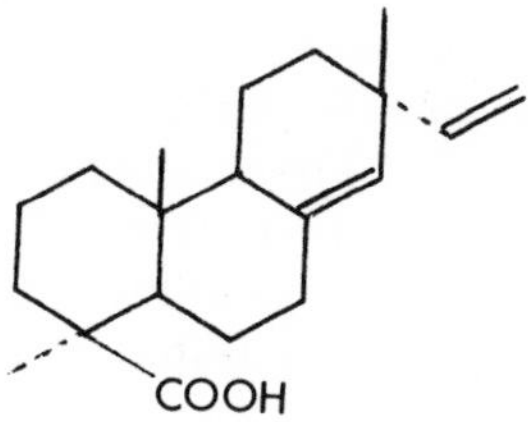

XXVI

been investigated, while gibberellin activity has been widely demonstrated in the subfamily Papilionoideae.

In the genus *Erythrophleum* at least ten alkaloids have been characterized in various species (Table 13). All alkaloids thus far isolated are esters of amino alcohols and derivatives of the tricyclic cassamic acid (XXVII) oxidized at C-7. Since the homogeneity of this group is recognized, it would be interesting to determine whether other differential characteristics, such as the

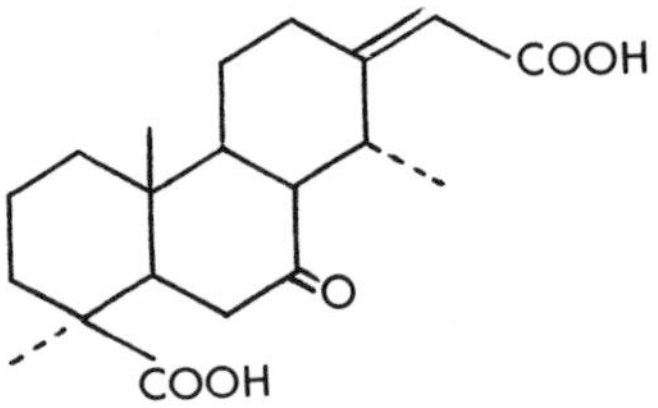

COOH

O

COOH

XXVII

position of oxidation, are of taxonomic significance at a subgeneric level.

Trees of the tribe of Amherstieae exude many resins or copals. These substances have provided investigators with rich collections of bicyclic diterpenes. Thus far over 15 alcohols, aldehydes, and acids possessing the so-called antipodal (10α-methyl) labdane skeleton have been characterized (Table 14). This high degree of homogeneity is, however, somewhat disturbed by the presence of (+)-hardwickic acid (10β-methyl) in *Copaifera officinalis* and (+)-cativic acid in *Prioria copaifera* (Table 15). There is apparently no sound taxonomic basis for questioning the placing of *Prioria* and *Copaifera* in the tribe, so that we cannot consider the absolute configuration of the labdane compounds as a tribal characteristic contrary to the skeleton itself.

With *Trachylobium* another difficulty arises: We have found that the trunk resin provides bicyclic diterpenes (Table 14), but the resin of seedpods is a mixture of tetra- and pentacyclic diterpenes derived from (–)-kaurene (XXXV) and (–)-trachylobane (XXVIII) (Table 16). Such an example is really quite important since it emphasizes once again the importance of publishing the precise nature of the plant organ studied.

It is quite interesting to consider the new trachylobane skeleton (XXVIII). It is a stable molecule analogous to the ion XXIX postulated (Wenkert, 1955) as a link between the ions XXX, XXXI, and XXXII from which the various tetracyclic diterpenes hibaene

TABLE 12

Comparison of Structures of Tri- and Diterpenes

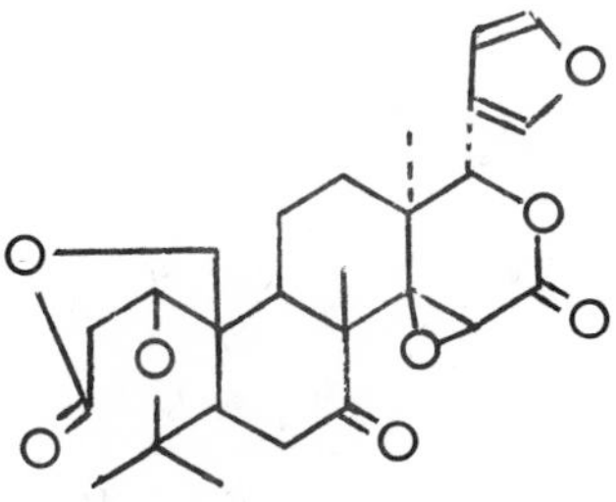

Limonin
(Rutaceae)
(Barton, 1961)

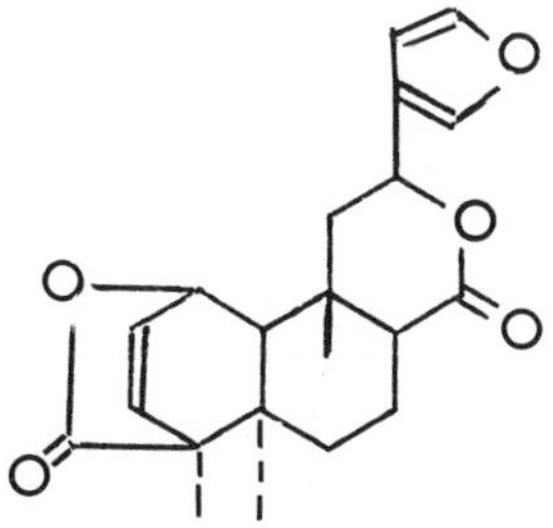

Columbin
(Menispermaceae)
(Overton et al., 1961)

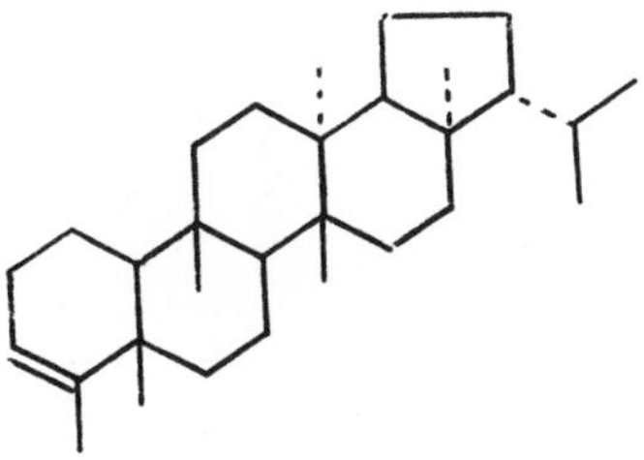

Filicene
(Fern)
(Ageta et al., 1964)

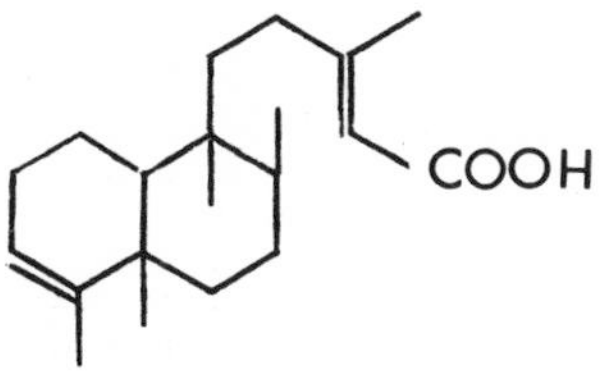

Kolaveol
(Leguminosae)
(Misra et al., 1964)

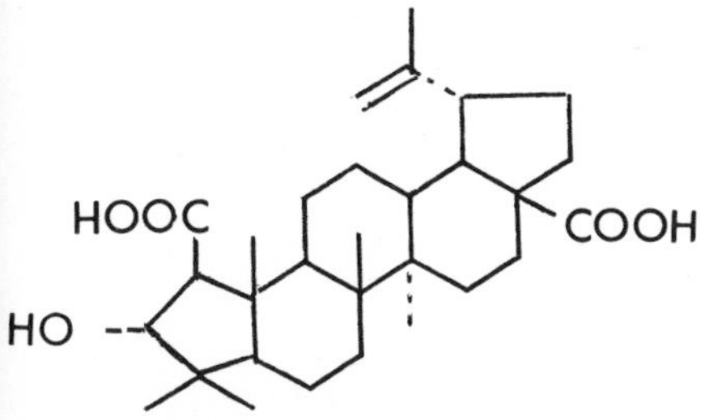

Ceanothic acid
(Rhamnaceae)
(de Mayo and Starrat, 1962)

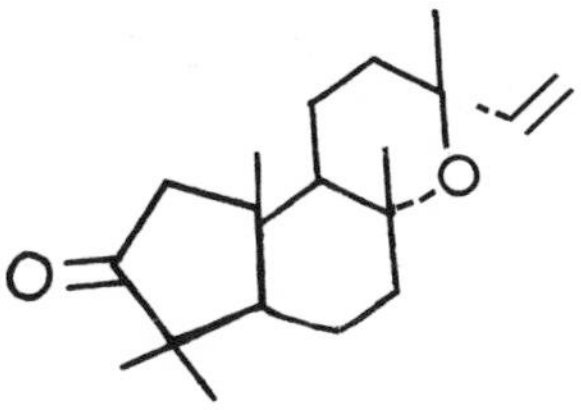

Colensenone
(Podocarpaceae)
(Grant and Carman, 1962)

TABLE 13

Alkaloids of *Erythrophleum*

Structure	R_1	R_2	R_3	Reference
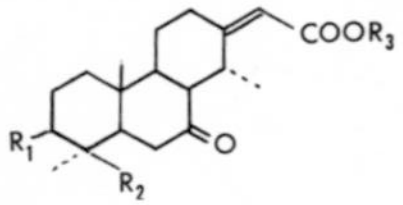 COOR₃ structure	H	$COOCH_3$	$CH_2CH_2NMe_2$	Arya and Engel (1961)
	OH	$COOCH_3$	$CH_2CH_2NMe_2$	Arya and Engel (1961)
	OH	CH_3	CH_2CH_2NHMe	Ottinger et al. (1965)
	OH	CH_3	Unknown amino alcohol	Gensler and Shermann (1959)
	OH	$COOCH_3$	Unknown amino alcohol	Arya (1961)
	OH	$COOCH_3$	Unknown amino alcohol	Chapman et al. (1963)
	H	COOMe	Unknown amino alcohol	Chapman et al. (1963)
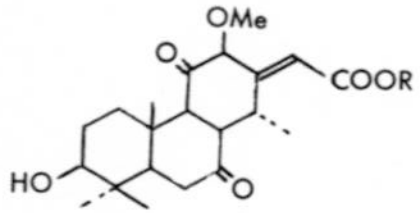 COOR structure (HO, OMe)	R Unknown			Tursch and Tursch (1964)
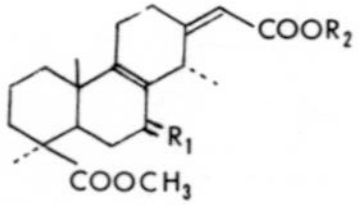 COOR₂ structure (COOCH₃)	R_I = O		R_2 Unknown amino alcohol	Chapman et al. (1965)
	R_I = OH, H		R_2 Unknown amino alcohol	Arya (1962)

TABLE 14

Some Acids of Amherstieae

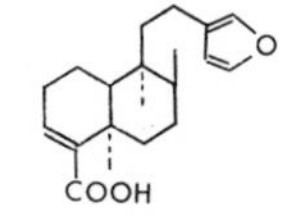

(−)-Hardwickic acid,
Hardwickia pinnata
(Misra et al., 1964)

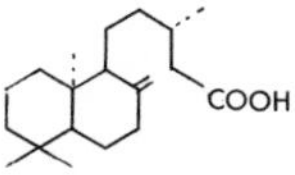

Eperuic acid,
Eperua falcata
(Blake and Jones, 1963)

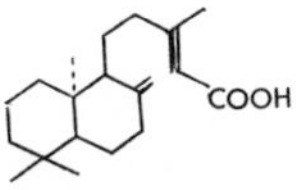

Daniellic acid,
Daniellia oliveri
(Haeuser et al., 1961)

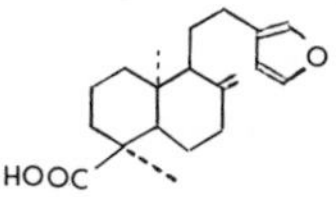

Copalic acid,
Hymenaca courbaril
(Nakano and Djerassi, 1961)

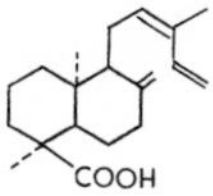

Ozic acid,
Daniella ogea
(Bevan et al., 1966)

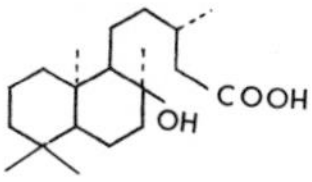

8β-Hydroxyeperuanic acid,
Trachylobium verrucosum
(Hugel et al., 1965a)

TABLE 15

Acids of Amherstieae in Normal Configuration

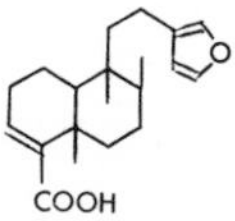

(+)-Hardwickic acid,
Copaifera officinalis
(Cocker et al., 1965)

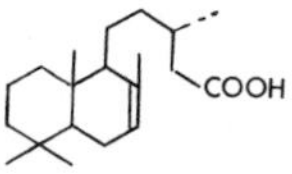

(+)-Cativic acid,
Prioria copaifera
(Grant, 1954)

TABLE 16

Diterpenes of Seed Pods of *Trachylobium*
(Hugel et al., 1965a)

Kauranol

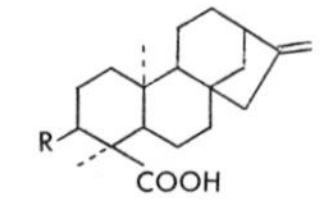

Kaurenic acid

R = H
R = OAc

Trachylobanic acid

R_I = H R R_2 = COOH
R_I = H R_2 = CH_2OH
R_I = OAc R_2 = COOH

XXVIII

XXIX

XXX

XXXI

XXXII

XXXIII

XXXIV

XXXV

(XXXIII), isoatisene (XXXIV), and kaurene (XXXV) may arise. It is in fact possible to isomerize trachylobane *in vitro* to (–)-hibaene (XXXIII), (–)-isoatisene (XXXIV), and (–)-kaurene (XXXV), three naturally occurring diterpenes (Hugel et al., 1965b). Thus the structure of trachylobane and its acidic cleavage are of a great biogenetic interest.

The derivatives of (–)-kaurene are present in many plants besides *Gibberella fujikuroi*, for example, fungi (Djerassi et al., 1961), lichens (Lehn and Huneck, 1965), conifers (Aplin et al., 1963), Ericaceae (Kakisawa et al., 1962), Euphorbiaceae (Henrick and Jefferies, 1965), Cesalpinieae (Hugel et al., 1965a), and Compositae (Dolder et al., 1960).

The wide occurrence of (–)-kaurene derivatives in nature is probably related to the fact that kaurene is a precursor of gib-

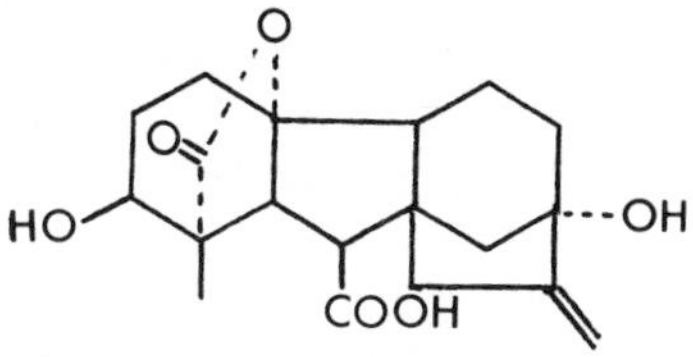

XXXVI

berellic acid (XXXVI) (Cross et al., 1964; Graebe et al., 1965), a compound which is known to be a plant growth hormone. It should be possible then, by a proper search, to find (–)-kaurene in many more plants than are known to contain it.

Most of the studies on the promotion of growth by gibberellins were carried out on the commercially important plants of the family Leguminosae, subfamily Papilionoidae (e.g., *Glycine*, *Trifolium*, *Pisum*, *Lathyrus*, *Phaseolus*). Moreover, rare examples of the characterization of gibberellins in higher plants are given by the isolation of gibberellin A5 (XXXVII) and gibberellin A6 (XXXVIII) from the bean *Phaseolus multiflorus* (MacMillan and Suter, 1958).

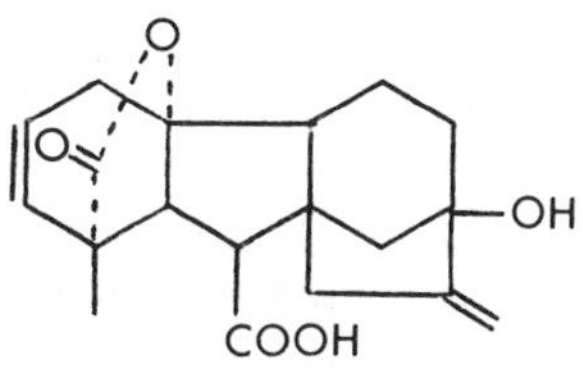

XXXVII 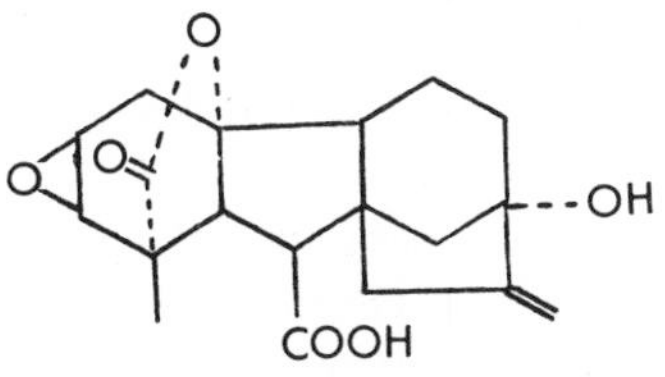   XXXVIII

Last, it should be noted that the biological interest of gibberellins goes beyond the theoretical studies of the systematists, since physiologists connect the activity of this group of compounds with the mechanism of growth, one of the most important problems in biology.

VII. CONCLUSIONS

An operational view of diterpenes is to assume that the principal, useful role of geranylgeranyl pyrophosphate is to form kaurene and the gibberellins, but that plants use the excess geranylgeraniol to generate the many diterpenes encountered in nature.

By analogy the tetra- and pentacyclic triterpenes would be the waste products of the biosynthesis of the steroids proceeding from squalene through lanosterol. The role of sterols in animals is well known but their precise functions in plants remain to be explained.

It seems that one can attach more and more taxonomic significance to polyterpenes as they become more distant from the simple, direct pathways. The further away, biogenetically, from squalene or from geranylgeraniol one goes, the less risk there is of an evolutionary convergence that might yield, in two widely separated botanical groups, the same complex modification of the parent substance.

ADDENDUM

Three communications about squalene cyclization have shown that this reaction is more complex than that originally considered. Lanostadiene (Barton and Moss, 1966) and 2,3-oxidosqualene (Van Tamelen et al., 1966; Corey et al., 1966) were recently shown to be intermediates in this reaction. It is noteworthy that in the former experiments (yeast enzymes) cyclization would precede oxidation at position 3, whereas the later experiments (rat liver homogenate) suggest initial epoxidation at positions 3 and 4.

The number of triterpenes characteristic of members of the Rutales is forever increasing. Among these new compounds, an example is melianone (Lavie et al., 1966) extracted from *Melia azedarach* (Meliaceae), whose structure is closely related to that of turraeanthin.

Cryptogams have afforded some more hydrocarbons with the hopan skeleton (Ageta and Iwata, 1966). The lack of oxygen function at position 3 was recently mentioned for the first time on a tetracyclic system: cycloartane and some of its homologues occur in *Polypodium vulgare* (Berti et al., 1967).

The biosynthesis and the possible catabolism of cycloartenol as a phytosterol precursor in vegetable tissues continue to be studied by two groups (Goad and Goodwin, 1966; Ehrhardt et al., 1967).

With regard to diterpenes, in the series of resin acids from the Cesalpinoideae, the first cooccurrence of both antipodal series (labdane and epurane) in the same species, *Oxystigma oxyphyllum*, was mentioned by Ekong and Okogun (1967).

REFERENCES

Ageta, H., and K. Iwata. 1966. Tetrahedron Lett.: 6069.

_____, K. Iwata, and S. Natori. 1964. Tetrahedron Lett.: 3413.

Akisanya, A., C. W. L. Bevan, T. G. Hirst, F. G. Halsall, and D. A. H. Taylor. 1960. J. Chem. Soc.: 3827.

Aplin, R. T., R. C. Cambie, and P. S. Rutledge. 1963. Phytochemistry, 2: 205.

Arya, V. P. 1961. J. Ind. Chem. Soc., 38: 829.

_____. 1962. J. Sci. Ind. Res. (India), 21: 342.

_____, and B. G. Engel. 1961. Helv. Chim. Acta, 44: 1650.

Asselineau, C., and J. Asselineau. 1957. Bull. Soc. Chim.: 1359.

Barton, D. H. R. 1961. Pure Appl. Chem., 551.

_____, and G. P. Moss. 1966. Chem. Commun.: 261.

_____, and E. Seoane. 1956. J. Chem. Soc.: 4158.

Berti, G., F. Bottari, A. Marsili, J. M. Lehn, P. Witz, and G. Ourisson. 1963. Tetrahedron Lett.: 1283.

_____, F. Bottari, A. Marsili, I. Morelli, and M. Polvani. 1967. Tetrahedron Lett.: 125.

Bevan, C. W. L., D. E. U. Ekong, T. G. Halsall, and P. Toft. 1965. Chem. Commun.: 636.

_____, D. E. U. Ekong, and J. I. Okogun. 1966. Chem. Commun: 44.

Birch, A. J., D. J. Collins, S. Muhammad, and J. P. Turnbull. 1963. J. Chem. Soc.: 2762.

Bissett, N. G., M. A. Diaz, C. Ehret, G. Ourisson, M. Palmade, F. Patil, P. Pesnelle, and J. Streith. 1966. Phytochemistry, 1865.

Blake, S., and G. Jones. 1963. J. Chem. Soc.: 430.

Boiteau, P., B. Pasich, and A. R. Ratsimamanga. 1964. Les Triterpénoides en Physiologie Végétale et Animale, Paris, Gauthier-Villars.
Bredenberg, J. B., and R. Gmelin. 1962. Acta Chem. Scand , 16: 1802
Chapman, G. T., B. Jacques, D. W. Mathieson, and V. P. Arya. 1963. J. Chem. Soc.: 4010.
_____, J. N. T. Gilbert, B. Jacques, and D. W. Mathieson. 1965. J. Chem. Soc.: 403.
Cocker, W., A. L. Moore, and A. C. Pratt. 1965. Tetrahedron Lett.: 1983.
Conolly, J. D., and K. H. Overton. 1964. Ann. Rep. LXI, 346.
Corey, E. J., W. E. Russey, and P. R. O. de Montenallo. 1966. J. Amer. Chem. Soc., 88: 4750.
Corsano, S., and E. Micione. 1965. Tetrahedron Lett.: 2377.
Cross, B. E., R. H. B. Galt, and J. R. Hanson. 1964. J. Chem. Soc.: 295.
Diaz, M. A., and G. Ourisson. Unpublished.
_____, G. Ourisson, and N. G. Bissett. 1966. Phytochemistry, 855.
Djerassi, C. 1957. Festschr. Arthur Stoll, 330, Basel, Birkhäuser.
_____, M. Cais, and L. A. Mitschler. 1959. J. Amer. Chem. Soc., 81: 2386.
_____, P. Quitt, E. Mossetig, R. C. Cambie, P. S. Rutledge, and L. H. Briggs. 1961. J. Amer. Chem. Soc., 83: 3720.
Dolder, F., H. Lichti, E. Mossetig, and P. Quitt. 1960. J. Amer Chem. Soc., 82: 246.
Ehrhardt, J. D., L. Hirth, and G. Ourisson. 1967. Phytochemistry, 815.
Ekong, D. E. U., and J. I. Okogun. 1967. Chem. Commun.: 72.
Engler, A. 1964. Syllabus der Pflanzenfamilien, Berlin, Gebrüder Borntraeger.
Enzell, C. R., and B. R. Thomas. 1965. Acta Chem. Scand., 19: 913.
Erdtmann, H. 1963. *In* Chemical Plant Taxonomy, Swain, T., ed., 89, London, Academic Press.
Eschenmoser, A., L. Ruzicka, O. Jeger, and D. Arigoni. 1955. Helv. Chim. Acta, 38: 1890.
Fischer, F., and N. Seiler. 1961. Liebig Ann. Chem., 644: 162.
Geissman, T. A., and G. A. Ellestad. 1962. Tetrahedron Lett.: 1083.
Gensler, W. J., and G. M. Shermann. 1959. J. Amer. Chem. Soc., 81: 5217.

Goad, L. J., and T. W. Goodwin. 1966. Biochem. J., 99: 735.
Goodwin, T. W. 1965. *In* Biosynthetic Pathways in Higher Plants, Swain, T., ed., 57, London, Academic Press.
Graebe, J. E., D. T. Dennis, C. D. Upper, and C. A. West. 1965. J. Biol. Chem., 240: 1847.
Grant, F. W. 1954. J. Amer. Chem. Soc., 76: 5001.
Grant, P. K., and R. M. Carman. 1962. J. Chem. Soc.: 3740.
Haeuser, J., R. Lombard, F. Lederer, and G. Ourisson. 1961. Tetrahedron, 12: 205.
Halsall, T. G., and R. T. Aplin. 1964. Progress in the Chemistry of Organic Natural Compounds, XXII: 153.
Henrick, C. A., and P. R. Jefferies. 1965. Aust. J. Chem., 18: 2005.
Hugel, G., L. Lods, J. M. Mellor, D. W. Theobald, and G. Ourisson. 1965a. Bull. Soc. Chim. France: 2882.
_____, L. Lods, J. M. Mellor, and G. Ourisson. 1965b. Bull. Soc. Chim. France: 2894.
Hutchinson, J. 1960. The Families of Flowering Plants, London, Oxford Univ. Press.
Kakisawa, H., M. Yanai, T. Kozima, K. Nakanishi, and H. Mishima. 1962. Tetrahedron Lett.: 215.
Kaufmann, H. P., and A. K. Sen Gupta. 1964a. Fette, Seifen, Anstrichmittel, 66: 461.
_____, and A. K. Sen Gupta. 1964b. Chem. Ber., 97: 2652.
Lavie, D., M. K. Jain, and I. Kirson. 1966. Tetrahedron Lett.: 2049.
Lehn, J. M., and S. Huneck. 1965. Z. Naturforsch. [B], 20: 10.
Lin, Y. Y., H. Kakisawa, Y. Shiobara, and K. Nakanishi. 1965. Chem. Pharm. Bull. (Tokyo), 13: 986.
Macmillan, J., and P. J. Suter. 1958. Naturwissenschaften, 45: 46.
Mallory, F. B., J. T. Gordon, and R. L. Conner. 1963. J. Amer. Chem. Soc., 85: 1362.
De Mayo, P., and A. N. Starrat. 1962. Canad. J. Chem., 40: 788.
Misra, R., R. C. Pandey, and Dev. Sukh. 1964. Tetrahedron Lett.: 3751.
Nakano, T., and C. Djerassi. 1961. J. Org. Chem., 26: 167.
Nicholas, H. J. 1964. Biochim. Biophys. Acta, 84: 80.
Ottinger, R., G. Chiurdoglu, and Vandendris. 1965. Bull. Soc. Chim. Belge, 74: 198.
Ourisson, G. 1963. Tetrahedron Lett.: 1283.
_____, P. Crabbe, and O. R. Rodig. 1964. Tetracyclic Triterpenes, Paris, Herrmann.

Overton, K. H., N. G. Weir, and A. Wylie. 1961. Proc. Chem. Soc.: 211.

Phillips, D. D., and A. W. Johnson. 1957. Chem. Industr.: 1211.

Polonski, J. 1964. Proc. Chem. Soc.: 292.

Rao and Rose. 1958. Trans. Bose Res. Inst. (Calcutta), 21: 23.

Rehm, S. 1960. Ergebn. Biol., XXII: 108.

_____, P. R. Enslin, A. D. J. Meeuse, and J. H. Wessels. 1957. J. Sci. Food Agric., 12: 679.

Richards, J. H., and J. B. Hendrickson. 1964. The Biosynthesis of Steroids, Terpenes and Acetogenins, New York, Benjamin.

Shibata, S., O. Tanaka, K. Sorma, Y. Ilida, T. Ando, and H. Nakamura. 1965. Tetrahedron Lett.: 207.

Taylor, D. A. H. 1960. J. Chem. Soc.: 3827.

Tsuda, Y., and K. Isobe. 1965. Tetrahedron Lett.: 3337.

Tursch, B., and E. Tursch. 1964. An. Brasil. Quim., 21: 23.

Van Tamelen, E. E., J. D. Willett, R. B. Clayton, and K. E. Lord. 1966. J. Amer. Chem. Soc., 88: 4752.

Warnhoff, E. W., and C. M. M. Halls. 1965. Canad. J. Chem., 43: 3311.

Wenkert, E. 1955. Chem. Industr.: 281.

Yosioka, I., and T. Nakanishi. 1963. Chem. Pharm. Bull. (Tokyo), 11: 1468.

_____, A. Matsuda, and I. Kitagawa. 1966. Tetrahedron Lett.: 613.

part IV

FLAVONOIDS

9

C-GLYCOSYL FLAVONOIDS

†Ralph E. Alston
Cell Research Institute and Department of Botany
The University of Texas, Austin

I. INTRODUCTION

The *C*-glycosyl flavonoids have not been neglected in recent reviews of flavonoid chemistry. Since the valuable comprehensive review of Seikel (1963), several additional reviews have appeared (Haynes, 1965; Chopin, 1966; Wagner, 1966). The latter group was concerned principally with newly discovered *C*-glycosyl compounds and with further analyses of structures that had been in dispute. The increasingly valuable technique of nuclear magnetic resonance spectroscopy was emphasized in these reviews especially as applied to analysis of the configuration and position of attachment of *C*-glycosylic groups. Historically, the nature of the *C*-glycosylic side chain has been the chief obstacle to the total structure determination of these flavonoids because of difficulties in cleaving the carbon-carbon bond to preserve intact the glycosyl moiety. Identity of the hexopyranosy nature of orientin and isoorientin was established, however, by Koeppen and Roux (1965a,b) through detection of *D*-glucose as a product of ferric chloride oxidation of the tetramethyl ethers.

The present treatment of *C*-glycosyl flavonoids will include only a summary of recent contributions to their structural chemistry. In addition, the compounds will be considered from a somewhat more biological viewpoint to the extent that a discussion of problems of their biosynthesis and their taxonomic distributions will be included. It is customary, when considering the biological implications of almost any type of distribution, to expect that the reader will be left with more problems at the conclusion than were recognized to exist in the beginning. This situation holds for the *C*-glycosyl flavonoids.

II. DISTRIBUTIONS OF *C*-GLYCOSYL FLAVONES IN THE PLANT KINGDOM

Tables 1 and 2 list the currently known *C*-glycosyl flavonoid and their taxonomic distributions. As suggested by G. H. N. Towers recently (see discussion following Seikel, 1963), *C*-glycosyl flavones and *O*-glycosyl flavones are apt to be confused on chromatograms, and the former may escape detection in screening for flavonoids by use of paper chromatography. In the solve systems that are often used in our laboratory, compounds such as isovitexin and isoorientin may be confused with 3-*O*-glycosides of the flavonols kaempferol and quercetin, on the basis of both R_f values and their colors in ultraviolet light in the presenc of ammonia. In contrast, on chromatograms *C*-glycosyls such a vitexin and orientin in these solvents run close to the flavone 7-*O*-glycosides derived from apigenin and luteolin, but which yield a more brilliant yellow color in ultraviolet light in the presence of ammonia. With more general knowledge of the chromatographic properties of *C*-glycosyl flavonoids, these compounds may be found to have an unexpectedly wide distribution. Experience in our laboratory will serve as an example. In no in

TABLE 1

Known C-Glycosyl Flavonoids

Flavones

1. Vitexin
2. Isovitexin (saponaretin)
2a. 7-Methylisovitexin (swertisin)
3. Saponarin type
4. Isosaponarin type
5. Vicenin types
6. Vitexin *O*-glycoside
7. Vicenin *O*-glycoside
8. Vitexin 4′-xyloside
9. Vitexin rhamnoside
10. Violanthin (a vicenin type, but different from vicenin)
11. Isovitexin arabinoside
12. Orientin (and epi-orientin)
12a. 5-Methylepi-orientin
12b. 5,7-Dimethylorientin
13. Isoorientin
13a. 7-Methylisoorientin (swertiajaponin)
14. Lutonarin type
15. Isolutonarin type
16. Lucenin types
17. Orientin *O*-glycoside
18. Lucenin *O*-glycoside
19. Isoorientin xyloside (adonivernoside)
19a. Isoorientin rhamnoside
20. Cytoside (4′-methylvitexin)
20a. Bayin (5-deoxyvitexin)
21. Scoparin
22. Scoparin 6-*C*-glycoside
23. Scoparin 6-*C*-glycoside *O*-glycoside
24. 3′-Methyllutonarin
25. Diosmetin 8-*C*-β-D-glucoside
25a. Diosmetin 6-*C*-β-D-glucoside

Flavanones

26. Hemiphloin
27. Isohemiphloin

Dihydrochalkones

28. Aspalathine

Isoflavones

29. Puerarine
30. Puerarine xyloside

Flavonol

31. Keyakinin

Flavanonol

32. Keyakinol

stance were our studies of *C*-glycosyl flavonoids begun deliberately because of an interest in these particular compounds. In each instance the flavonoid chemistry of the plants was being investigated as part of a genetic or taxonomic problem, and the presence of glycoflavonoids was coincidental.

Our first contact with *C*-glycosyl flavonoids came in a taxonomic survey of the family Lemnaceae. The compounds are of widespread occurrence in the family. They are present in all known species of *Lemna*, *Spirodela*, and *Wolffia* (McClure and Alston, 1966). Additionally, Mr. David Ockendon, who was work-

TABLE 2

Distribution of C-Glycosyl Flavonoids

Plant Source	Type of *C*-Glycosyl Compounds	Reference
Mosses		
Mnium affine	1, 5, 6, 7, 12, 16, 17, 18	Melchert and Alston (1965)
Ferns		
Cyathea faujei		Ueno et al. (1963)
Sphenomeris chusana		Ueno et al. (1963)
Gymnosperms no reports		
Angiosperms		
Gramineae	3, 14, 24	Seikel and Geissman (1957)
Hordeum vulgare		Seikel and Bushnell (1959)
Avena sativa	9	Harborne and Hall (1964)
Avena sp. (6 others)	11	Harborne and Hall (1964)
Avena sp. (4 others)	9	
Triticum monococcum	3	
Triticum dicoccum	13, (?), 19a	
Triticum polonicum	3	
Agrostic canina	12, 13	Harborne and Hall (1964)
Stipa calamagrostic	12, 13	Harborne and Hall (1964)
Poa compressa	12, 13	Harborne and Hall (1964)
Oryza sativa	1, 2, 13, 13	Harborne and Hall (1964)
Briza media	1, 2	Harborne and Hall (1964)
Lemnaceae	(additional different but unidentified *C*-glycosyl types occur among these species)	McClure (1965)
Spirodela intermedia	1, 12, 4	McClure and Alston (1966)
Spirodela polyrhiza	1, 12, 13, 14	McClure and Alston (1966)

TABLE 2 (continued)

Plant Source	Type of *C*-Glycosyl Compounds	Reference
Spirodela bipertorata	1, 12, 13, 14	McClure and Alston (1966)
Spirodela oligorhiza	1, 12, 4, 13, 14, 16, 5, 2	Jurd et al. (1957)
Lemna minor	1, 12, 4, 13, 14, 16, 5, 2	McClure and Alston (1966)
Lemna gibba	1, 12, 13, 2	McClure and Alston (1966)
Lemna obscura	13, 16, 5	McClure and Alston (1966)
Lemna trisulca	1, 12, 13, 16, 5	McClure and Alston (1966)
Lemna perpusilla	5	McClure and Alston (1966)
Lemna trinervis	16, 5	McClure and Alston (1966)
Lemna valdaviana	16, 5	McClure and Alston (1966)
Lemna minima	12, 13, 5	McClure and Alston (1966)
Wolffia arrhiza	12, 13, 5, 2	McClure and Alston (1966)
Wolffia columbiana	1, 12, 13, 5, 2	McClure and Alston (1966)
Iridaceae		
Iris chrysophylla	1, 2, 13, 5, 12, 16	Carter (unpublished)
Iris tenax	2, 13	Carter (unpublished)
Caryophyllaceae		
Saponaria officinalis	3	Barger (1906)
Combretaceae		
Combretum micranthum	1, 2	Jentzsch et al. (1962)
Compositae		
Helichrysum bracteatum	12, 13	Rimpler et al. (1963)
Tragopogon dubius		Krochewsky (1967)
Tragopogon pratense		Krochewsky (1967)

TABLE 2 (continued)

Plant Source	Type of C-Glycosyl Compounds	Reference
Tragopogon porrifolius		Krochewsky (1967)
Cruciferae		
Alliaria officinalis		Paris and Delaveau (1962)
Fagaceae		
Castanosperum australe	20a	Eade et al. (1962)
Nothofagus fusca	(possibly 3′ or 5′ C-glycosyl of dihydrachalcone)	Hillis (unpublished)
Gentianaceae		
Swertia japonica	2a, 13, 13a	Komatsu and Tomimori (1966)
Leguminosae		
Aspalathus linearis	12, 13, 28	Koeppen and Roux (1965a,b)
Cytisus laburnum	20	Paris (1957), Stambouli and Paris (1961)
Lespedeza capitata	13	Paris and Charles (1962)
Parkinsonia aculeata	12, 12a, 12b	Bhatia et al. (1965)
Psoralea spp. (all of 30 tested)	1, 2, 12, 13, 5, 16, 21, 22, 7, 18, 23	Ockendon et al. (1966)
Pueraria thunbergiana	29, 30	Shibata et al. (1962)
Sarothamnus scoparium	1, 12, 21	Hörhammer et al. (1962)
Spartium junceum	12	Hörhammer et al. (1960)
Tamarindus indica	1, 12	Lewis and Neelakantan (1964); Bhatia et al. (1964)
Malvaceae		
Hibiscus syriacus	3	Nakaoki (1944)
Myrtaceae		
Eucalyptus hemopholia	26, 27	Hillis and Carle (1963)

TABLE 2 (continued)

Plant Source	Type of *C*-Glycosyl Compounds	Reference
Oxalidaceae		
Oxalis cernua	12	Shimokoriyama and Geissman (1962)
Polygonaceae		
Polygonum orientale	1, 2, 12, 13	Hörhammer et al. (1962)
Ranunculaceae		
Adonis vernalis	12, 19	Hörhammer et al. (1960)
Trollius europaeus	1, 12	Sachs (1963)
Rosaceae		
Crataegus divers	8	Hrugasiewicz (1964)
Crataegus oxyacantha	8	Fiedler (1955)
Rutaceae		
Citrus limonum	25, 25a (+ derivatives of apigenin and luteolin)	Horowitz (1966) Chopin et al. (1964)
Verbenaceae		
Vitex lucens	1, 2, 5, 8, 12, 13, 16	Seikel et al. (1966)
Vitex peduncularis	1	Rao and Venkateswarlu (1956)
Vitex megapotamica	12, 13	Hänsel et al. (1965)
Vitex agnuscastus	Orientin type	Hänsel et al. (1965)
Vitex negundo	Not identified	Hänsel et al. (1965)
Vitex trifolia	Not identified	Hänsel et al. (1965)
Violaceae		
Viola tricolor	10 (possibly 5, 16; personal communication)	Hörhammer et al. (1965)
Ulmaceae		
Zelkora serrata	31, 32	Funaoka and Tanaka (1957)

ing on a monograph of a section of the genus *Psoralea* (family Leguminosae), found that all species of *Psoralea* that were available for this study produced a number of different *C*-glycosyl flavones (Ockendon et al., 1966). Again, in a collaborative study of the Compositae genus *Tragopogon* (Brehm and Ownbey, 1965), Krochewsky (1967) found that the major flavonoids of this genus were *C*-glycosyl flavones. Next, in the study of interspecific hybridization in Iris (family *Iridaceae*), Miss Lynne Carter, in our laboratory, found that the major flavonoids of both *Iris chrysophylla* and *Iris tenax* were *C*-glycosyl flavones. Finally, in collaboration with Dr. T. Melchert, who was interested in the identities of some presumed flavonoids in the moss *Mnium*, these compounds were also identified as *C*-glycosyl flavones (Melchert and Alston, 1965). From this experience it is easy to become convinced that *C*-glycosyl flavonoids have a far more widespread distribution than is presently recognized. Previously, Harborne and Hall (1964) noticed that *C*-glycosyl flavonoids were common among a number of grasses tested. In early work, the types of *C*-glycosyl flavonoids encountered were repeatedly, almost monotonously, those related to the vitexin-orientin series of flavone However, the parent flavonoids, apigenin (from vitexin) and luteolin (from orientin) are extremely common themselves and it is likely that a much greater variety of *C*-glycosyl flavonoids exists Indeed, examination of Table 1 discloses that several different subtypes of flavonoids are now known to be represented among *C*-glycosyl compounds, for example, flavones, isoflavones, flavanones, dihydrochalkones, flavonols, and flavanonols. Because of the fact that flavonols, which are extremely common flavonoids, have so far yielded only a single *C*-glycosyl type of compound (keyakinin), the validity of the structure of this compound is of considerable academic interest. However, keyakinin has been restudied quite recently by Hillis and Horn (1966) and its structure has been further confirmed as a flavonol, with the total structure as indicated in Fig. 1.

It should be possible to propose a large number of theories concerning the taxonomic implications of the distributions of

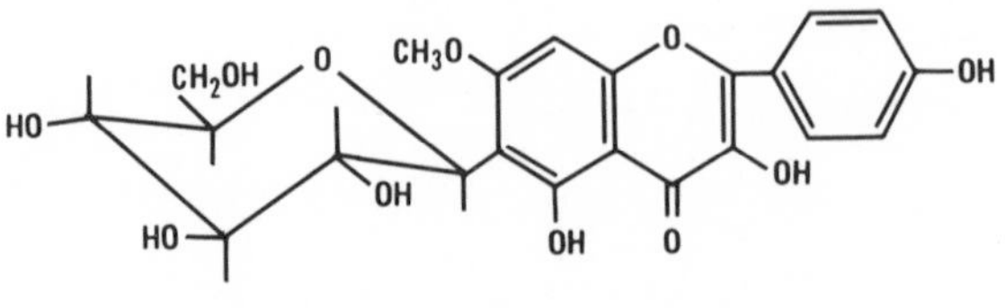

Fig. 1. Structure of keyakinin.

C-glycosyl flavonoids. Yet, discretion is said to be preferred, in some quarters, to valor, and only conservative general statements will be risked on this occasion. It is, of course, noteworthy that a number of different *C*-glycosyl flavonoids occur in the moss *Mnium*. In addition, Hegnauer referred to a previous report by Molisch of the occurrence of saponarin in the moss *Madotheca platyphylla*. Further confirmation of the suspected occurrence of *C*-glycosyl flavonoids in mosses has come from work by Harborne (1966). The presence of *C*-glycosyl flavonoids in mosses, ferns, and higher plants is definitely of taxonomic interest as is the presence of the rather unusual anthocyanidin types, apigeninidin and luteolinidin, in the moss *Bryum* (Bendz and Martensson, 1964) and in ferns (Harborne, 1966). Indeed, if the implications of this fragmentary chemical data were accepted at face value, we might reconsider the question of which types of flavonoids are "primitive" as opposed to "advanced." Although this question of chemical primitiveness among flavonoids has not apparently been considered comprehensively on biogenetic grounds, it seems to be generally regarded that anthocyanidins such as cyanidin, and *O*-glycosides of flavonoids are representatives of more primitive flavonoids. Primitive, as used here, signifies that such compounds may have appeared early among the flavonoids following the appearance of the characteristic C_{15} structural unit.

The presence of flavonoids in mosses is of further interest because of previous speculation (Alston and Turner, 1963) that the origin of flavonoids per se was dependent on a well-developed pattern of lignin biosynthesis. Mosses, of course, lack typical lignin. Finally, certain morphological features of bryophytes in general suggest the possibility that these plants might be ancestral to the vascular plants, and the copresence of flavonoids in vascular plants and in bryophytes provides some further support for such a viewpoint. A question that is only indirectly in the domain of evolution concerns the implications of certain common patterns of associations of *C*-glycosyl flavonoids in mosses and in angiosperms. So far as is known at present, the group of *C*-glycosyl flavonoids characteristic of *Mnium* is basically the same group that occurs most abundantly in the angiosperms as a whole. If further work discloses parallelisms between the types of *C*-glycosyl flavonoids characteristic of these two groups of plants, then one is offered greater encouragement to search for a functional explanation, either involving mechanisms of biosynthesis or the role of these groups of flavonoids in the plant.

It is of course premature to draw conclusions from the lack of reports of *C*-glycosyl flavonoids in gymnosperms. However, a useful project would be that of screening a representative group of gymnosperms for such compounds. As will be noted later, chromatographic screening of *C*-glycosyl flavonoids is quite efficient insofar as detection of the more common *C*-glycosyl flavonoids is the sole objective.

The question of the importance of *C*-glycosyl flavonoids in the taxonomy of lower categories is of less general interest, and in fact, from a theoretical consideration, the same criteria and limitations apply to these compounds as apply to any other group of secondary compounds. For the record, however, it is noteworthy that *C*-glycosyl flavonoids are of great importance in the recognition of certain of the Lemnaceae, a family of minute aquatic plants (the smallest angiosperms), many species of which do not flower and which often resemble each other quite closely. *C*-Glycosyl flavonoid chemistry made a major contribution to the taxonomy of the family Lemnaceae (McClure, 1964), and if all of the minor flavonoid components of the various species of Lemnaceae become characterized these compounds may provide considerable insight to evolutionary patterns within this family. In contrast to the duckweeds (Lemnaceae), 30 different species of *Psoralea* (family Leguminosae), including species with wide differences in their external appearances, have a closely similar distribution of *C*-glycosyl flavonoids, at least in their leaves. In this situation one might consider that *C*-glycosyl flavonoids serve as a valuable generic character, although not all of the closely related genera have been compared with *Psoralea*. In other instances, it seems that *C*-glycosyl flavonoids may be rare in a genus. For example, Hillis (personal communication), who surveyed leaves of over 330 *Eucalyptus* species for polyphenols, detected a compound with chromatographic properties of hemiphloin in only about five species.

No genetic studies have been conducted on the inheritance of specific *C*-glycosyl flavonoid types. Work in progress on *Tragopogon* is promising in this respect (Ownbey and Brehm, 1965). Also, interspecific hybrids of chemovars of *Mnium affine* offer excellent possibilities (Melchert and Alston, 1965). Genetic evidence on how vitexin, isovitexin, and vicenin syntheses are interrelated would be of potential value in considering the general question of the sequence and mode of *C*-glycosylation.

III. CHEMICAL STRUCTURES OF *C*-GLYCOSYL FLAVONES

As noted by Haynes (1965), recent work on the structures of vitexin, isovitexin (saponaretin), orientin, and isoorientin (homoorientin) has firmly established their identities as indicated in Fig. 2. Thus vitexin (and orientin) is 8-*C*-β-*D*-glucopyranosyl-apigenin (or luteolin), and isovitexin (and isoorientin) is 6-*C*-β-*D*-glucopyranosyl-apigenin (or luteolin). The notable contributions to the establishment of the structures described above were those of Horowitz and Gentili for vitexin and isovitexin, and Koeppen and Roux (1965a,b) for orientin and isoorientin. The names isoorientin and isovitexin suggested by Haynes (1965) are in accord with the established interrelationships of these four compounds. Unfortunately, in the same paper, Haynes retained the name of hemiphloin for the flavanone analogous to isovitexin and isohemiphloin for the flavanone analogous to vitexin. It is suggested that these names should be transposed. Hemiphloin has been oxidized to yield isovitexin (Hillis and Carle, 1963), and a comparison of the shifts of protons at C-6 and C-8 of hemiphloin and isohemiphloin in NMR analyses of these compounds and naringenin establishes the *C*-glycosyl position in isohemiphloin (Hillis and Horn, 1965), so the identities of these *C*-glycosyl flavonones appear to be on a firm basis.

Recently, the di-*C*-glycosyl substances similar to vicenin and lucenin have been investigated by the method of NMR spectroscopy (Seikel and Mabry, 1965; Hörhammer et al., 1965). The compounds are considered to be 6,8-di-*C*-glycosyl derivatives of

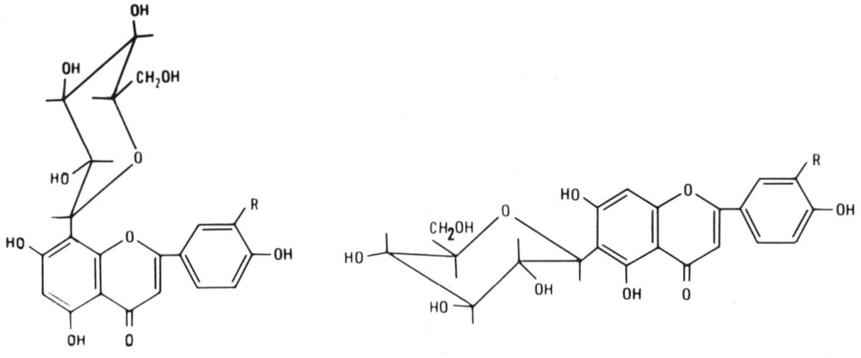

Fig. 2. Common C-glycosyl flavones. Left side: vitexin (R = H); and orientin (R = OH). Right side: isovitexin (R = H) and isoorientin (R = OH).

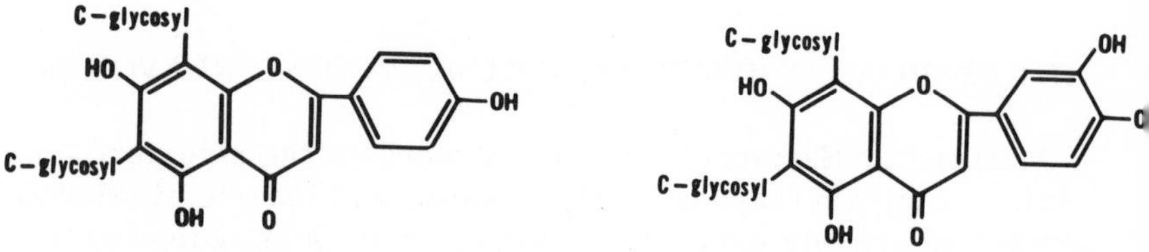

Fig. 3. Left side: violanthin (also, probably represents vicenin-1). Right side: lucenin-1.

apigenin (vicenin-1 and violanthin) and luteolin (lucenin-1), as indicated in Fig. 3. Complete structural details of the glycosidic substituents have not yet been published for these compounds.

It is not surprising that detailed analyses of mixed *C*- and *O*-glycosyl flavonoids have been relatively few in comparison with the wide variety of attacks on the simpler *C*-monosides. Recently, however, Horowitz and Gentili (1966a,b) have reported the structures *O*-*D*-xylosylvitexin from *Vitex lucens* and *Citrus sinensis*, and *p*-hydroxybenzoylvitexin from *Vitex lucens*. The structure of the former compound is interesting from the point of view of previous ideas concerning the site of *O*-glycosylation in *O*-glycosyl derivatives of isovitexin (saponaretin) in *Vitex lucens* For example, saponarin has been considered to be isovitexin 4′-*O*-glucoside. Two other compounds, namely lutonarin and isolutonarin, represent an analogous series derived from isoorientin.

In *O*-*D*-xylosylvitexin the xylose is attached to vitexin as a 2′-*O*-β-*D*-xylopyranoside (Fig. 4) (presumably the first appearance of a 2-*O*-xylosyl derivative of glucose). Horowitz and

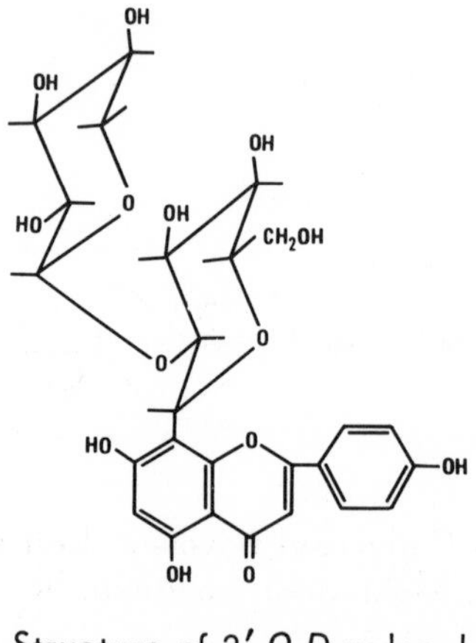

Fig. 4. Structure of 2′-*O*-*D*-xylosylvitexin.

Gentili used the NMR data to assign the above structure. Additionally the ultraviolet spectra of the xyloside in ethanol in the presence of sodium acetate, in sodium ethylate, and in aluminum chloride were identical with those of vitexin in these reagents, indicating that glycosylation did not involve a phenolic hydroxyl. This recent work suggests that a careful reappraisal of all assignments of the positions of *O*-glycosyl substituents of *C*-glycosyl flavonoids is necessary. However, there is no reason to consider that mixed *C*- and *O*-glycosides of a type involving a phenolic hydroxyl do not occur. Developments in this area of *C*-glycosyl flavonoid chemistry will be of great interest.

The second compound determined by Horowitz and Gentili, an acylated *C*-glycosyl flavonoid, is the first of its type to be characterized, although McClure (1964) suspected that acylated *C*-glycosyl flavonoids occurred in a number of species of Lemnaceae (e.g., in *Lemna perpusilla*). The fact that only small amounts of plant material can be provided in the types of sterile culture conditions used by McClure limited the methods of analysis to chromatographic techniques, ultraviolet spectroscopy, and controlled hydrolysis. According to McClure (personal communication), Seikel confirmed through similar analytical techniques the presence of acyl derivatives in material sent to her by him.

NMR spectra of derivatives of *p*-hydroxybenzoylvitexin indicate that again the substituent group is at the 2 position of the sugar.

Seikel, in particular, has called attention to the fact that several different types of vicenin-like and lucenin-like compounds occur, and this situation has been especially complicated in *Vitex lucens* wherein a total of eight compounds that seem to be 6,8-di-*C*-glycosyl derivatives of either apigenin or luteolin are present. Some of each flavone series are also interconvertible (Seikel et al., 1966). In the lucenin series (Table 3), a total of four compounds is separable by two-dimensional paper chromatography in *n*-butanol-acetic acid-water (4: 1: 5) and 3 percent NaCl. Lucenin-1 has been the most thoroughly investigated (Seikel and Mabry, 1965). This compound and lucenin-3 occur in *Vitex lucens* in about equal amounts. Lucenin-2 and lucenin-4 are minor components of *Vitex lucens*. A similar series based on vicenin occurs in this species, and similarly, vicenin-1 and vicenin-3 are major components while vicenin-2 and vicenin-4 are minor components. According to Seikel and co-workers all of the lucenin types have ultraviolet spectra closely similar to those of orientin (in the presence of the usual diagnostic reagents such as sodium

TABLE 3

Interconversion of Lucenin and Vicenin Types by Acid Hydrolysis[a]
[a]Modified from Seikel et al. (1966).

	Derivatives							
	Lucenin				Vicenin			
Compound Treated	1	2	3	4	1	2	3	4
Lucenin 1	+		+	+				
Lucenin 2		+						
Lucenin 3	+		+	+				
Lucenin 4	+		+	+				
Vicenin 1					+		+	+
Vicenin 2						+		
Vicenin 3					+		+	+
Vicenin 4					+		+	+

acetate, sodium ethylate, and aluminum chloride), and all of the vicenin types have ultraviolet spectra closely similar to those of vitexin. All available evidence indicates that the entire series of eight compounds is of the 6,8-di-*C*-glycosyl type. Acid hydrolysi (1 *N* HCl in 50 percent methanol) yields the results summarized i Table 3. Since vitexin or isovitexin yields an equilibrium mixtur in hot HCl (as do the analogous orientin and isoorientin), presumably by opening of ring C and recyclization, Seikel has used this known interconversion to postulate a basis for interpreting the results of acid treatment of vicenins and lucenins. If R_1 is the 6-*C*-glycosyl substituent and R_2 is the 8-*C*-glycosyl substituent, the following conclusions hold. If $R_1 = R_2$, no new derivatives should appear following acid treatment (this situation is typical of vicenin-2 and lucenin-2). If $R_1 \neq R_2$, then opening of ring C followed by recyclization through the alternative linkage would yield a different product. This situation would account for the interconvertibility of lucenin-1 and lucenin-3, for example, but would not readily explain the origin of lucenin-4.

It is noteworthy that Seikel reported that minor components, chromatographically identical with lucenin-2 and vicenin-2, were present in a crude extract containing violanthin provided by Hörhammer (see Hörhammer et al., 1965). Violanthin itself was not identical with any of the other vicenin types of substances. A sample of vicenin obtained from McClure, derived from *Lemna obscura*, was apparently identical with vicenin-2.

In our own chromatographic studies we have observed compounds having chromatographic and spectral properties of lucenin and vicenin in mosses, in *Psoralea*, in *Tragopogon*, and in *Iris*, in addition to the Lemnaceae. All of these compounds show an additional spot when chromatographed after acid treatment. This is true even for vicenin of *Lemna obscura*, which Seikel believes to be equivalent to vicenin-2. Furthermore, compounds of *Psoralea rigida* considered to be *O*-glucosidic derivatives of vicenin and lucenin (on the basis of the fact that they yield vicenin or lucenin upon hydrolysis with glucosidase) do not exhibit this double spotting, but still yield the typical di-*C*-glycoside upon hydrolysis. All of these results provide a somewhat confusing body of data about the nature of these compounds. It is possible that some type of isomerism involving the *C*-glycosyl substituents forms the basis for much of the conflicting data.

IV. BIOSYNTHETIC STUDIES

One of the chief interests in the biosynthesis of *C*-glycosyl flavonoids concerns the stage at which *C*-glycosylation occurs, and, ultimately, the details of the mechanism itself. A corollary of this problem is the sequence and mode of attachment of the sugars in di-*C*-glycosyl compounds such as vicenin and lucenin, and the time and mechanism of *O*-glycosylation in the mixed *C*- and *O*-glycosides.

Although there has been some speculation on the questions above concerning *C*-glycosylation, until recently little or no experimental work had been carried out relevant to these problems. Hörhammer and Wagner (1961) raised the question of whether *C*-glycosylation occurred before the synthesis of the aglycone, but they considered that no strong theoretical argument favored this alternative over *C*-glycosylation of the aglycone itself. Whalley (1961) considered *C*-glycosylation as most probably an ultimate step in biosynthesis but did not discuss the rationale of the prediction. Autoradiographic studies by Wallace (1967) (Wallace and Alston, 1966) have provided strong support for *C*-glycosylation prior to the formation of the flavone aglycone in orientin. This and related work will not be summarized.

The chemosystematic study of McClure (1964) showed that duckweeds (family Lemnaceae) were potentially quite favorable subjects for biosynthetic studies of *C*-glycosyl and other flavonoids. A notable advantage was the fact that each species had its own complement of flavonoids, permitting one to synthesize one type of radioactive flavonoid in one species and then isolate the compound for feeding experiments with another species which synthesizes a possible derivative of the first compound.

Van Dyke (1965) was the first to develop a system for the efficient incorporation of ^{14}C into *C*-glycosyl flavonoids of a member of the Lemnaceae. He added ^{14}C-labeled aromatic amino acids to the medium in which *Lemna perpusilla* was grown aseptically for seven days. By means of paper chromatography and autoradiography, the major *C*-glycosylflavonoids (vicenin and two presumed acylated derivatives of vicenin) were shown to be highly radioactive following exposure to ^{14}C-phenylalanine. This work showed that feeding experiments with these rootless l-plants were feasible.

The experiment relevant to the question of the time of *C*-glycosylation was carried out utilizing the system of Van Dyke with ^{14}C-labeled phenylalanine and *Spirodela polyrhiza.* This species produces two anthocyanins and at least four flavones. The major flavone components are orientin and luteolin 7-β-*D*-glucoside. When *S. polyrhiza* plants are grown under sterile conditions with ^{14}C-phenylalanine for seven days, all flavonoids are heavily labeled (Fig. 5). The specific activities of the four major flavones were approximately the same (Table 4). These results indicate that the uptake of phenylalanine into both *O*- and *C*-glycosides occurs with about equivalent efficiency—probably into

TABLE 4

Specific Activity × 10^{-3} μ C/mM

Orientin	1.49
Luteolin[a]	1.63
Vitexin	2.03
Apigenin[a]	2.47

[a]Obtained by enzymatic hydrolysis of the corresponding 7-β-*D*-glycosides in order to facilitate separation from the *C*-glycosyl types.

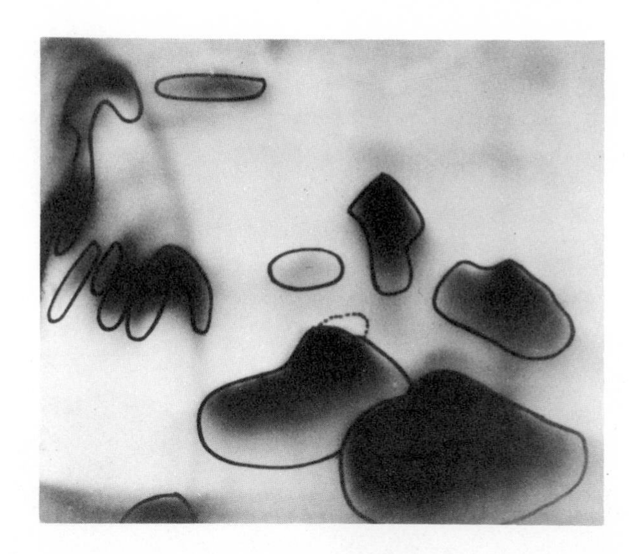

Fig. 5. Autoradiograph of chromatogram of extract from fronds of *Spirodela polyrhiza* clone 7003 grown in presence of ^{14}C phenylalanine. All major flavonoids are heavily labeled. (Courtesy of James Wallace.)

ring B. Radioactive luteolin was obtained from *S. polyrhiza* by enzymatic hydrolysis of luteolin 7-*β*-*D*-glucoside; the luteolin was then added to the medium as before, and a new culture of *S. polyrhiza* was grown in this medium for seven days. Flavonoids of this culture were extracted as before and chromatographed. Although the same array of flavonoids was produced, luteolin 7-*β*-*D*-glucoside was the only glycoside showing any definite radioactivity (Fig. 6). These results suggest that *O*-glycosylation, but not *C*-glycosylation, of the fully cyclized flavonoid occurs. These results do not indicate when *C*-glycosylation occurs, but it is possible that the chalcone analogue of luteolin would serve as a precursor. An analogous experiment utilizing ^{14}C-labeled apigenin gave generally similar results. An encouraging aspect of this work was the discovery that flavone aglycones could be absorbed by the living plant from the medium.

Preliminary results are available from other experiments involving (1) *O*-glycosylation of *C*-glycosyl compounds; (2) possible conversion of 8- to the 6-*C* glycosyl analogue, vitexin to isovitexin; and (3) *O*-methylation of *C*-glycosyl compounds such as isoorientin. The success of these experiments requires first that the glycosides enter the cell. For positive experiments successful conversion of the labeled precursor into another flavonoid

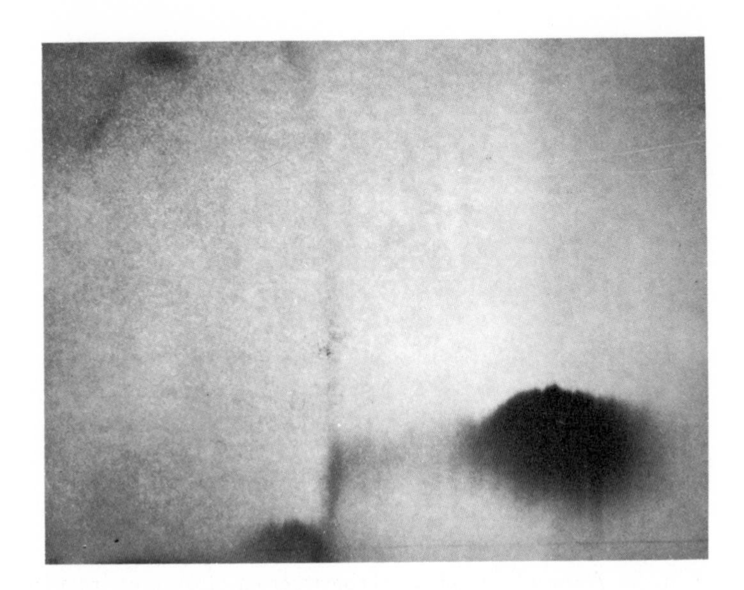

Fig. 6. Autoradiograph of chromatogram of extract from fronds of *Spirodela polyrhiza* grown in the presence of ^{14}C-luteolin in the culture medium (31 days exposure). Notice that the only glycoside labeled is luteolin-7-β-*D*-glucoside. Labeled orientin is not present, or at least it cannot be detected by this method.

is adequate. For negative experiments the radioactive substance must be shown to accumulate in the plant. Unequivocal proof of this accumulation may be rather difficult, but carefully washed plant material exposed to the medium for various periods of time indicated that considerable uptake of compounds such as orientin and saponaretin occurred.

For the study of possible *O*-glycosylation of a *C*-glycosyl flavonoid, labeled isovitexin was fed to a culture of *Lemna minor*, a species known to form both saponaretin and isosaponarin (and probably saponarin). In this experiment, labeled isovitexin was detected on autoradiographs given one month of exposure. These results are not considered to be unequivocal proof that *O*-glycosylation of the *C*-glycosyl flavone does not occur.

To test the possibility that a second *C*-glycosylation of a compound such as orientin could occur, labeled orientin was provided to a culture of *Lemna minor* (which makes lucenin). In this experiment there was no labeled lucenin detected on the autoradiograph after an exposure of one month.

It should be emphasized that the results of these experiments are not always considered to be conclusive, especially those that have given negative results. However, it is evident that the duckweeds offer a promising system to work with, which should provide further interesting results in the future.

V. PATTERNS OF ASSOCIATIONS OF *C*-GLYCOSYL FLAVONOIDS AND OTHER FLAVONOIDS

While only little can be deduced about *C*-glycosyl flavonoid chemistry from a consideration of the associations of these compounds in the plant, the topic is nevertheless an interesting one. Most observations will be based on direct knowledge of the flavonoid content of species that have been the objects of systematic investigations. *C*-Glycosyl flavonoids occur frequently alone (or with anthocyanins), or with *O*-glycosylated flavones or flavonols or both. When mixtures of *C*- and *O*-type flavone glycosides are present in a plant, the ring-B substitution patterns are similar. An excellent example is *Spirodela polyrhiza* discussed earlier. In nearly every instance when flavonols are also present, the latter are quite minor components. These generalizations may not prove to be valid when more extensive data are available.

Di-*C*-glycosyl types such as vicenin and lucenin are quite common and frequently these are the major flavonoids present. Next, are the 8-*C*-glycosyl types such as vitexin and orientin. Only in *Tragopogon* species have we found the 6-*C*-glycosyl types to predominate.

ADDENDUM

Chopin et al. (1966) have recently reported the synthesis, by means of 6-*C*-glucosylation of corresponding flavones, of several *C*-glucosyl flavonoids (e.g., 6-*C*-glucosyl derivatives of acacetin, benzyl-4′-apigenin, apigenin, dibenzyl-3′,4′-luteolin, and luteolin). The reaction was effected in the presence of acetobromglucose with a yield of *C*-glycosyl of less than 1 percent (in contrast to a large quantity of the *O*-glucosidic derivative). A small amount of *C*-8 glucoside accompanied the *C*-6 glucoside, but elsewhere these authors (Chopin et al., 1965) considered that the *C*-glucosylation, as well as the *C*-methylation, of acacetin using acetobromglucose was effected perhaps exclusively at position 6, with the subsequent formation of isocytisin (8-*C*-glucosyl acacetin) by isomerization.

The total structure of bayin (from the wood of *Castanospermum australe*) has been reported as 8-*C*-β-*D*-glucopyranosyl-7,4′-dihydroxyflavone (Eade et al., 1966). The 8-*C*-β-*D*-glucopyranosyl structure was confirmed previously by NMR evidence (Hillis and Horn, 1965) and Eade and co-workers established the remainder of the structure, notable as lacking the 5-oxygen function. It is also noteworthy that this species produces two isoflavones (formononetin and afromosin) which also lack the 5-hydroxyl group.

Kroschewsky, Markham and Mabry (unpublished work) at The University of Texas have isolated 7-*O*-methylisovitexin (swertisin) from *Tragopogon dubius* (Compositae), and the compound is known to occur in the species *T. pratensis*, *T. mirus*, and *T. miscellus*.

ACKNOWLEDGMENTS

The author is indebted to Prof. J. Chopin, Dr. W. E. Hillis, Dr. R. M. Horowitz, and Prof. H. Wagner for receipt of their papers prior to publication, and to Mr. James W. Wallace for use of unpublished data. National Institutes of Health Grant GM-11111 and NSF Grant 5448X are acknowledged with appreciation.

REFERENCES

Alston, R. E., and B. L. Turner. 1963. Biochemical Systematics, Englewood Cliffs, N. J., Prentice-Hall.

Barger, J. 1906. J. Chem. Soc., 89: 1210.

Bhatia, V. K., S. R. Gupta, and T. R. Seshadri. 1964. Curr. Sci. (India), 33: 581.

_____, S. R. Gupta, and T. R. Seshadri. 1965. Curr. Sci. (India), 22: 634.

_____, S. R. Gupta, and T. R. Seshadri. 1966. Tetrahedron, 22: 1147.

Brehm, B. G., and M. Ownbey. 1965. Amer. J. Bot., 52: 811.
Chopin, J. 1966. Les C-glycoflavonoides. *In* Mentzer, Ch., and Fatianoff, O., eds., Actualités de Phytochemie, Paris, Masson, in press.
_____, M. Bouillant, and A. Durix. 1965. C. R. Acad. Sci. (Paris), 260: 4850.
_____, A. Durix, and M. Bouillant. 1966. Tetrahedron Lett., 31: 3657.
_____, B. Roux, and M. A. Durix. 1964. C. R. Acad. Sci. (Paris), 13: 259.
Eade, R. E., I. Salasoo, and J. J. H. Simes. 1962. Chem. Industr.: 1720.
_____, I. Salasoo, and H. H. Simes. 1966. Aust. J. Chem., 19: 1717.
Fiedler, U. 1955. Arzneimittelforsch., 5: 609.
Funaoka, K., and M. Tanaka. 1957. Nippon Kokuzai Gakkaishi, 3: 144, 173, 218.
Hänsel, R., C. Leuckert, H. Rimpler, and K. D. Schaff. 1965. Phytochemistry, 4: 19.
Harborne, J. B. 1966. *In* Swain, T., ed, Comparative Phytochemistry, 271-295.
_____, and E. Hall. 1964. Phytochemistry, 3: 421.
Haynes, L. J. 1965. *In* Advances in Carbohydrate Chemistry, vol. 20, 357, New York, Academic Press.
Hillis, W. E., and H. Carle. 1963. Aust. J. Chem., 16: 147.
_____, and D. H. S. Horn. 1965. Aust. J. Chem., 18: 531.
_____, and D. H. S. Horn. 1966. Aust. J. Chem., 19: 705.
Hörhammer, L., and H. Wagner. 1961. *In* Ollis, W. D., ed., Chemistry of Natural Phenolic Compounds, London, Pergamon.
_____, H. Wagner, and W. Leeb. 1960. Arch. Pharm. (Weinheim), 293/65: 264.
_____, H. Wagner, and P. Beyersdorff. 1962. Naturwissenschaften, 49: 392B.
_____, H. Wagner, R. Rosprim, T. J. Mabry, and H. Rösler. 1965. Tetrahedron Lett., 22: 1707.
Horowitz, R. M. 1966. Unpublished.
_____, and B. Gentili. 1964. Chem. Industr.: 498.
_____, and B. Gentili. 1966a. Chem. Industr.: 625.
_____, and B. Gentili. 1966b. Abstract Sixth Annual Meeting of the Plant Phenolics Group (now Phytochemical Society) of North America, April 6-8, Austin, Texas.
Hrugasiewicz, K. 1964. Chem. Abstr., 60: 13094.

Jentzsch, K., P. Spiegel, and L. Fuchs. 1962. Planta Medica, 10: 1.
Jurd, L., T. A. Geissman, and M. Seikel. 1957. Arch. Biochem. Biophys., 67: 284.
Koeppen, B. H. 1963. J. Lab. Clin. Med., 000: 141.
_____, and D. G. Roux. 1965a. Tetrahedron Lett., 39: 3497.
_____, and D. G. Rous. 1965b. Biochem. J., 97: 444.
_____, C. J. B. Smit, and D. G. Roux. 1962. Biochem. J., 83: 507.
Komatsu, M., and T. Tomimori. 1966. Tetrahedron Lett., 16: 1611.
Kroschewsky, J. R. 1967. Ph.D. Dissertation. University of Texas, Austin.
Lewis, Y. S., and S. Neelakantan. 1964. Curr. Sci. (India), 33: 460.
McClure, J. B. 1964. Ph.D. Dissertation, University of Texas, Austin.
_____, and R. E. Alston. 1966. Amer. J. Bot., 53: 849.
Melchert, T. E., and R. E. Alston. 1965. Science, 150: 1170.
Nakaoki, T. 1944. J. Pharm. (Japan), 64: (11A): 57.
Ockendon, D., R. E. Alston, and K. Naifeh. 1966. Phytochemistry, 5:601.
Owenbey, M., and B. G. Brehm. 1965. Science, 150: 381.
_____, and A. Charles. 1962. C. R. Acad. Sci. (Paris), 254: 352.
_____, and P. Delaveau. 1962. C. R. Acad. Sci. (Paris), 254: 928.
Rao, C. B., and V. Venkateswarlu. 1956. Curr. Sci. (India), 25: 328.
Rimpler, H., L. Langhammer, and H. J. Frenzel. 1963. Planta Medica, 11: 325.
Sachs, H. 1963. Thesis, München.
Seikel, M. K. 1963. *In* Runeckles, V. C., ed., Aspects of Plant Phenolic Chemistry, 31, Toronto.
_____, and A. J. Bushnell. 1959. J. Org. Chem., 24: 1995.
_____, and T. A. Geissman. 1957. Arch. Biochem. Biophys., 71: 17.
_____, and T. J. Mabry. 1965. Tetrahedron Lett., 16: 1105.
_____, J. H. S. Chow, and L. Feldman. 1966. Phytochemistry, 5, in press.
Shibata, S., T. Murakami, Y. Nishikawa, and W. Budidarmo. 1962. Chem. Abstr., 56: 3564.
Shimokoriyama, M., and T. A. Geissman. 1962. Recent Prog. Chem. Nat. Symth. Colouring Matters, 255.

Stambouli, A., and R. Paris. 1961. Ann. Pharm. France., 19: 435.

Ueno, A., N. Oguri, K. Hori, Y. Saiki, and T. Harada. 1963. J. Pharm. Soc. Jap., 83: 420; 82: 1486.

Van Dyke, G. W. 1965. M. A. Thesis. University of Texas, Austin.

Wagner, H. 1966. *In* Swain, T., ed., Comparative Phytochemistry, 309, New York, Academic Press.

Wallace, J. W. 1967. PhD. Dissertation. University of Texas, Austin.

____, and R. E. Alston. 1966. Plant and Cell Physiology, 7:699.

Whalley, W. B. 1961. *In* Ollis, W. D., ed., Chemistry of Natural Phenolic Compounds, London, Pergamon.

10

NEW STRUCTURAL VARIANTS AMONG THE ISOFLAVONOID AND NEOFLAVANOID CLASSES*

W. D. OLLIS
Department of Chemistry
University of Sheffield
England

*This Chapter was submitted in June, 1966, and papers published since that date are not included in this review.

I. INTRODUCTION

The diversity of structural type associated with the flavonoids is considerable, and natural representatives of this group include flavones, flavonols, flavanones, flavanonols, anthocyanidins, catechins, and leucoanthocyanidins. It is also customary to include within the flavonoid class, the chalkones and the aurones. In contrast, the isoflavonoid group (Ollis, 1962) has until recently been restricted to a few variants such as the isoflavones, isoflavanones, rotenoids, pterocarpans (coumaranochromans), coumestones (coumaronocoumarins), 3-arylcoumarins, and angolensin (see Fig. 1).

The naturally occurring isoflavones and isoflavanones have been well reviewed by Dean (1963), but, although the first naturally occurring isoflavonoid, ononin, was described as early as 1855, by 1962 (Ollis) the number of known natural isoflavones was only twenty-five. Recently this number has been increased, but new isoflavones will only be discussed in this review where their co-occurrence with other isoflavonoids is of interest. Natural isoflavanones were first recognized in 1952 (Ollis, 1962) and their relation to the rotenoids (Crombie, 1963) is such that the rotenoids are usually regarded as members of the isoflavonoid group. For many years only two natural pterocarpans were known, but recently there have been some interesting discoveries among this class of natural products (see Section IV). As suggested by Harper, Kemp and Underwood (1965a, 1965b), the name pterocarpan has been adopted to replace coumaranochroman. In this connection the proposal (Livingston, Witt, Lundin, and Bickoff, 1964, 1965) of the name coumestan for the coumaronocoumarins is particularly unfortunate, and if the name coumaronocoumarin should be replaced, then we suggest that the term coumestone should be adopted. The continued use of the terms pterocarpan and coumestan to refer to two different isoflavonoid types (see Fig. 1) will surely cause confusion. The sole representative of the 3-arylcoumarins known to occur nat-

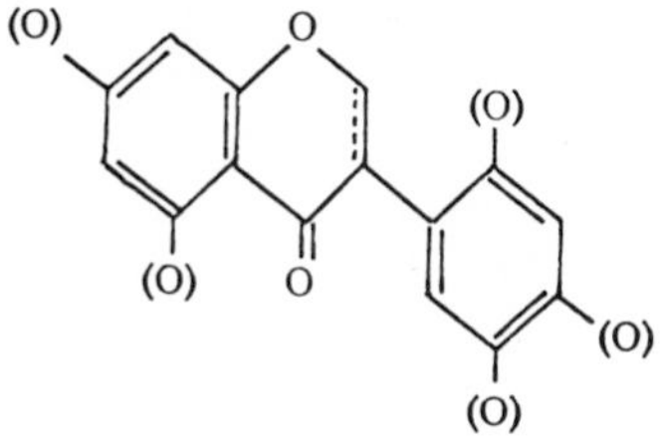

Isoflavones
Isoflavanones

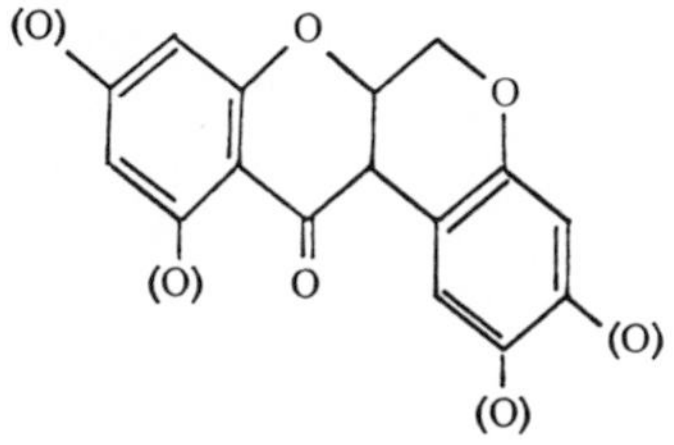

Rotenoids

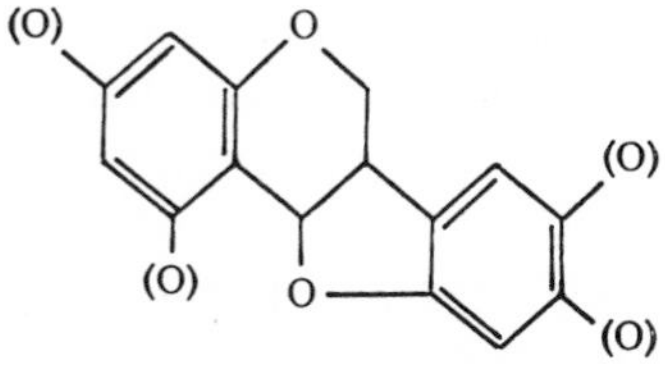

Pterocarpans
[Coumaranochromans]

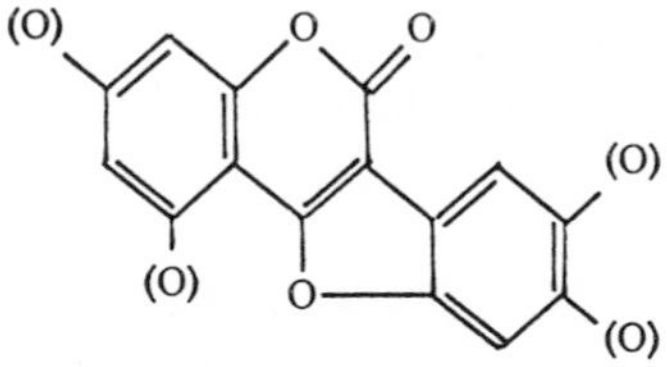

Coumestones
[Coumaronocoumarins]
[Coumestans]

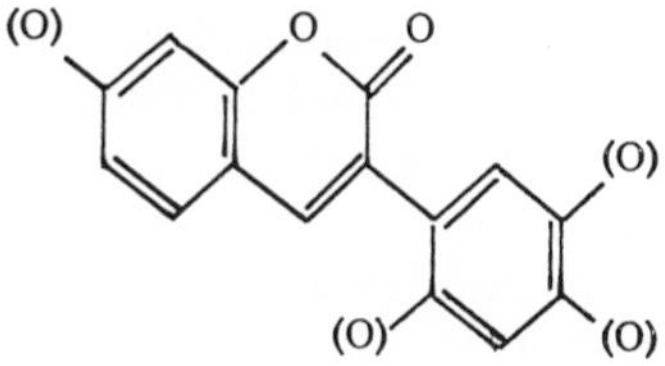

3-Arylcoumarins

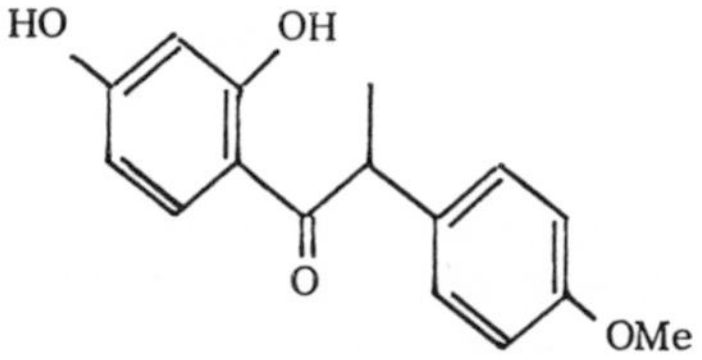

Angolensin

Fig. 1. Structural types among the isoflavonoid class of natural products (Ollis, 1962).

urally is pachyrrhizin (Simonitsch, Frei, and Schmid, 1957), and the only α-methyldeoxybenzoin at present known as a natural product is angolensin (Ollis, Ramsay, and Sutherland, 1965a).

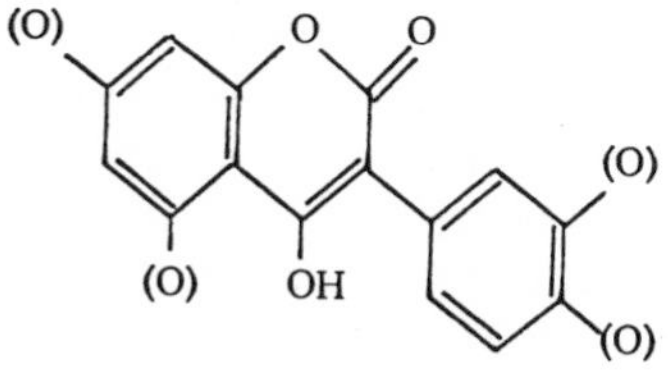

I. 3-Aryl-4-hydroxycoumarins

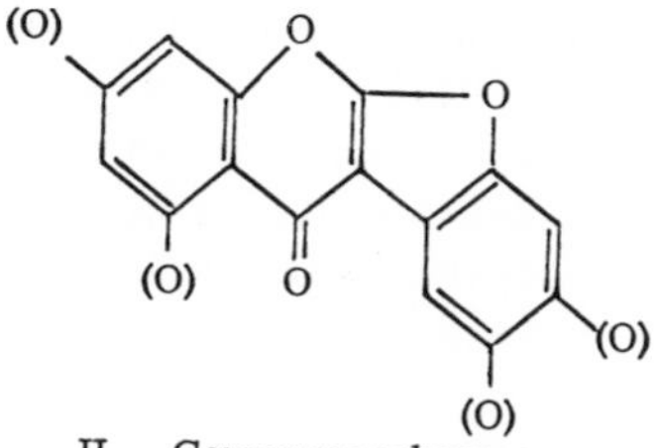

II. Coumaronochromone

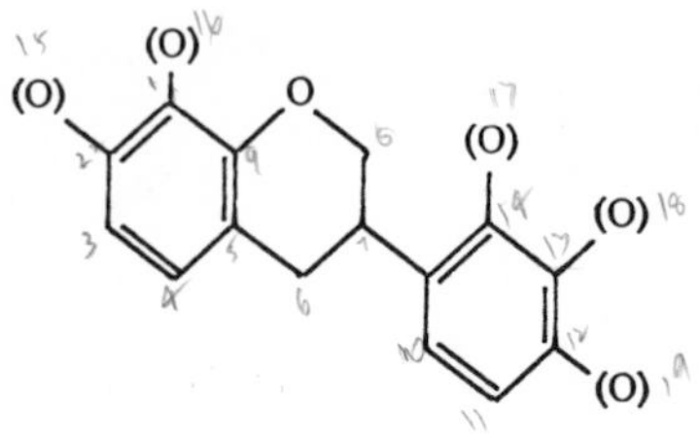

(O) (O) O (O) (O) (O)

III. Isoflavans

Recently discovered structural variants that may be included in the isoflavonoid class are the 3-aryl-4-hydroxycoumarins (I), a coumaronochromone (II), and several new isoflavans (III).

II. THE NATURAL OCCURRENCE OF 3-ARYL-4-HYDROXYCOUMARINS

Our interest in this class of natural products was a consequence of making an examination of the compound robustic acid. This compound was first isolated from *Derris robusta* in 1942 (Krishna and Ghose, 1942; Harper, 1942), but little information regarding its constitution was provided by subsequent structural studies (Clark, 1943; Subba Rao and Seshadri, 1946b; Sreerama Murti, Subba Rao, and Seshadri, 1948). We therefore undertook a further examination of this problem (East, 1963; Wheeler, 1963

The root material of *Derris robusta* yielded a complex mixture from which nine pure compounds (DR-1, -2, -3, -4, -5, -6, -7, -8, -9) were isolated by chromatography. Ultraviolet and infrared spectroscopic studies (Jurd, 1962; Ollis, 1962) indicated that DR-1, -2, -3, and -4 were related isoflavones and the NMR spectrum of DR-4 immediately identified its constitution as the known 5,7-dimethoxy-3′,4′-methylenedioxyisoflavone (IV) (Iyers, Shah, and Venkataraman, 1951).

The association of isoprenoid residues with the structures of many natural phenolic compounds is well known (Ollis and Sutherland, 1961), and the examination of NMR spectra (Burrows Ollis, and Jackman, 1960; Ollis, Ramsay, Sutherland, and Mongkolsuk, 1965b) often easily reveals the presence of isoprenoid units such as a $\gamma\gamma$-dimethylallyl group ($Me_2C{=}CH{-}CH_2{-}$) and a 2,2-dimethylchromene residue ($Me_2C(0){-}CH{=}CH{-}$). The NMR spectra of DR-1, -2, -3, and -4 indicated that they could be represented by the partial structures of Figure 2.

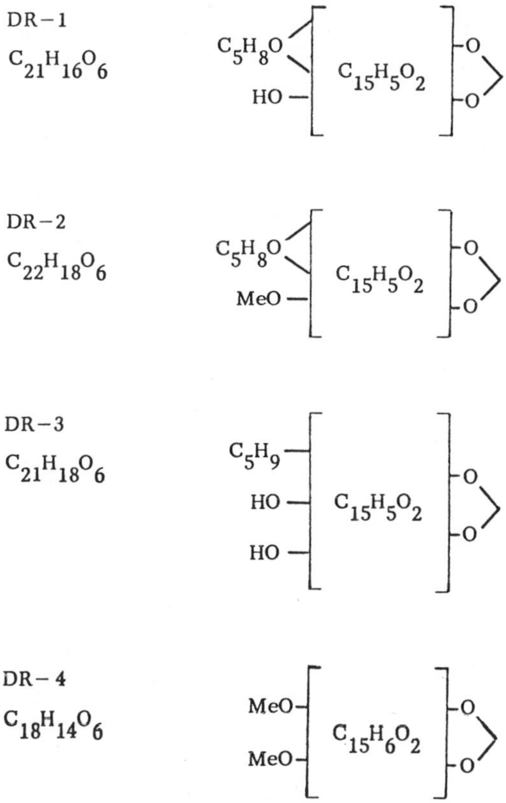

DR-1
$C_{21}H_{16}O_6$
C_5H_8O, HO — [$C_{15}H_5O_2$] —O, —O (methylenedioxy)

DR-2
$C_{22}H_{18}O_6$
C_5H_8O, MeO — [$C_{15}H_5O_2$] —O, —O (methylenedioxy)

DR-3
$C_{21}H_{18}O_6$
C_5H_9 —, HO —, HO — [$C_{15}H_5O_2$] —O, —O (methylenedioxy)

DR-4
$C_{18}H_{14}O_6$
MeO—, MeO— [$C_{15}H_6O_2$] —O, —O (methylenedioxy)

Fig. 2. The partial structures of four isoflavones from *Derris robusta*.

The constitutions of DR-1, -2, and -3 were easily established when the structural features shown in Figure 2 were recognised from their NMR spectra. The belief that DR-3 was an isoflavone related to its congener (IV) was fully supported by the NMR spectra of DR-3 diacetate and DR-3 dimethyl ether, which showed, *inter alia*, signals characteristic of a piperonyl [ABX system (3 protons) and a singlet (2 protons)] and a $\gamma\gamma$-dimethylallyl (C_5H_9) group. Assuming that DR-3 and DR-4 had similar oxygenation patterns, two possible structures for DR-3, with the $\gamma\gamma$-dimethylallyl group in either position 6 or position 8, could be considered. The former structure (V) for DR-3 was established since mild acid-catalysed cyclisation of DR-3 gave a mixture of isomers; these isomers, (VI and VIIa), were clearly differentiated spectroscopically. Alkaline hydrogen peroxide oxidation of the methyl ether (VIIb) gave piperonylic acid and a salicylic acid, $C_{13}H_{16}O_5$, which was undoubtedly the compound (IX). Methylation of DR-1 gave DR-2 and reduction of DR-2 gave the methyl ether VIIb. These results established the constitutions of four new isoflavones isolated from *Derris robusta* as DR-1 (VIIIa), DR-2 (VIIIb), DR-3 (V), and DR-4 (IV).

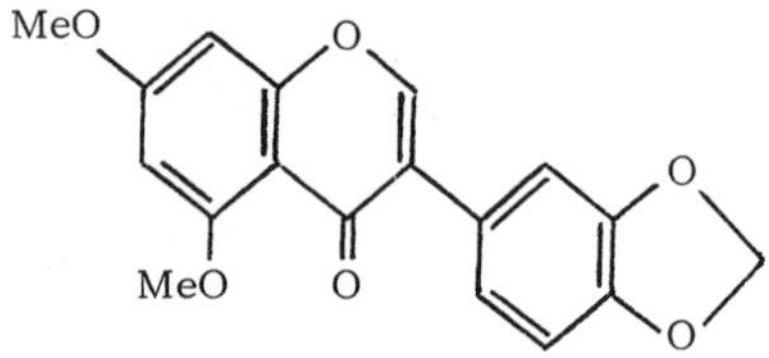

MeO
MeO O O O

IV. DR–4

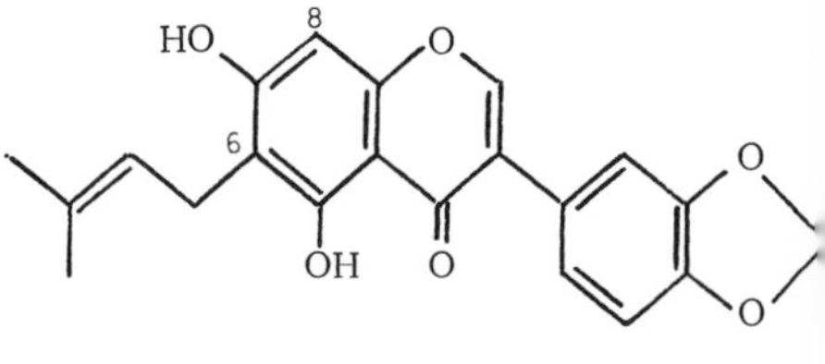

HO 8 O 6 O
OH O O

V. DR–3

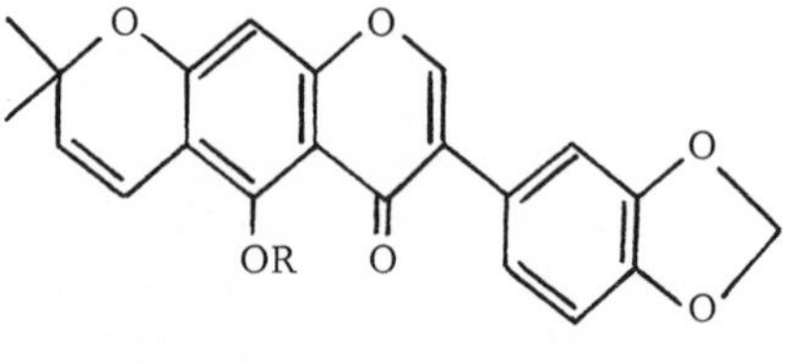

O O O
OR O O

VIIIa. R = H; DR–1

VIIIb. R = Me; DR–2

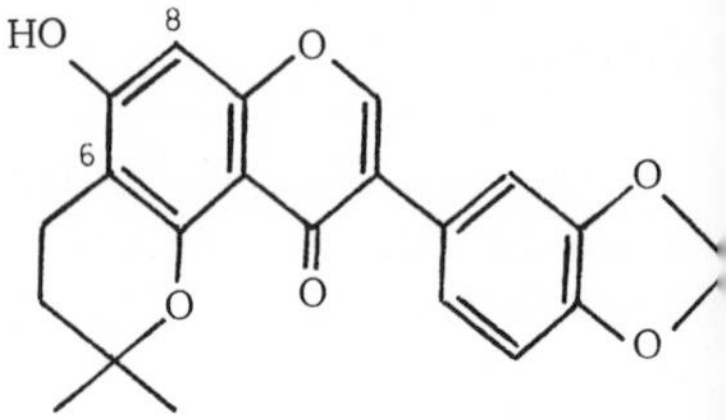

HO 8 O 6 O
O O O

VI

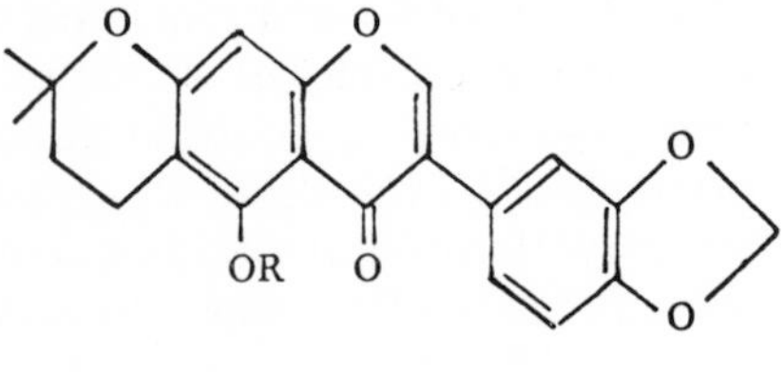

O O O
OR O O

VIIa. R = H

VIIb. R = Me

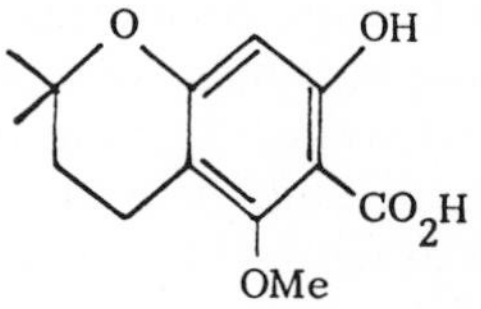

O OH
CO2H
OMe

IX

Having settled the constitutions of the four isoflavones, it was now possible to consider the nature of the five congeners, DR-5, -6, -7, -8, and -9. These five compounds all showed similar absorption ($\nu_{C=O}$ 1705–1715 cm^{-1}) in the carbonyl region of their infrared spectra and the partial structures of DR-5, -6, and -7 summarised in Figure 3 followed from their NMR spectra. The amounts of DR-8 and DR-9 that were available at this stage precluded examination of their NMR spectra.

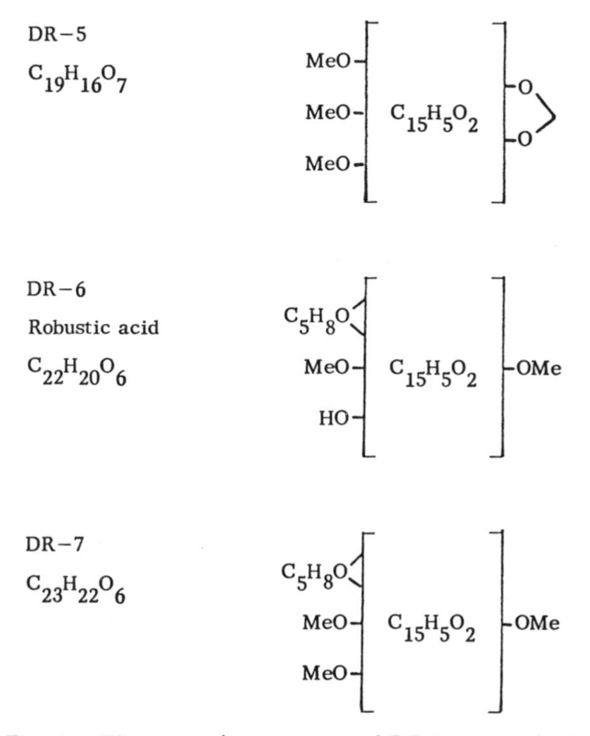

Fig. 3. The partial structures of DR-5, -6, and -7.

The infrared spectral characteristics of DR-5 suggested that it was a coumarin derivative and this, and the presence of many common features in the NMR spectra of DR-4 (IV) and DR-5 (see Fig. 2), suggested that DR-5 had the structure (XII). This was confirmed by partial synthesis. The deoxybenzoin (X) obtained from DR-4 (IV) by mild alkaline hydrolysis gave with methoxycarbonyl chloride (Gilbert, McGookin, and Robertson, 1957) the 3-aryl-4-hydroxycoumarin XI, which on methylation gave DR-5 (XII).

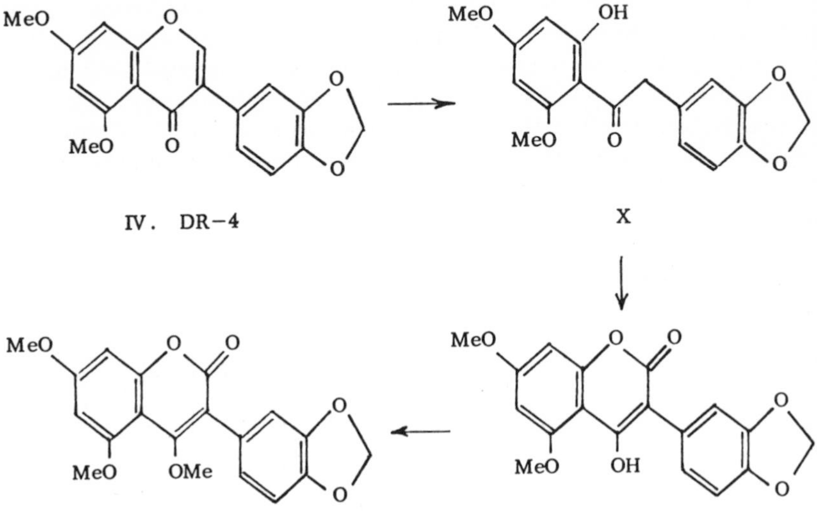

IV. DR–4

X

XI

XII. DR–5

The constitutional relation of the other compounds to the proven structure of DR-5 (XII) was now considered. pK_a' determination established that robustic acid (DR-6) had similar acidity properties to the 4-hydroxycoumarin (XI) and this evidence, coupled with the NMR spectra (see Fig. 3) of its acetate and methyl ether, suggested that robustic acid had the structure (XIII). This proposal was fully confirmed by the degradation of the dihydro derivative of robustic acid (XV) by alkaline hydrogen peroxide, which yielded the acid (IX) and *p*-anisic acid. Methylation of robustic acid (DR-6, XIII) gave DR-7, thus settling its constitution (XIV).

The compounds DR-8 and DR-9 were available only in very small amounts, so that initially our opinion regarding their possible structures was dictated by biogenetic analogy and their ultraviolet and infrared spectra. These suggested that DR-8 and DR-9 were related to robustic acid (XIII) and its methyl ether (XIV), so the structures (XVI and XVII) were considered. This hypothesis was confirmed by partial synthesis. Alkaline hydrolysis of DR-2 (VIIIb) gave the deoxybenzoin (XVIII), and this with methoxycarbonyl chloride gave DR-8 (XVI). Methylation of DR-8 gave DR-9 (XVII).

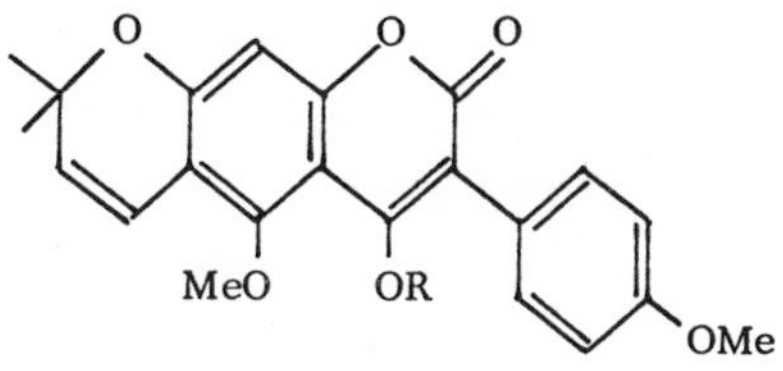

XIII. R = H; DR-6 Robustic acid

XIV. R = Me; DR-7 Robustic acid methyl ether

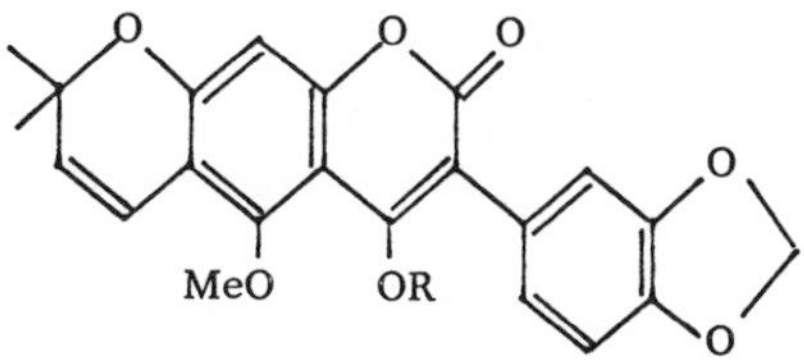

XVI. R = H; DR-8

XVII. R = Me; DR-9

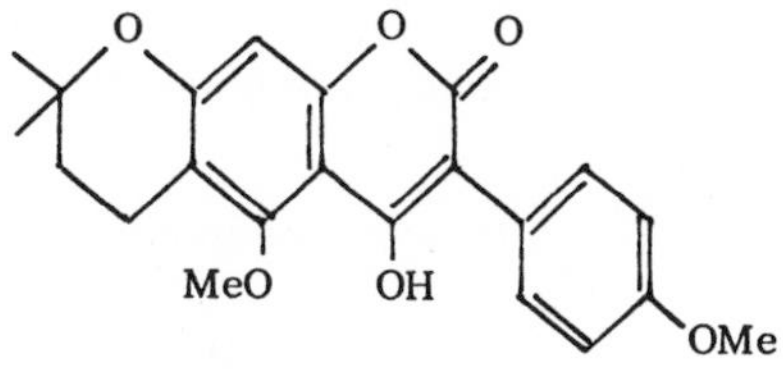

XV

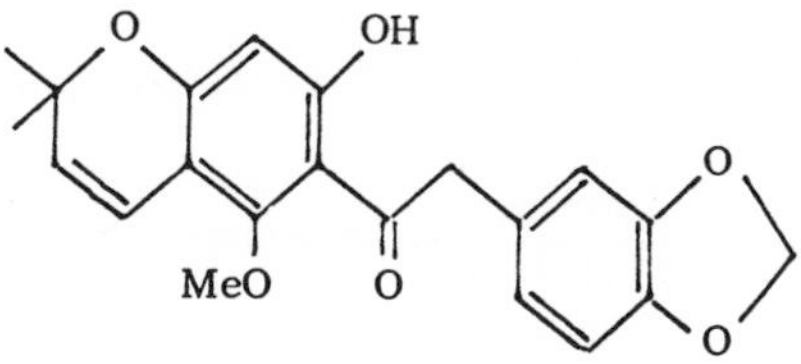

XVIII

This investigation of the nine natural products isolated from *Derris robusta* emphasises the way in which the application of modern physical methods permits the selection of chemical reactions that may be expected to provide the maximum amount of structural information. It is also clear that structure determination is considerably facilitated when a number of structurally related natural products is isolated from a single plant source. This task of isolation has been made much easier by modern chromatographic methods. Fourteen chemical reactions (Table 1) were used to establish the constitutions of the nine natural products isolated from *Derris robusta*.

There is a satisfying relation between the structures of the four isoflavones and the five 3-aryl-4-hydroxy(methoxy)coumarins isolated from *Derris robusta*. This relation suggests that there may well be common features in the biosynthetic programs (see Fig. 4) leading to these two types of natural product.

Lonchocarpic acid and scandenin are two natural products that have been known for a number of years (Jones, 1934; Jones and Haller, 1943; Clark, 1943), but, although a structural similarity between these two compounds and robustic acid was suspected (Subba Rao and Seshadri, 1946a; Sreerama Murti, Subba Rao, and Seshadri, 1948), little progress had been made (Subba Rao and

TABLE 1

Chemical Transformations Used to Interrelate the Nine Natural Products Isolated from *Derris Robusta*

DR-1 —Methylation→ DR-2

DR-2 ↓ Hydrogenation

Dihydro-DR-2 —Oxidation→ $C_{13}H_{16}O_5$ (IX)

↑ Methylation

DR-3 —Acid→ Low melting point isomer + High melting point isome

DR-4 —Alkaline hydrolysis→ Deoxybenzoin —$Cl \cdot CO_2Me$→ 4-Hydroxycoumarin —Methylation→ DR-5

Robustic acid (DR-6) —Methylation→ DR-7 —Hydrogenation→ Dihydro-DR-7 —Oxidation→ $C_{13}H_{16}O_5$ (IX)

DR-2 —Alkaline hydrolysis→ Deoxybenzoin —$Cl \cdot CO_2Me$→ DR-8 —Methylation→ DR-

Khan, 1961, 1963) toward the definition of their full structures. The recognition (See Fig. 4) of 3-aryl-4-hydroxycoumarins as a new structure variant among the isoflavonoid group encouraged us to consider the interpretation of these earlier studies.

Lonchocarpic acid and scandenin were known to be isomers, $C_{26}H_{26}O_6$, and since they formed diacetates, dimethyl ethers, and monomethyl ethers, they could be represented by the partial structure $C_{25}H_{21}O_3(OH)_2(OMe)$. The tetrahydro derivatives of scandenin and lonchocarpic acid also yielded similar derivatives The chemical behaviour and spectroscopic properties of scanden and lonchocarpic acid that had been reported emphasised their structural relation to robustic acid. Oxidation of scandenin and lonchocarpic acid had been shown to yield *p*-hydroxybenzoic acid The structures (XIX) and (XX) could therefore be given serious

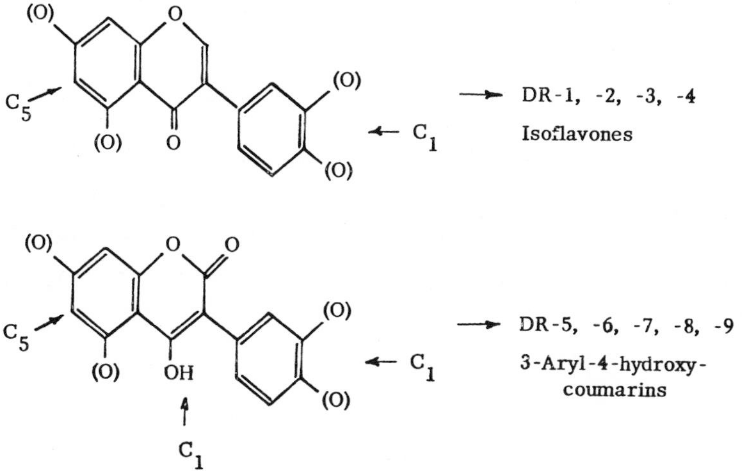

→ DR-1, -2, -3, -4
Isoflavones

→ DR-5, -6, -7, -8, -9
3-Aryl-4-hydroxy-coumarins

Fig. 4. The related biosynthetic programs leading to isoflavones and 3-aryl-4-hydroxycoumarins.

consideration immediately for lonchocarpic acid and scandenin, but no evidence had been published that permitted final assignments. Thus, our starting point for the consideration of the constitutions of lonchocarpic acid and scandenin was the structures (XIX) and (XX), which were in accord with previously published evidence and the biogenetically acceptable assumption (Ollis and Sutherland, 1961) that they both contained a $\gamma\gamma$-dimethylallyl substituent in association with a robustic acid (XIII) type of structure.

The problem of discriminating between the two structures (XIX and XX) for lonchocarpic acid and scandenin was similar in form to the situation that we had already encountered in our structural study of sericetin (XXI) isolated from *Mundulea sericea* (Burrows, Ollis, and Jackman, 1960; Ollis, 1961).

Our structural investigation of the natural products isolated from *Derris robusta* was made much easier by opportunities for interrelation (see Table 1) provided by the isolation of nine structurally related natural products, and it was therefore decided to make a detailed study of *Derris scandens* in the hope of isolating some isoflavones as well as 3-aryl-4-hydroxycoumarins. In this endeavour we derived considerable assistance from Professor N. V. Subba Rao (Osmania University, Hyderabad, India), and the results of this collaboration are now reported.

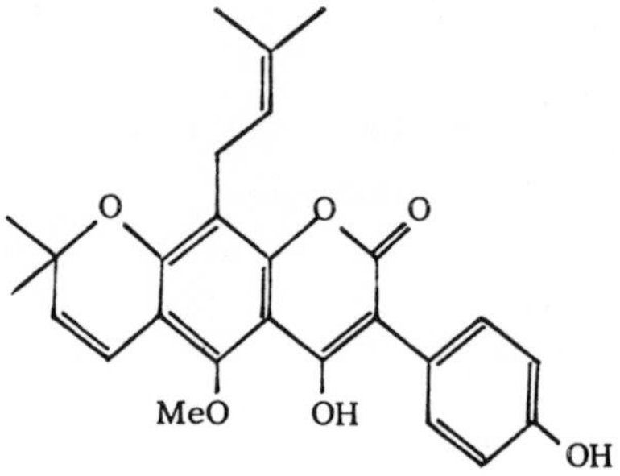

XIX

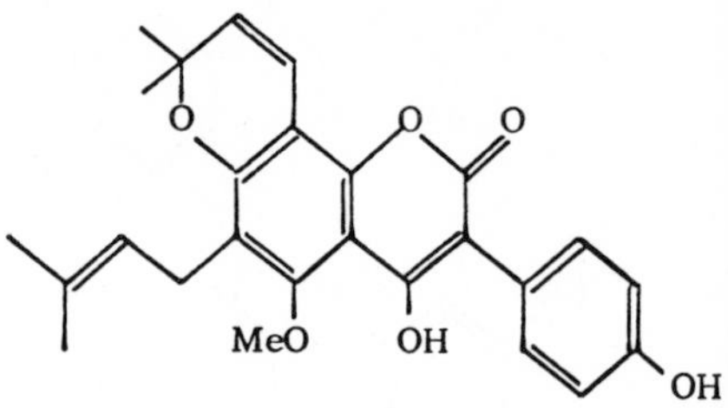

XX

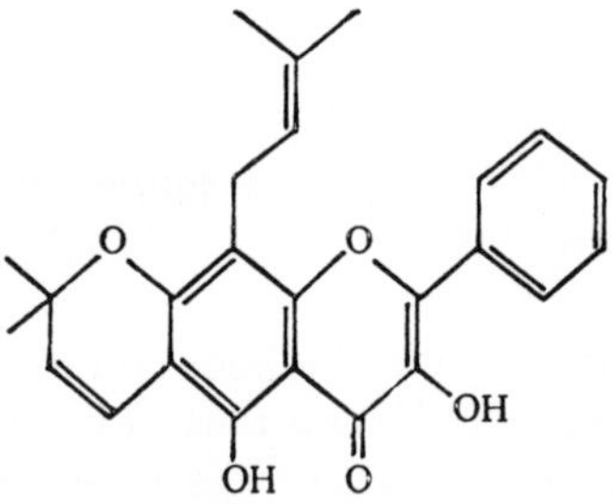

XXI Sericetin

Two varieties of *Derris scandens* were examined. One was collected from the Warangal region and the other from the Chanda District (Central Provinces). The substances isolated from these two sources are listed in Table 2.

TABLE 2

Natural Products Isolated from *Derris scandens*

Ch. and W.	Scandenin	$C_{26}H_{26}O_6$
Ch. and W.	Lonchocarpic acid	$C_{26}H_{26}O_6$
Ch.	Compound DS-1	$C_{27}H_{2} O_6$
W.	Warangalone	$C_{25}H_{24}O_5$
Ch.	Chandalone	$C_{25}H_{24}O_5$

(Ch. = Chanda variety; W = Warangal variety)

Our investigation (Harmer, 1965) of the five natural products isolated from *Derris scandens* (Table 2) may be summarized as follows. By the application of the standard spectroscopic methods, including particularly NMR spectroscopy and high resolution mass spectrometry (Falshaw, 1965), the substances warangalone and chandalone were recognised as isomeric isoflavones, $C_{25}H_{24}O_5$, whereas scandenin, lonchocarpic acid, and DS-1 were obviously derivatives of 3-aryl-4-hydroxycoumarin. It was decided to try to settle the constitution of the isoflavones first and then take advantage of the standard methods used in the *Derris robusta* investigation to convert these isoflavones into the corresponding 3-aryl-4-hydroxycoumarins. It was soon established that warangalone and chandalone could be represented by the partial structure (XXII) and scandenin and lonchocarpic acid by

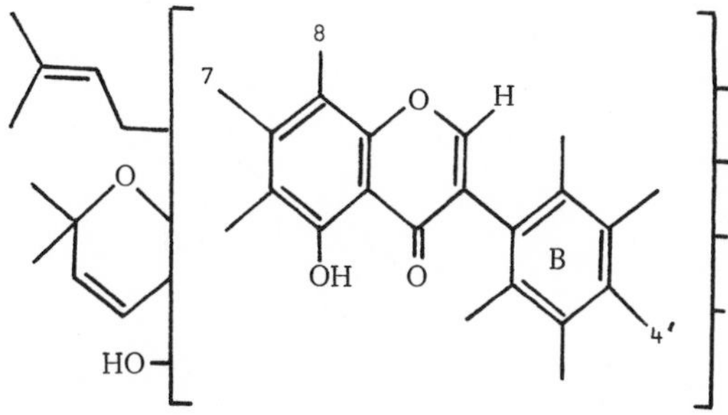

XXII. Partial structure of chandalone and warangalone

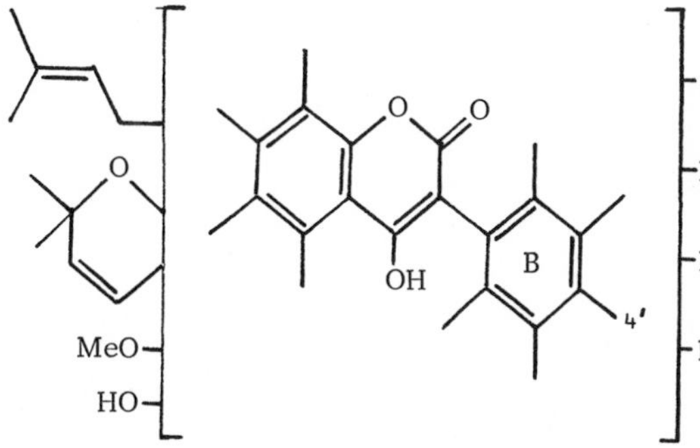

XXIII. Partial structure of scandenin and lonchocarpic acid

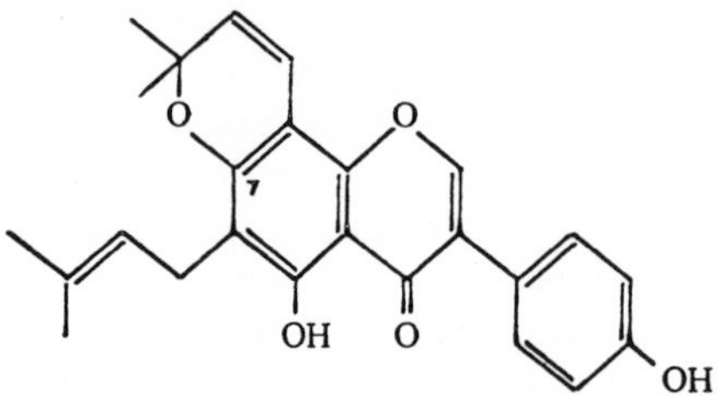

XXIV. Osajin

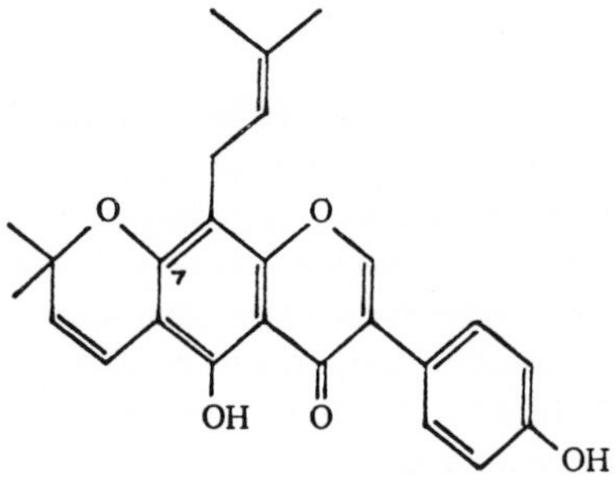

XXV. Warangalone

the partial structure (XXIII). DS-1 was available in such small amounts that detailed investigation was not possible at this stage.

The derivation of the full structure of warangalone involved the following argument. The NMR spectrum of warangalone and its derivatives required that the four aromatic protons be placed on ring B; they formed an A_2B_2 system and this and their chem-

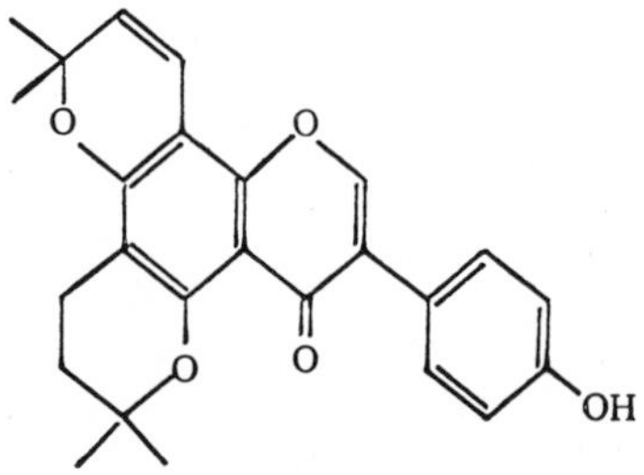

XXVI

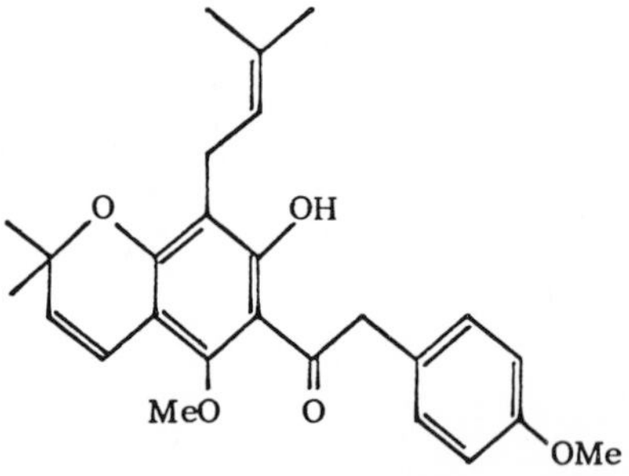

XXVII

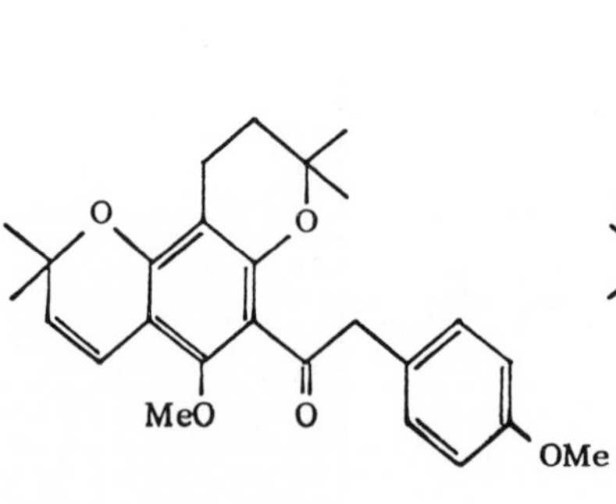

XXVIII

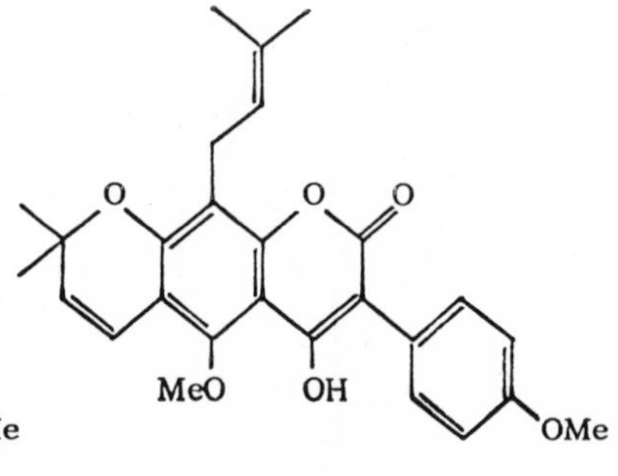

XXIX. DS-1

ical shifts required placing the phenolic hydroxyl in position 4′. Thus, subject to the biogenetic restriction that the 2,2-dimethylchromene oxygen was placed in position 7, two structures (XXIV or XXV) required consideration for warangalone. One of these could be immediately excluded since the structure (XXIV) was already allocated to osajin. One of the classical investigations in natural product chemistry is the study by Wolfrom and his colleagues (Wolfrom and Wildi, 1951) of osajin (XXIV) and pomiferin isolated from *Maclura pomifera*. These studies extended over 1939–1951 and it is clear how the availability of physical methods has completely altered the character and tempo of such studies.

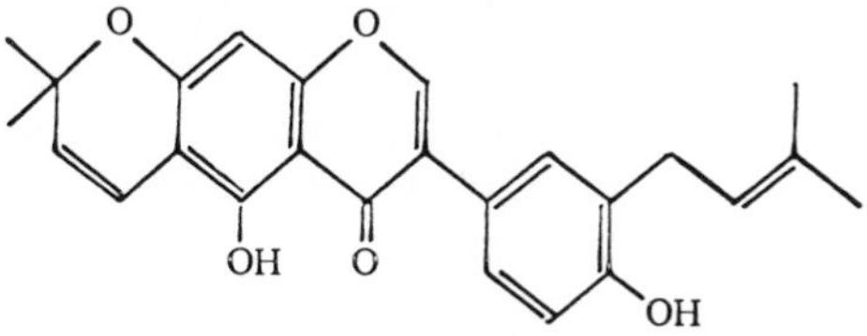

O O OH O OH

XXX. Chandalone

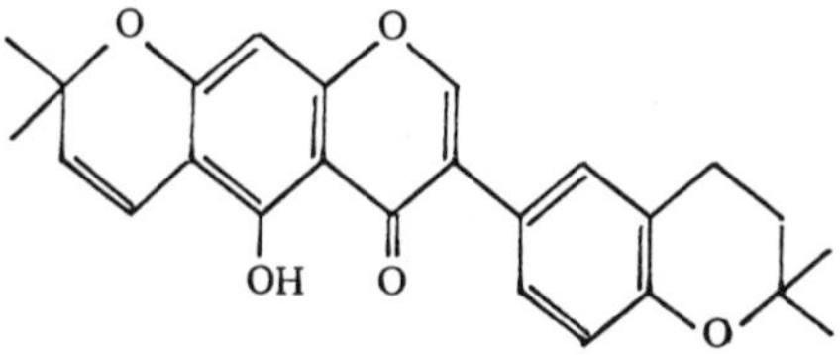

O O OH O O

XXXI

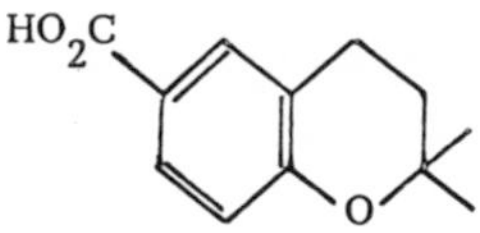

HO_2C O

XXXII

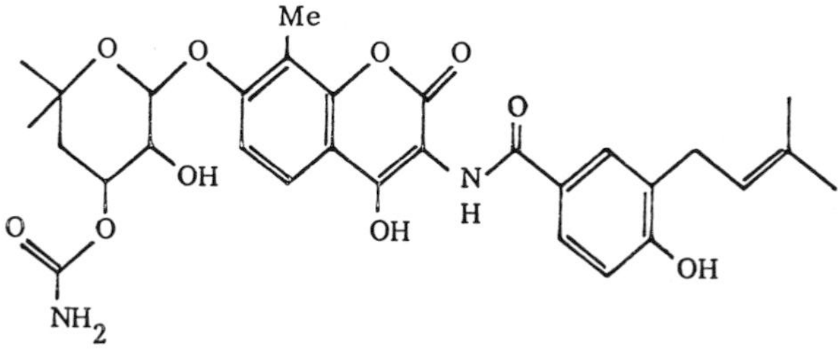

Me O O O O O OH N H OH OH O O NH_2

XXXIII. Novobiocin

The constitution of osajin is compatible with its acid-catalyzed transformation into the isomer (XXVI), whereas under similar conditions warangalone was unaffected. The NMR spectrum of osajin (Burrows, Ollis, and Jackman, 1960) was very similar to that of warangalone and the structure of warangalone (XXV) was fully confirmed by the alkaline hydrolysis of warangalone dimethyl ether to the deoxybenzoin (XXVII), which was cyclized by acid to the ketone (XXVIII).

The reaction of the deoxybenzoin (XXVII) with methoxycarbonyl chloride provided an unexpected bonus in that the product (XXIX) was identical with the natural product, DS-1, which was available only in trace amounts from *Derris scandens.*

Although the general features of the NMR spectra of chandalone and its derivatives were similar to those of warangalone (XXV), there were significant differences. Thus, the four aromatic protons (see partial structure XXII) of chandalone constituted an ABX system plus a singlet. This singlet in chandalone and its derivatives had chemical shift characteristics that were very similar to those of DR-1 (VIIIa) and its derivatives. Thus, the structure (XXX) could be confidently considered for chandalone. This was confirmed by its acid-catalyzed isomerization, giving the 2,2-dimethylchroman XXXI, which, with alkaline hydrogen peroxide, gave an acid, $C_{12}H_{14}O_3$. This acid was shown to be identical with the 2,2-dimethylchroman-6-carboxylic acid (XXXII) previously obtained during investigations of the antibiotic novobiocin (XXXIII) (Hinman, Caron, and Hocksema, 1957).

The availability of warangalone of proven structure (XXV) now made it possible to comment on the possible structures (XIX and XX) of lonchocarpic acid and scandenin. The transformation of warangalone (XXV) to DS-1 (XXIX) has already been described and this partially synthetic DS-1 gave, on methylation, a monomethyl ether identical with lonchocarpic acid dimethyl ether. This, in association with the evidence already discussed,

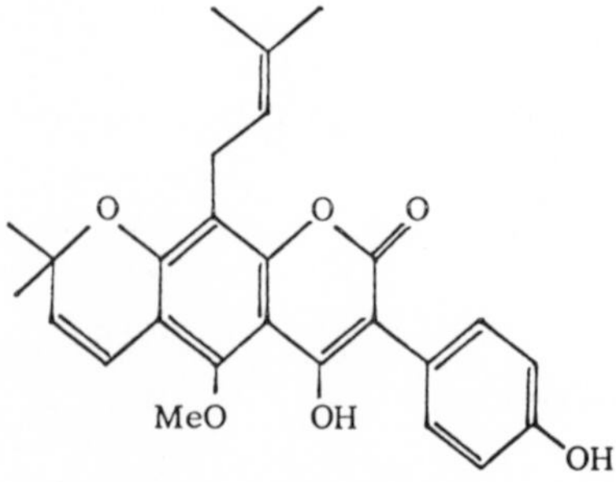

XIX. Lonchocarpic acid

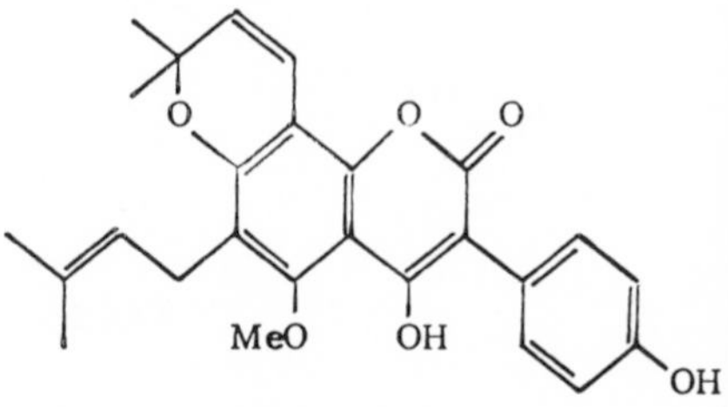

XX. Scandenin

established the constitution of lonchocarpic acid as (XIX), from which it follows that scandenin has the constitution (XX).

The constitutions of lonchocarpic acid, scandenin, and robustic acid have also been examined by Pelter and his colleagues (Johnson, Pelter, and Stainton, 1966). Their structural arguments are in a number of respects different but complementary to ours, and they reach the same structural conclusions regarding these three 3-aryl-4-hydroxycoumarins. They have not, however, reported the natural co-occurrence of the 3-aryl-4-hydroxycoumarins with the structurally corresponding isoflavones. This co-occurrence is certainly of biogenetic interest and there can be little doubt that the biosyntheses of isoflavones and 3-aryl-4-hydroxycoumarins follow similar pathways (see Fig. 4).

III. THE BIOGENETIC RELATIONS BETWEEN ISOFLAVONES, COUMARONOCHROMONES, AND 3-ARYL-4-HYDROXYCOUMARINS

The extractives of *Piscidia erythrina* provide an interesting group of structurally related natural products including piscidic acid, four isoflavones, four rotenoids, and a new type of isoflavonoid called lisetin (Falshaw, Ollis, Moore, and Magnus, 1966). The isoflavones are jamaicin (XXXIV; Stamm, Schmid, and Büchi, 1958), ichthynone (XXXV; Schwarz, Cohen, Ollis, Kaczka, and Jackman, 1964; Dyke, Ollis, Saintsbury, and Schwarz, 1964), and two new members of this group, piscerythrone (XXXVI) and piscidone (XXXVII) (Falshaw and Ollis, 1966; Falshaw, Ollis, Moore and Magnus, 1966). The rotenoids are rotenone (XXXVIII), sumatrol (XXXIX), millettone (XL), and isomillettone (XLI). This plant also yields the 6a,12a-dehydrorotenoid dehydromillettone (XLII) directly.

In addition to these compounds (XXXIV–XLII), *Piscidia erythrina* also yields lisetin (XLIII), which represents a new type of isoflavonoid, the coumaronochromones (II) (Falshaw and Ollis, 1966). Lisetin is structurally related to the isoflavone piscerythrone (XXXVI), and this relationship is emphasised by the transformation of piscerythrone to lisetin (XLIII) in good yield by oxidation with alkaline potassium ferricyanide. This transformation may be represented as an oxidative coupling process involving the diradical (XLIV) and the tautomeric cyclohexadienone (XLV), giving (XLV, arrows) lisetin (XLIII).

Lisetin (XLIII) may be transformed into the 3-aryl-4-hydroxycoumarins (XLVI), and (XLVII) by the following processes.

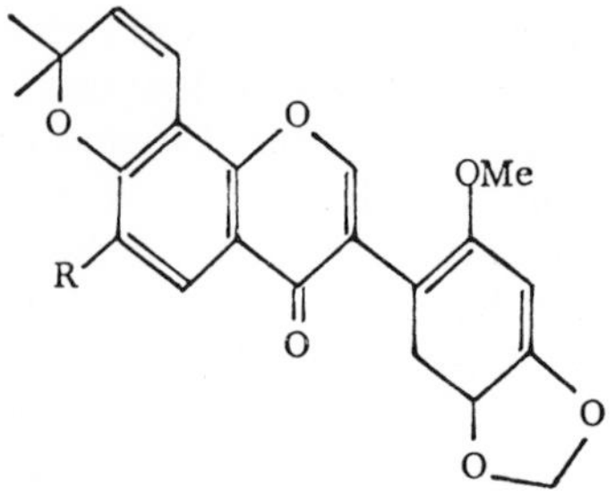

XXXIV. R = H Jamaicin
XXXV. R = OMe Ichthynone

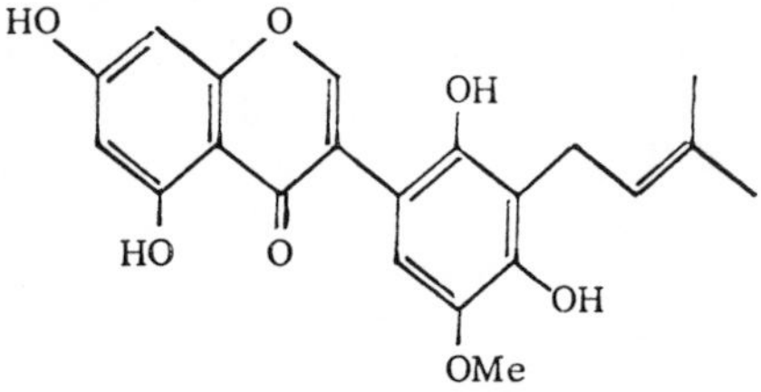

XXXVI. Piscerythrone

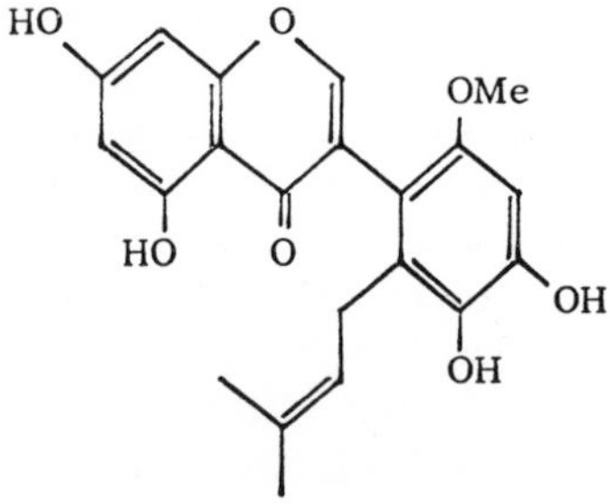

XXXVII. Piscidone

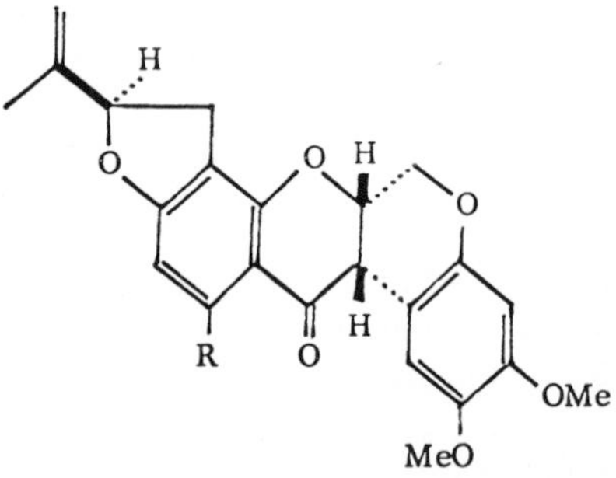

XXXVIII. R = H Rotenone
XXXIX. R = OH Sumatrol

Alkaline hydrolysis of lisetin trimethyl ether gives the 4-hydroxycoumarin (XLVI) and similarly isolisetin dimethyl ether gives the compound (XLVII). Isolisetin is formed from lisetin by acid-catalysed cyclisation.

The formation of lisetin by oxidation of piscerythrone and the transformation of lisetin into derivatives of 3-aryl-4-hydroxycoumarins raises the interesting question of whether these chemical processes may be regarded as models for biosynthetic processes.

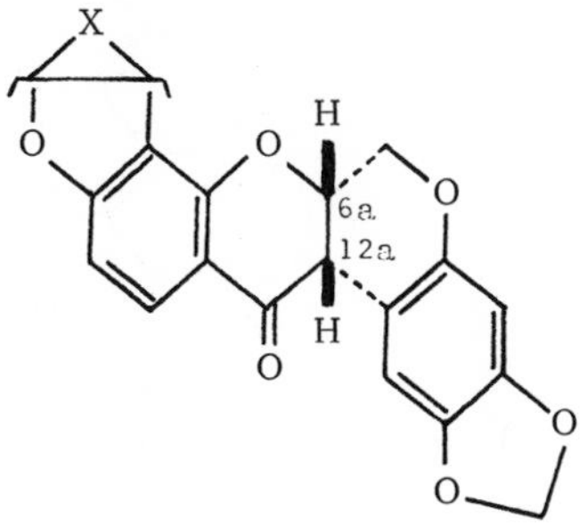

XL. X = $Me_2C-CH=CH$. Millettone

XLI. X = $CH_2=CMe-CH-CH_2$. Isomillettone

XLII. cf. XL with $C_{6a}-C_{12a}$ double bond. Dehydromillettone

Our studies on *Derris robusta* and *Derris scandens* demonstrate the natural co-occurrence of isoflavones and 3-aryl-4-hydroxycoumarins of corresponding structural type. It is possi-

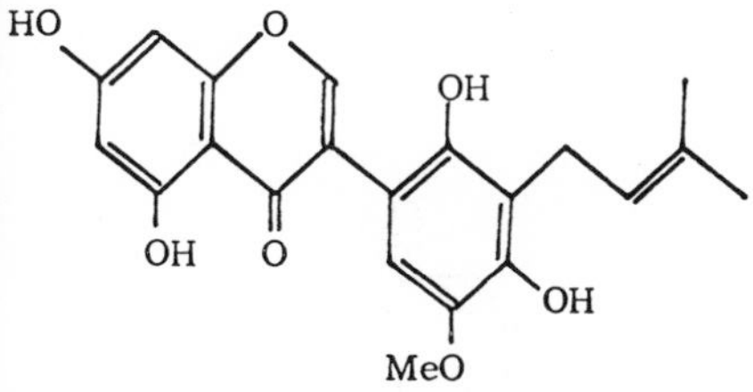

XXXVI. Piscerythrone

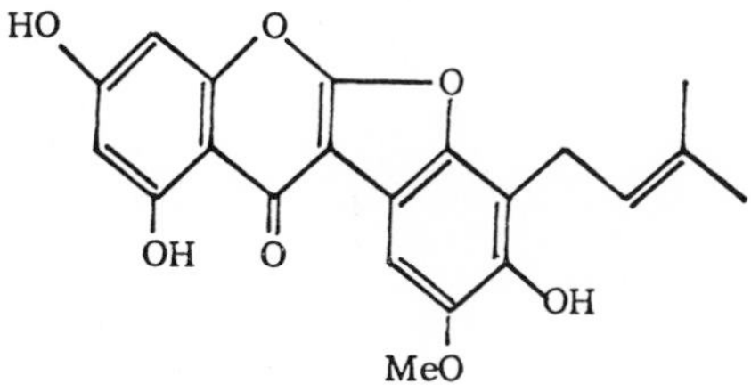

XLIII. Lisetin

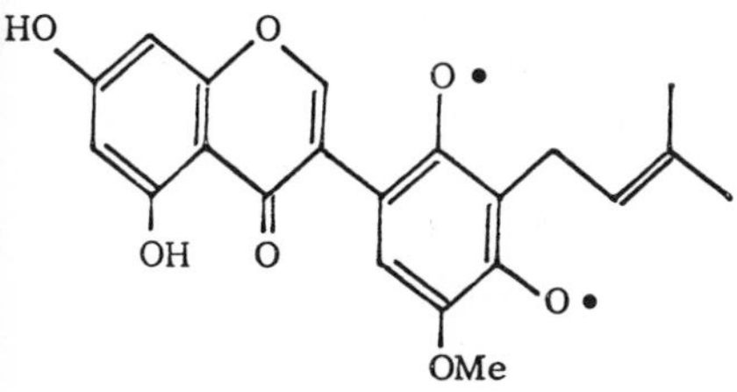

XLIV

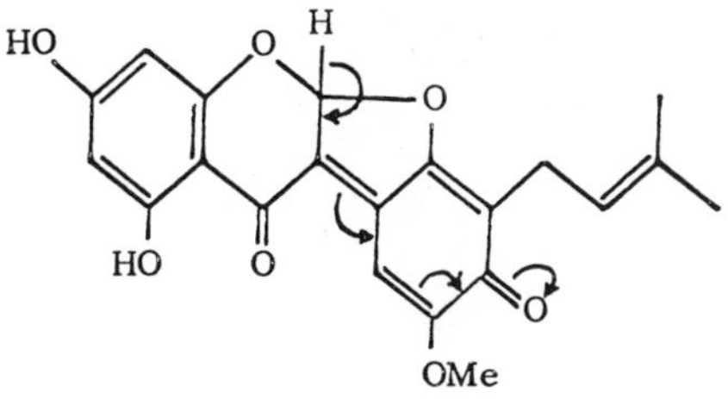

XLV

Scheme A.

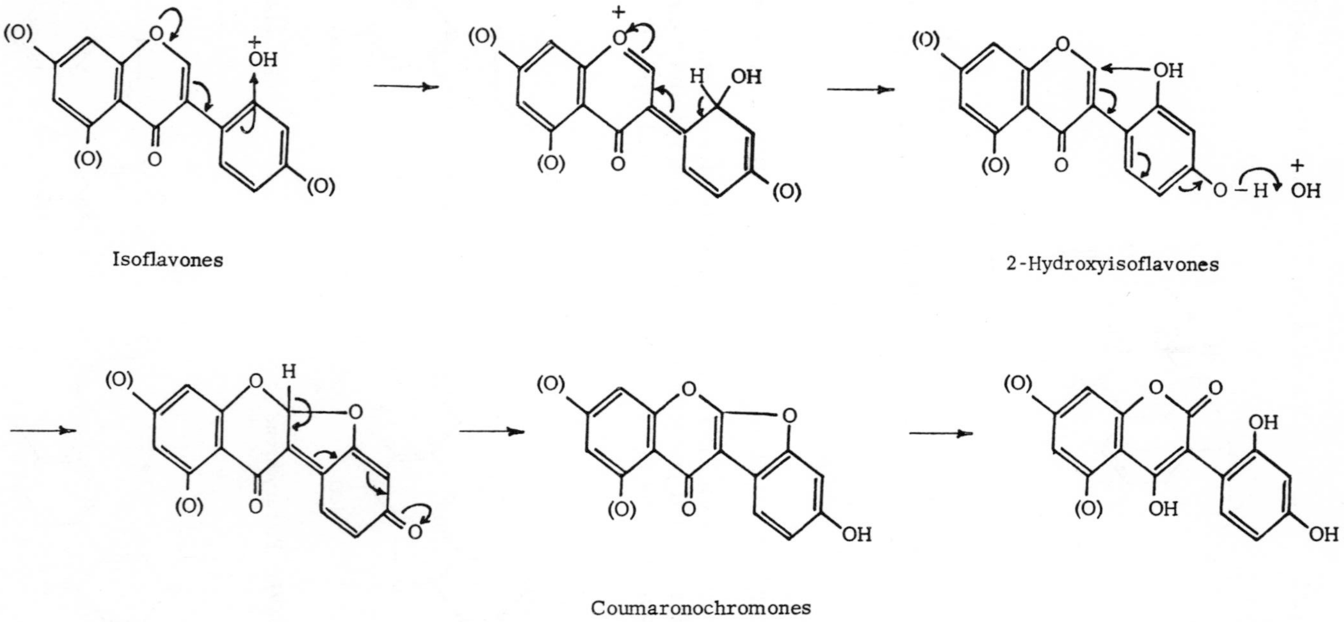

Isoflavones

2-Hydroxyisoflavones

Coumaronochromones

Scheme B.

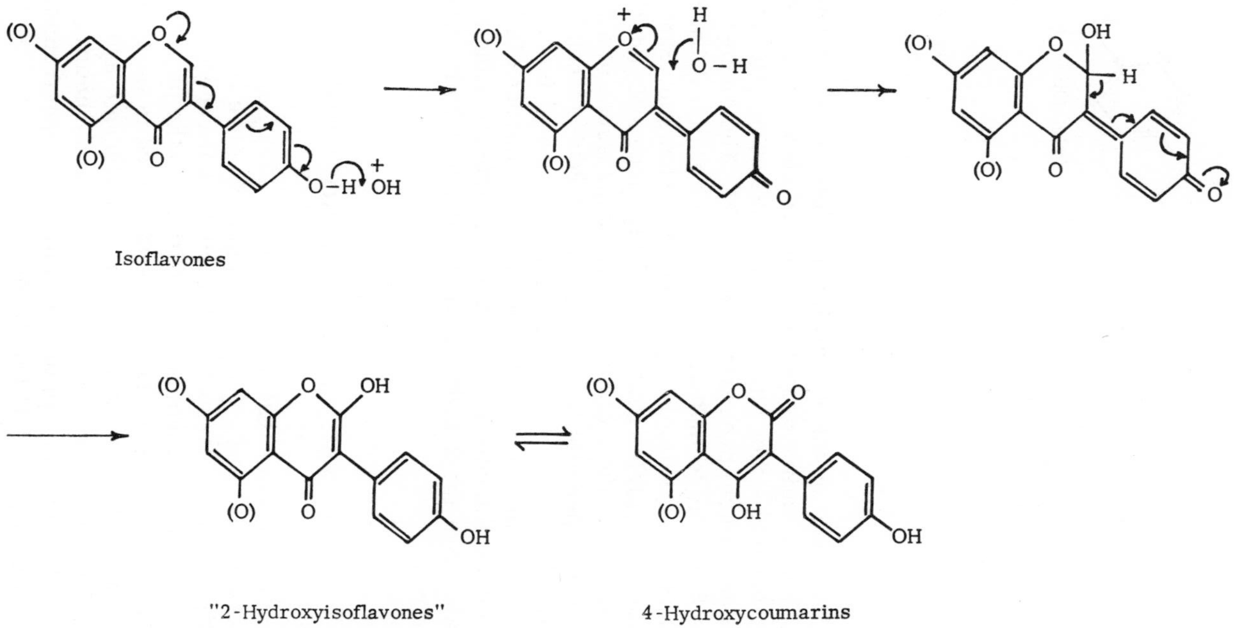

Isoflavones

"2-Hydroxyisoflavones"

4-Hydroxycoumarins

Fig. 5. Possible biogenetic relationships between isoflavones, coumaronochromones, and 3-aryl-4-hydroxycoumarins.

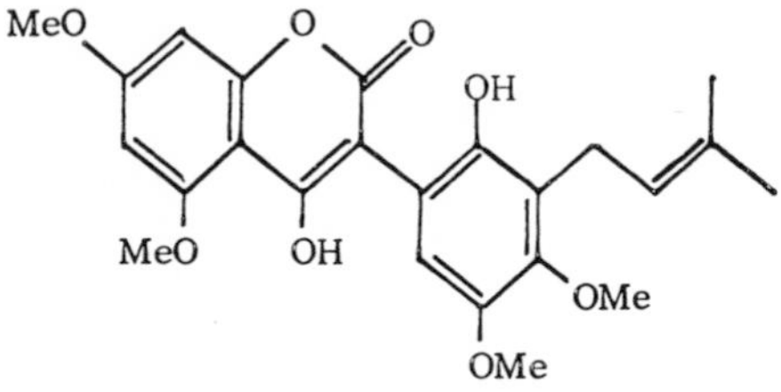

MeO O O OH MeO OH OMe OMe

XLVI

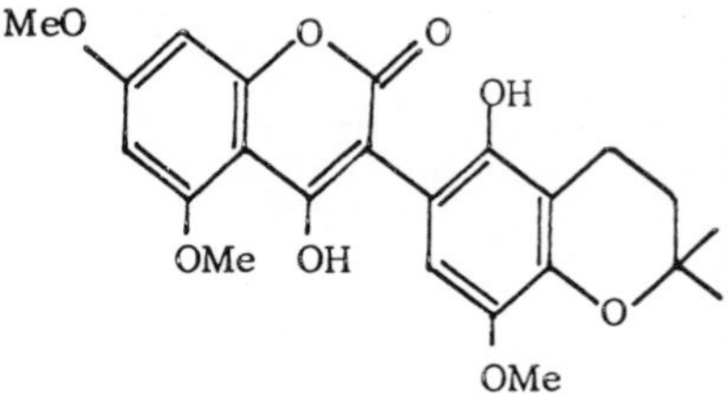

MeO O O OH OMe OH O OMe

XLVII

ble to comment on these structural relations as indicated in Figure 5. The oxidative processes that are shown in Figure 5 could be represented either as ionic processes using ^{+}OH or as the equivalent radical processes (see XLIV). The former is used in Figure 5 merely to indicate reasonable structural transformations and does not imply a preferred mechanism. Scheme A indicates a possible reason for the high frequency of 2′-oxygenation among the isoflavones and the presence of a 2′-hydroxyl group permits the formation of a coumaronochromone (see A, Fig. 5). In the absence of a 2′-hydroxyl function 4-hydroxycoumarins could be formed from isoflavones in the sense indicated by scheme B. Scheme B also indicates the tautomeric equivalence of the 2-hydroxyisoflavone and 4-hydroxycoumarin structures.

IV. THE NATURAL PTEROCARPANS

For many years (Ollis, 1962; Dean, 1963), only (–)-homopterocarpin (LIV) and (–)-pterocarpin (LX) were the known representatives of this class of isoflavonoids, but recently there has been a considerable number of additions to the pterocarpan group (see Table 3). Some of the compounds listed in the table have been provided by our investigations of *Dalbergia* and *Machaerium* species and considerable assistance in our structural examination of pterocarpans was provided by mass spectrometry (Cook, 1965; Falshaw, 1965; Ollis, 1966).

Pterocarpin was originally formulated incorrectly, but examination of its NMR spectrum established that it had the oxygenation pattern shown in formula (LX) (Bredenberg and Shoolery, 1961). Subsequently a *cis*-B/C ring fusion (XLVIII or XLIX) was proposed (Suginome and Iwadare, 1962) on the basis of the coupling constant observed for the 6a and 11a protons in (±)-homopterocarpin (LVI). The absolute configuration of optically active pterocarpans has only been recently established (Itô, Fujise, and Mori, 1965). This involved an ingenious correlation of (−)-trifolirhizin (LXX) with (−)-paraconic acid, whose absolute configuration was deduced as a consequence of its derivation from aucubin. This led to the proposal that (−)-pterocarpans had the 6a-R, 11a-R configuration, which is in accord with the proposals based on optical rotatory dispersion studies that were made by Clark-Lewis (1962) for the absolute configurations of (−)-pterocarpin (LX) and (−)-homopterocarpin (LIV). The configurations for the pterocarpans that we have isolated (see Table 3) are based on specific rotation measurements and optical rotatory dispersion studies.

The natural pterocarpans provide an unusual stereochemical circumstance. The majority of the natural pterocarpans are laevorotatory and have the absolute configuration (XLVIII), but several (+)-pterocarpans have been discovered recently (see Table 3), thus showing that these natural products are produced in both antipodal forms. In this connection the natural occurrence of racemic pterocarpans is of interest and examples include (±)-demethylhomopterocarpin (LIII) from *Dalbergia cearensis*, *Dalbergia spruceana*, and *Dalbergia variabilis* (Cook, 1965; Kurosawa, 1965), and (±)-maackiain (LIX) from *Sophora japonica* (Shibata and Nishikawa, 1963) and *Dalbergia spruceana* (Cook, 1965). The natural (±)-demethylhomopterocarpin (LIII) was shown to be identical with the compound prepared by synthesis (Cocker, McMurry, and Staniland, 1965; Cook, 1965).

Several pterocarpans have been prepared recently, by either partial or total synthesis. The first synthesis of the pterocarpan skeleton by Suginome (1962) using sodium borohydride reduction of the corresponding 2′-hydroxyisoflavone led to (±)-homopterocarpin (LVI) and the same general method has been used to synthesize (±)-demethylhomopterocarpin (LIII; Cocker, McMurry, and Staniland, 1965; Cook, 1965) and (±)-4-hydroxy-8,9-methyl-

TABLE 3

The Constitutions of Natural Pterocarpans

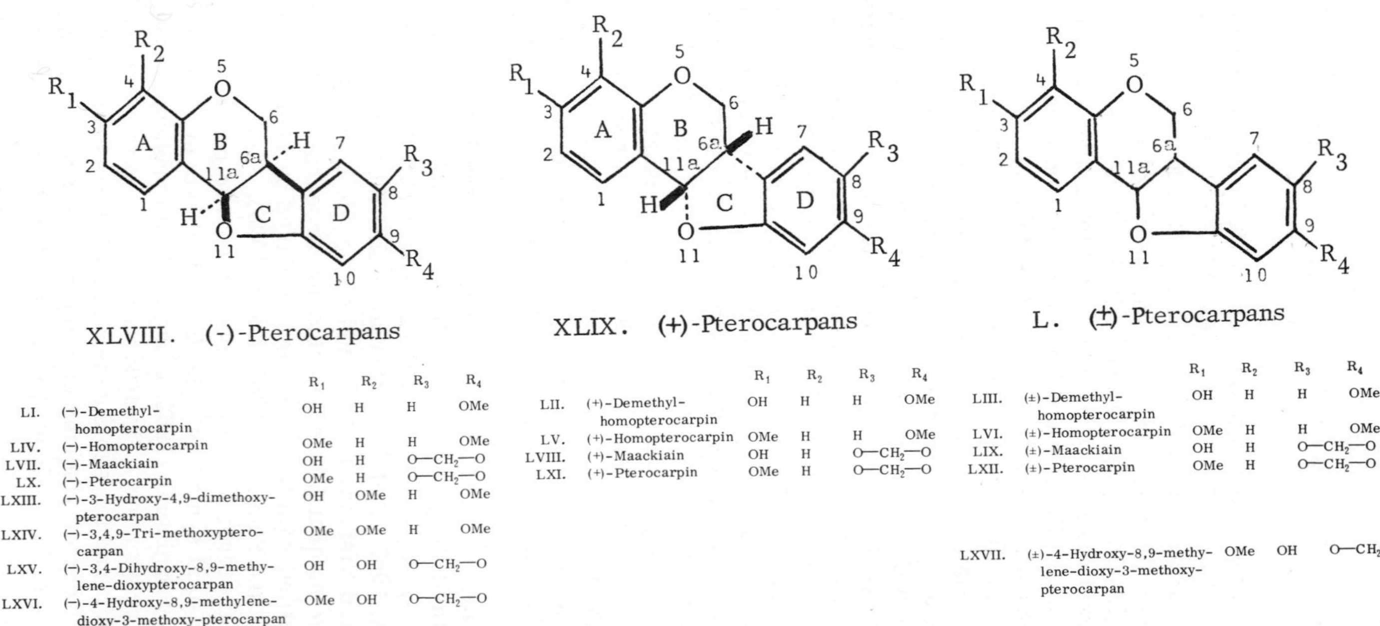

XLVIII. (-)-Pterocarpans

		R_1	R_2	R_3	R_4
LI.	(−)-Demethyl-homopterocarpin	OH	H	H	OMe
LIV.	(−)-Homopterocarpin	OMe	H	H	OMe
LVII.	(−)-Maackiain	OH	H	O—CH_2—O	
LX.	(−)-Pterocarpin	OMe	H	O—CH_2—O	
LXIII.	(−)-3-Hydroxy-4,9-dimethoxy-pterocarpan	OH	OMe	H	OMe
LXIV.	(−)-3,4,9-Tri-methoxyptero-carpan	OMe	OMe	H	OMe
LXV.	(−)-3,4-Dihydroxy-8,9-methy-lene-dioxypterocarpan	OH	OH	O—CH_2—O	
LXVI.	(−)-4-Hydroxy-8,9-methylene-dioxy-3-methoxy-pterocarpan	OMe	OH	O—CH_2—O	
LXVIII.	(−)-3-Hydroxy-8,9-methylene-dioxy-4-methoxy-pterocarpan	OH	OMe	O—CH_2—O	
LXIX.	(−)-8,9-Methylene-dioxy-3,4-dimethoxy-pterocarpan	OMe	OMe	O—CH_2—O	
LXX.	Trifolirhizin	β-D-glucosidyloxy	H	O—CH_2—O	

XLIX. (+)-Pterocarpans

		R_1	R_2	R_3	R_4
LII.	(+)-Demethyl-homopterocarpin	OH	H	H	OMe
LV.	(+)-Homopterocarpin	OMe	H	H	OMe
LVIII.	(+)-Maackiain	OH	H	O—CH_2—O	
LXI.	(+)-Pterocarpin	OMe	H	O—CH_2—O	
LXXI.	Sophorajaponicin	β-D-glucosidyloxy	H	O—CH_2—O	

L. (±)-Pterocarpans

		R_1	R_2	R_3	R_4
LIII.	(±)-Demethyl-homopterocarpin	OH	H	H	OMe
LVI.	(±)-Homopterocarpin	OMe	H	H	OMe
LIX.	(±)-Maackiain	OH	H	O—CH_2—O	
LXII.	(±)-Pterocarpin	OMe	H	O—CH_2—O	
LXVII.	(±)-4-Hydroxy-8,9-methy-lene-dioxy-3-methoxy-pterocarpan	OMe	OH	O—CH_2—O	

enedioxy-3-methoxypterocarpan (LXVII); the synthesis of this (±) -pterocarpan (LXVII; Cook, 1965) was required in order to settle the constitutions of the three (−)-pterocarpans (LXV, LXVI, LXVIII) that we isolated from *Dalbergia spruceana.* It is interesting to note that these three laevorotatory pterocarpans (LXV, LXVI, LXVIII) occur in *Dalbergia spruceana* with the racemic pterocarpans (LIII) and (LIX). Another route to pterocarpans is exemplified by the recent synthesis of (±)-homopterocarpin (LVI) from di-*O*-methylcoumestrol (Bowyer, Chatterjea, Dhoubadel, Handford, and Whalley, 1964).

(+)-Maackiain (LVIII) and (+)-pterocarpin (LXI) have not yet been detected as natural products, but they have been obtained by partial synthesis from sophorajaponicin (LXXI)(Shibata and Nishikawa, 1963). Hydrolysis of sophorajaponicin (LXXI) with emulsin gave (+)-maackiain (LVIII), which on methylation gave (+)-pterocarpin (LXI).

The natural products (−)-phaseolin (LXXII) from fungus-infected *Phaseolus vulgaris* (D. R. Perrin, 1964) and (−)-neodulin (LXXIII) from *Neorautanenia edulis* (Van Duuren, 1961; Crombie and Whiting, 1963) provide interesting recent additions to the pterocarpan class.

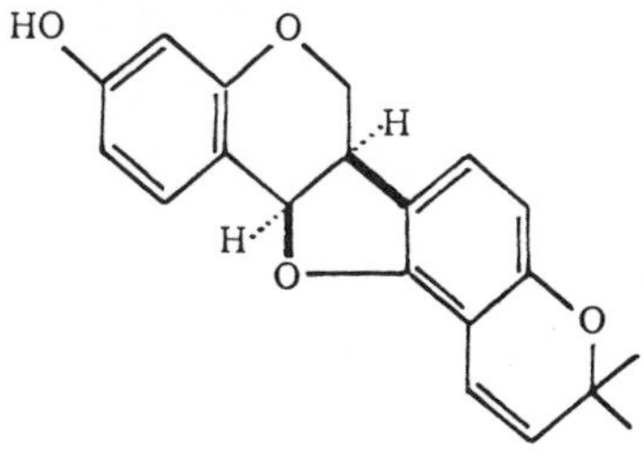

LXXII. Phaseolin

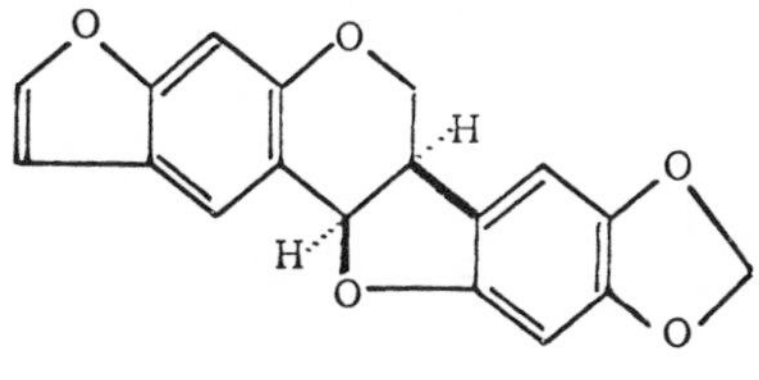

LXXIII. Neodulin (Edulin)

A. Pterocarpan Derivatives as Phytoalexins

Recently there has been considerable interest in the phytoalexins, the name given to a group of protective agents that are synthesised in plants in response to fungal attack (Cruickshank and Perrin, 1964). These antifungal agents include the natural phenolic compounds trifolirhizin (LXX)(Bredenberg and Hietala, 1961), an isocoumarin (Condon and Kuc, 1962), pisatin (LXXIV; D. R. Perrin and Bottomley, 1962; Perrin and Perrin, 1962), orchinol (Hardegger, Schellenbaum, and Corvodi, 1963), hircinol

TABLE 4

The Natural Pterocarpans

Name and Synonym	Formula	Plant Source	Reference
(−)-Demethylhomopterocarpin	LI	*Andira inermis*	Cocker et al. (1965)
		Swartzia madagascarensis	Harper et al. (1965)
(+)-Demethylhomopterocarpin	LII	*Dalbergia variabilis*	Kurosawa (1965)
		Machaerium kuhlmannii	Roberts (1966)
		Machaerium vestitum	Redman (1966)
(±)-Demethylhomopterocarpin	LIII	*Dalbergia cearensis*	Cook (1965)
		Dalbergia spruceana	Cook (1965)
		Dalbergia variabilis	Kurosawa (1965)
(−)-Homopterocarpin	LIV	*Pterocarpus santalinus*	Späth and Schläger (1940)
		Pterocarpus sojauxii	King et al. (1953)
		Pterocarpus osun	Akisanya et al. (1959)
		Pterocarpus indicus	Brooks (1911); Cooke and Rae (1964)
[Baphinitone]		*Baphia nitida*	Ryan and Fitzgerald (1913)
		Swartzia madagascarensis[a]	Harper et al. (1965)
(+)-Homopterocarpin	LV	*Machaerium villosum*	Gottlieb et al. (1965)
(−)-Maackiain	LVII	*Maackia amurensis*	Suginome (1962)
(−)-Inermin		*Andira inermis*	Cocker et al. (1962)
(−)-Demethylpterocarpin		*Swartzia madagascarensis*[a]	Harper et al. (1965)
(±)-Maackiain	LIX	*Sophora japonica*	Shibata and Nishikawa (1963)
		Dalbergia spruceana	Cook (1965)
(−)-Pterocarpin	LX	*Pterocarpus santalinus*	Späth and Schläger (1940)
		Pterocarpus dalbergioides	King et al. (1953)
		Pterocarpus macrocarpus	King et al. (1953)

		Pterocarpus indicus	Brooks (1911), Cooke and Rae (1964)
		Baphia nitida	McGookin et al. (1940); Robertson and Whalley (1954); Bredenberg and Shoolery (1961)
		Sophora subprostrata	Shibata and Nishikawa (1963)
		Swartzia madagascarensis[a]	Harper et al. (1965)
(−)-3-Hydroxy-4,9-dimethoxy-pterocarpan	LXIII	*Swartzia madagascarensis*	Harper et al. (1965)
(−)-3,4,9-Trimethoxyptero-carpan	LXIV	*Swartzia madagascarensis*	Harper et al. (1965)
(−)-3,4-Dihydroxy-8,9-methylenedioxypterocarpan	LXV	*Dalbergia spruceana*	Cook (1965)
(−)-4-Hydroxy-8,9-methylene-dioxy-3-methoxypterocarpan	LXVI	*Dalbergia spruceana*	Cook (1965)
(−)-3-Hydroxy-8,9-methylene-dioxy-4-methoxypterocarpan	LXVIII	*Swartzia madagascarensis*	Harper et al. (1965)
		Dalbergia spruceana	Cook (1965)
(−)-8,9-Methylenedioxy-3,4-dimethoxypterocarpan	LXIX	*Swartzia madagascarensis*	Harper et al. (1965)
Trifolirhizin [(−)-Maackiain β-D-glucoside]	LXX	*Trifolium pratense*	Bredenberg and Hietala (1961)
		Sophora subprostrata	Shibata and Nishikawa (1963)
		Sophora flavescens	Shibata and Nishikawa (1963)
Sophorajaponicin [(+)-Maackiain β-D-glucoside]	LXXI	*Sophora japonica*	Shibata and Nishikawa (1963)

[a]It is assumed that these three compounds are (−)-pterocarpans. This is implied in the preliminary communications, but the physical data have not yet been published.

(Urech, Fechtig, Nuesch, and Vischer, 1963), phaseolin (LXXII) (D. R. Perrin, 1964), and the terpenoid ipomeamarone (Kubota and Matsuura, 1953; Uritani, Akazawa, and Uritani, 1954).

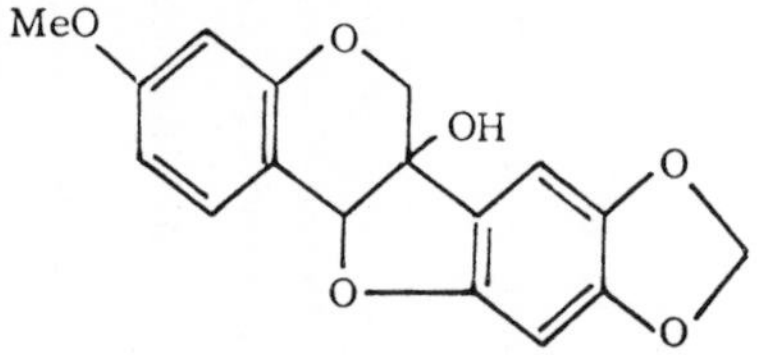

MeO O OH O O O

LXXIV. Pisatin

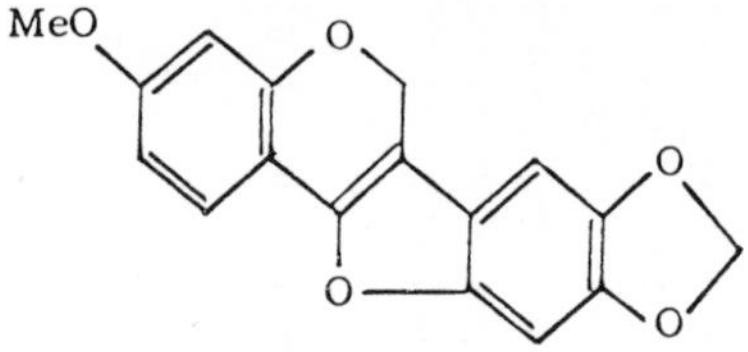

MeO O O O O

LXXV. Anhydropisatin

In connection with the antifungal activity of the pterocarpan trifolirhizin (LXX), it has been shown that (–)-homopterocarpin (LIV) is almost as active as pisatin (LXXIV). (+)-Pisatin (LXXIV) is produced when the seed pods of *Pisum sativum* are infected with *Monilinia fructicola*. The absolute configuration of (+)-pisatin is not known. Mild acid treatment of pisatin (LXXIV) gave

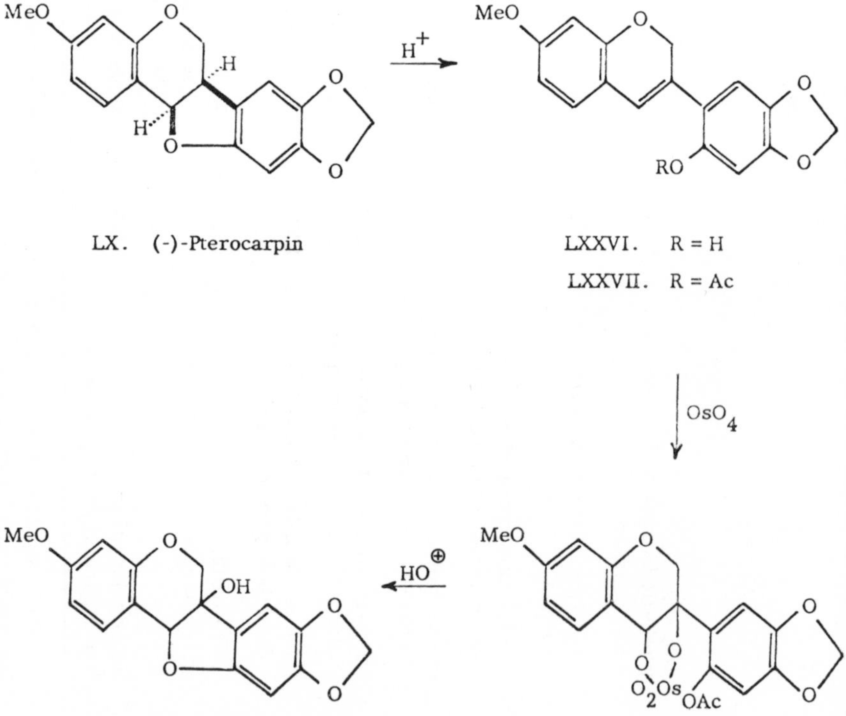

MeO O H H O O O

LX. (-)-Pterocarpin

$\xrightarrow{H^+}$

MeO O O O RO

LXXVI. R = H

LXXVII. R = Ac

$\downarrow OsO_4$

MeO O O O O_2Os OAc O O

$\xleftarrow{HO^{\oplus}}$

MeO O OH O O O

LXV. Pisatin

anhydropisatin (LXXV) and this instability had to be taken into account in the design of the partial synthesis of (±)-pisatin (LXV) from (−)-pterocarpin (LX) (Bevan, Birch, Moore, and Mukerjee, 1964).

We have recently isolated from *Dalbergia variabilis* (Kurosawa, 1965) a compound (+)-variabilin (LXXVIII; $[\alpha]_D^{20} = +211°$ in methanol), which is easily dehydrated to anhydrovariabilin (LXXIX). The racemate (LXXVIII) had previously been obtained from homopterocarpin (Bevan, Birch, Moore, and Mukerjee, 1964) and dehydration gave anhydrovariabilin (LXXIX). The isolation of anhydrovariabilin from *Swartzia madagascarensis* has recently been reported (Harper, Kemp, and Underwood, 1965a, 1965b), but it is possible that it is an artefact produced from variabilin (LXXVIII).

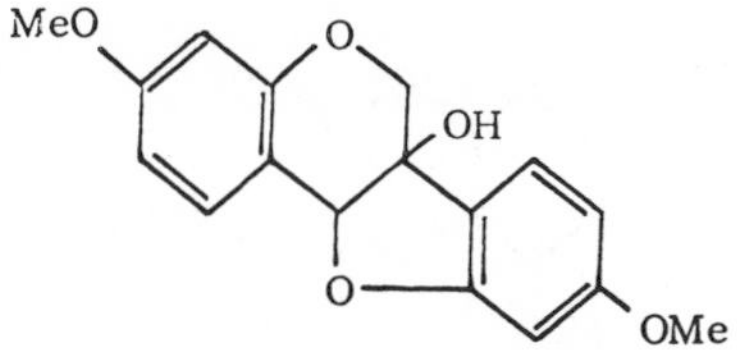

LXXVIII. Variabilin 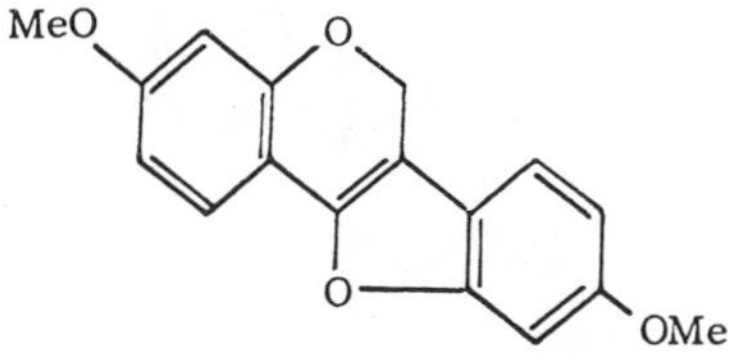 LXXIX. Anhydrovariabilin

V. THE NATURAL OCCURRENCE OF ISOFLAVANS

For many years the existence of only one natural isoflavan was recognized. This was the compound equol (LXXX), but as it had been isolated only from the urine of mares, stallions, cows, goats, and hens (Marrian and Beall, 1935; Wessely, Hirschel, Schögl-Petziwal, and Prillinger, 1938; Anderson and Marrian, 1939; Wessely and Prillinger, 1939; Dirscherl and Schodder, 1940; Suemitsu, Huira, and Makajima, 1955; Klyne and Wright, 1957, 1959; MacRae, Dale, and Common, 1960; Hertelendy and Common, 1964). The description of equol as a natural product, in the usual sense of the term, was therefore questionable. However, we have recently discovered three isoflavans (LXXXI, LXXXII, LXXXIII) during our examination of *Dalbergia* and *Machaerium* species.

The isoflavan nature of vestitol (LXXXI), mucronulatol (LXXXII), and duartin (LXXXIII) was first indicated by their

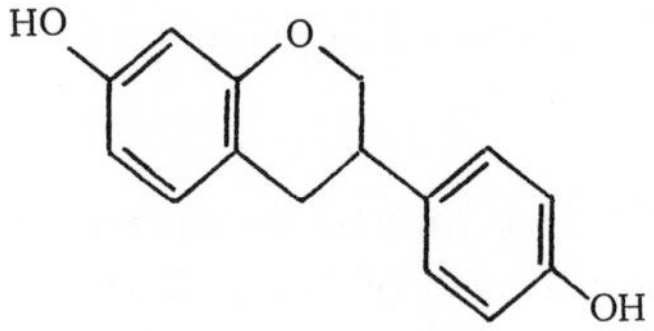

LXXX. Equol

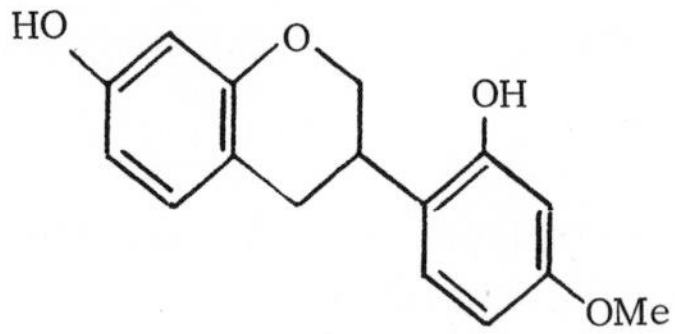

LXXXI. Vestitol

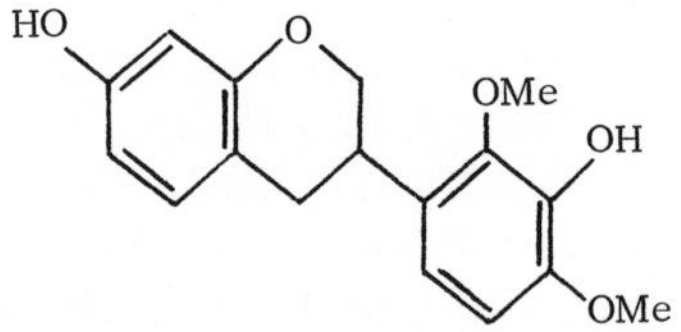

LXXXII. Mucronulatol

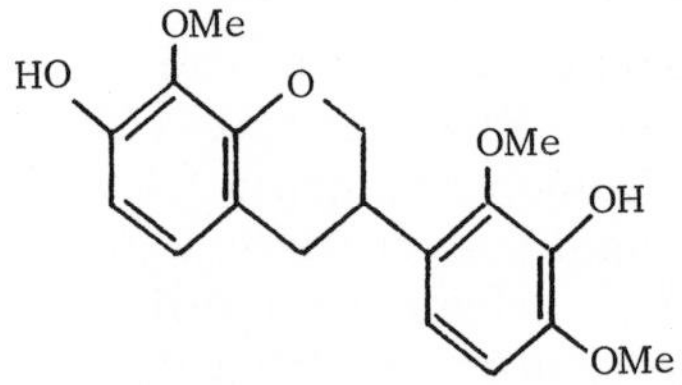

LXXXIII. Duartin

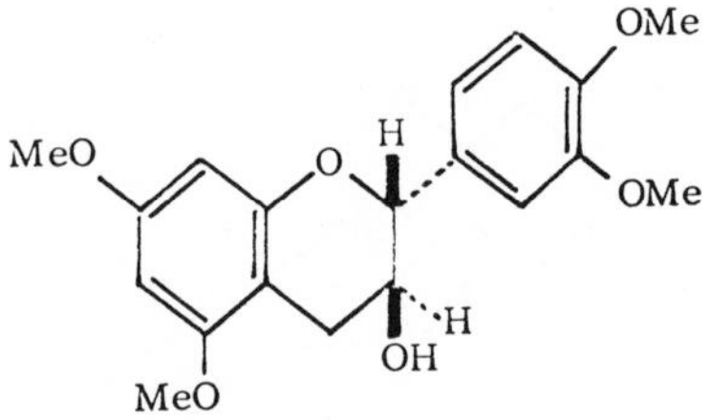

LXXXIX. (+)-Catechin tetramethyl ether

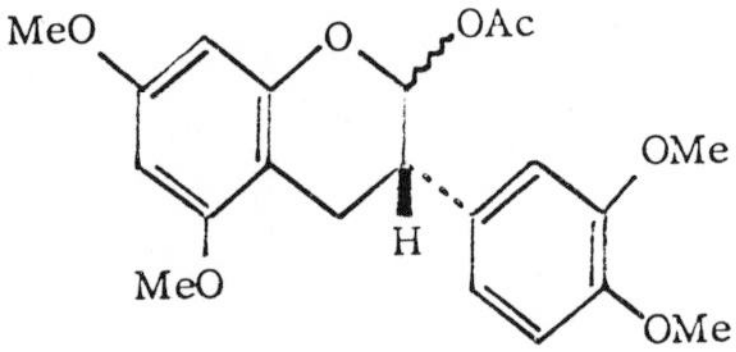

XC.

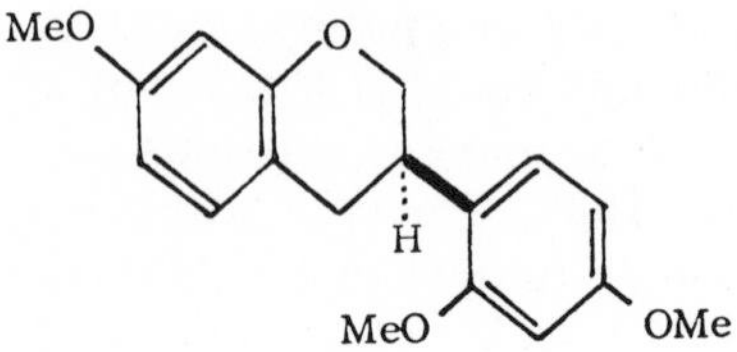

XCI.

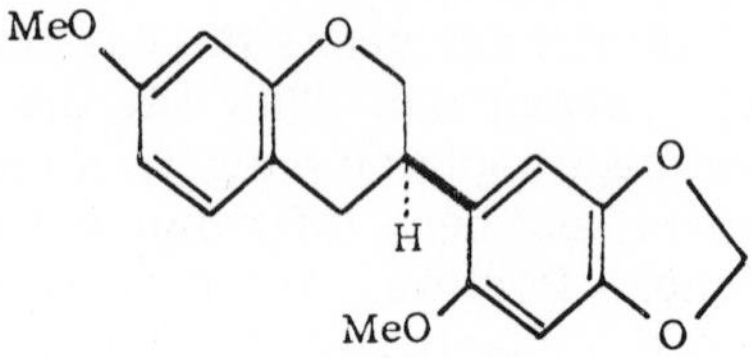

XCII.

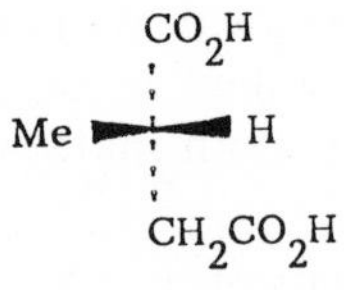

XCIII.

NMR and mass spectra (Falshaw, 1965) and was fully confirmed by standard methods involving their degradation via isoflavanone derivatives and by synthesis (Kurosawa, 1965; Magalhaẽs Alves, 1965; Redman, 1966). Vestitol, mucronulatol, and duartin have all been isolated in optically active forms and the definition of their absolute configuration involved the following argument.

(+)-Catechin tetramethyl ether of established absolute configuration LXXXIX has been stereospecifically transformed into the tetramethoxyisoflavan XC, which shows a positive plain O.R.D. curve (Clark-Lewis, 1962; Clark-Lewis, Dainis, and Ramsay, 1965). In contrast the isoflavans (XCI and XCII) derived from (−)-homopterocarpin (LIV) and (−)-pterocarpin (LX) both show negative plain curves (Clark-Lewis, 1962). Thus, examination of their optical rotatory dispersion curves may be used to define the absolute configuration of isoflavans and these assignments have been independently confirmed by Weinge's transformation of the diastereoisomeric 2-acetoxyisoflavans (XC) to (−)-methylsuccinic acid (XCIII) (Weinges, 1965; Weinges and Paulus, 1965).

The stereochemical arguments outlined above are in complete accord with the independent deductions of the absolute configurations of the pterocarpans (Itô, Fujise, and Mori, 1965) discussed in Section IV.

The absolute configurations of the isoflavans that we have isolated are summarised in Figure 6.

Examination of Figure 6 shows that isoflavans are produced in both R and S forms in that both antipodal forms of mucronulatol have been isolated from *Machaerium mucronulatum*, but the (−)-isomer (XCV) is apparently present in excess.

No definitive experiments have yet been carried out on the biosynthesis of pterocarpans and isoflavans, but it may be noted that Redman (1966) has made the interesting observation that (+)-demethylhomopterocarpin (LII)(see Table 3) and (+)-vestitol (XCIV)(see Fig. 6) of corresponding absolute stereochemistry have both been isolated from *Machaerium vestitum*. The suggestion that the biosynthesis of pterocarpans and isoflavans is similar to that of other isoflavonoids (Grisebach, Chap. 11 of this volume) is supported by the isolation of violanone (XCVII) from *Dalbergia violacea* (Gregson, 1965). The structure of this new isoflavanone corresponds to that of (±)-mucronulatol (LXXXII).

The isolation of these new isoflavans (see Fig. 6) is in some respects complementary to the recently discovered flavan (XCVIII) isolated after methylation from *Xanthorrhoea preissii* (Birch and Salahuddin, 1964). Although the hydroxylated flavans such as the catechins and leucoanthocyanidins are well known, this is the first simple flavan to be isolated from natural sources.

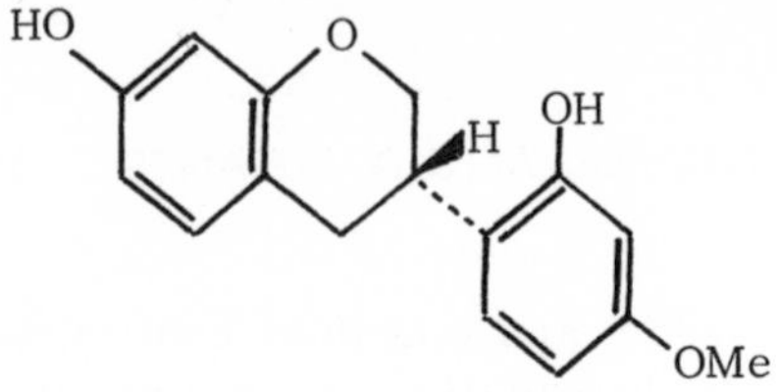

XCIV. (+)-Vestitol — *Machaerium vestitum*

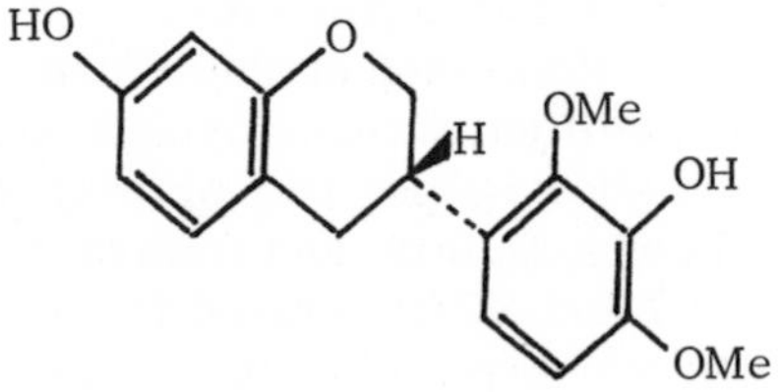

XCV. (-)-Mucronulatol — *Machaerium mucronulatum*, *Machaerium opacum*, *Machaerium vestitum*, *Machaerium villosum*

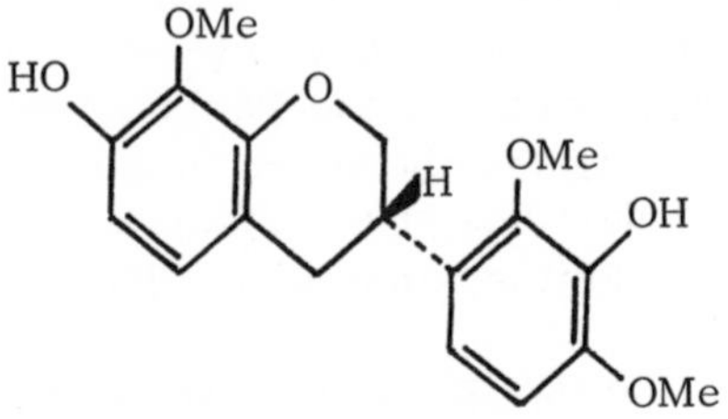

XCVI. (-)-Duartin — *Machaerium mucronulatum*, *Machaerium opacum*, *Machaerium villosum*

LXXXII. (±)-Mucronulatol — *Machaerium mucronulatum*

Fig. 6. The natural occurrence of isoflavans (Kurosawa, 1965; Magalhaẽs Alves, 1965; Gottlieb et al., 1965- Redman, 1966).

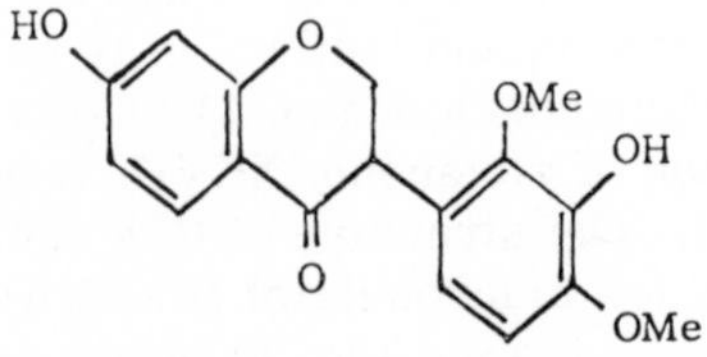

XCVII. Violanone

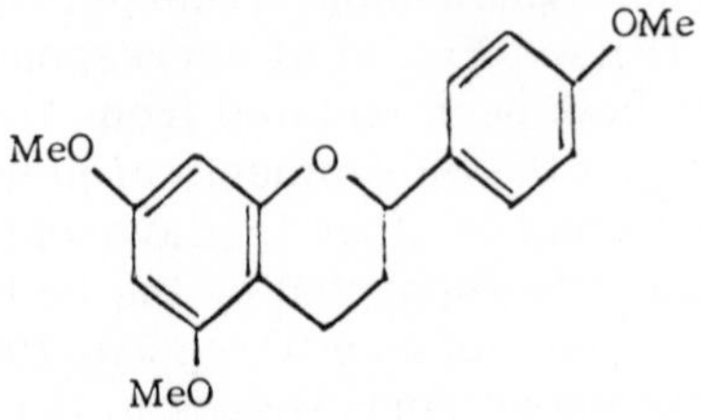

XCVIII

VI. THE NEOFLAVANOID CLASS OF NATURAL PRODUCTS

The term neoflavanoid was recently proposed (Eyton, Ollis, Sutherland, Gottlieb, Taveira Magalhaẽs, and Jackman, (1965a) to describe a newly recognised group of structurally related natural products with a C_{15} skeleton. The relationship of the neoflavanoids to the flavanoid and isoflavanoid types is indicated in Figure 7. The presence of a heterocyclic ring is not a condition for the use of these terms and, just as chalkones are included among the flavanoids (or flavonoids) and angolensin is regarded as an isoflavanoid (or isoflavonoid), so the dalbergiones and 4-arylcoumarins are included in the neoflavanoid group.

The proposal (Eyton, Ollis, Sutherland, Jackman, Gottlieb, and Taveira Magalhaẽs, 1962) that the dalbergiones and dalbergins (4-arylcoumarins; Fig. 8) were biogenetically related has been independently made (Seshadri, 1963). Studies among this group of natural products have been recently reviewed (Ollis, 1966), so the present account will be primarily concerned with our recent unpublished work on neoflavanoids with particular reference to the structural types indicated in Figure 8.

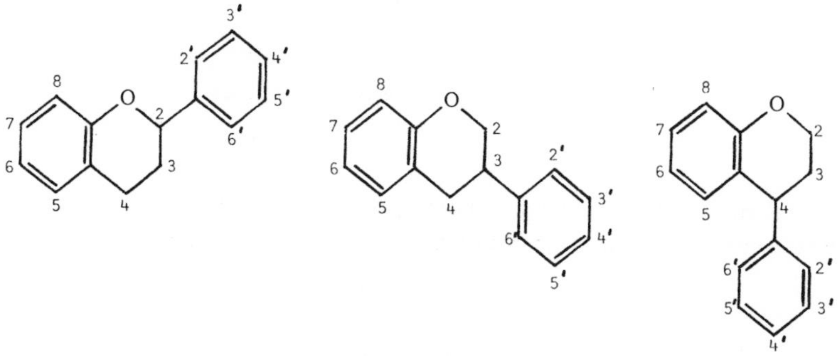

Flavanoid Isoflavanoid Neoflavanoid

Fig. 7. Type formulae of the flavanoids, isoflavanoids, and neoflavanoids.

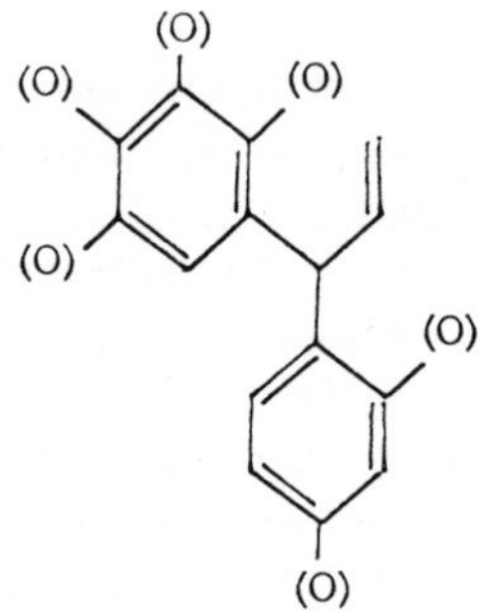

Dalbergiquinols

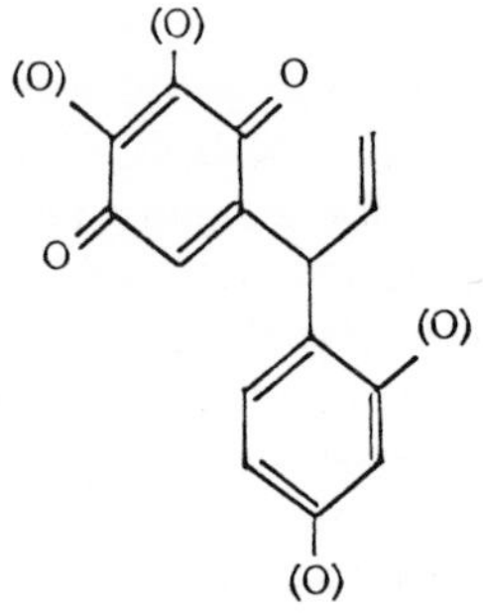

Dalbergiones

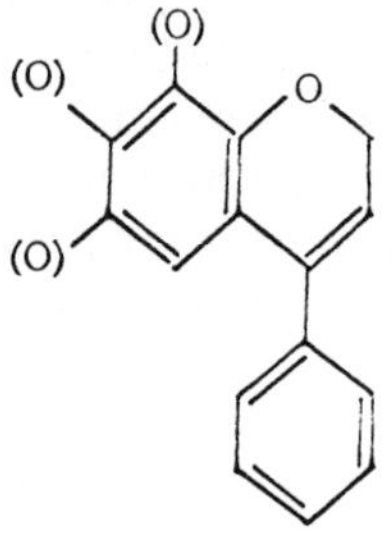

Neoflavenes

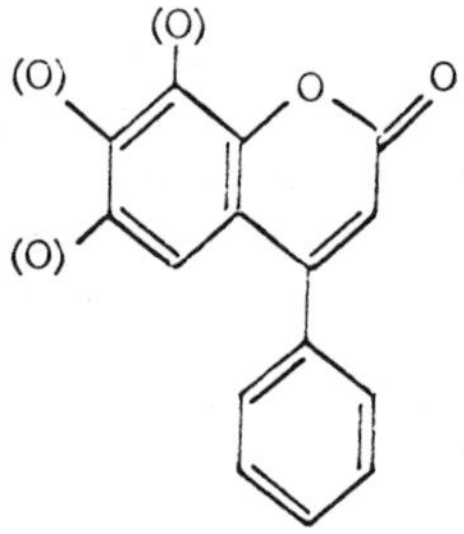

Dalbergins
(4-Arylcoumarins)

Fig. 8. Structural types belonging to the neoflavanoid group. (Positions of oxygenation that have been encountered are indicated.)

A. Dalbergiones

Since our recognition of the dalbergiones as a new class of natural quinones (Eyton, Ollis, Sutherland, Jackman, Gottlieb, and Taviera Magalhaẽs, 1962), these substances have been investigated by several groups. So far the dalbergiones have been shown to be restricted to the *Dalbergia* and *Machaerium* genera. The current phytochemical situation is summarised in Figure 9.

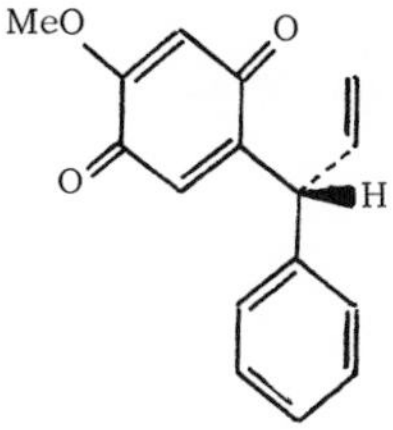

XCIX. R-4-Methoxydalbergione

Dalbergia latifolia [a,b]

Dalbergia nigra [c,d]

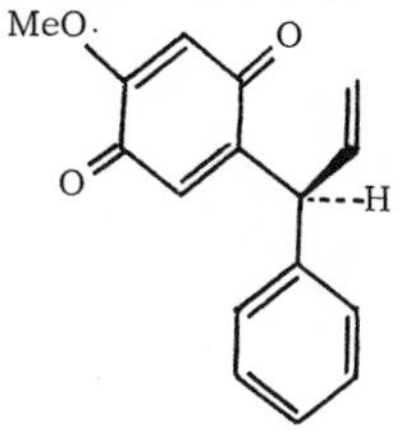

C. S-4-Methoxydalbergione

Dalbergia baronii [e]

Dalbergia spruceana [f]

Dalbergia miscolobium [d,g]

(syn. D. violacea)

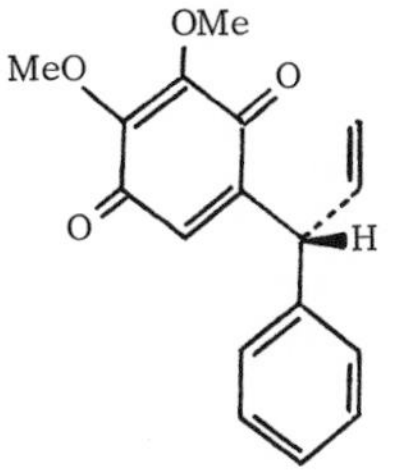

CI. R-3,4-Dimethoxydalbergione

Machaerium kuhlmannii [h]

Machaerium scleroxylon [i]

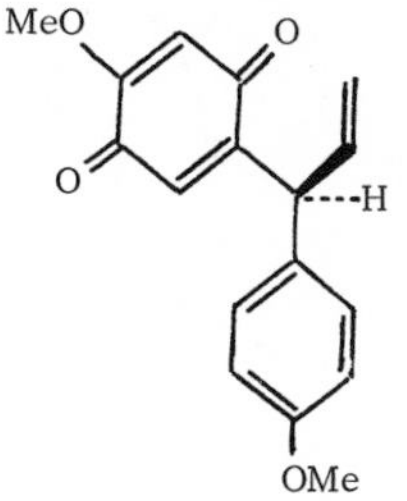

CII. S-4,4'-Dimethoxydalbergione

Dalbergia nigra [c,d]

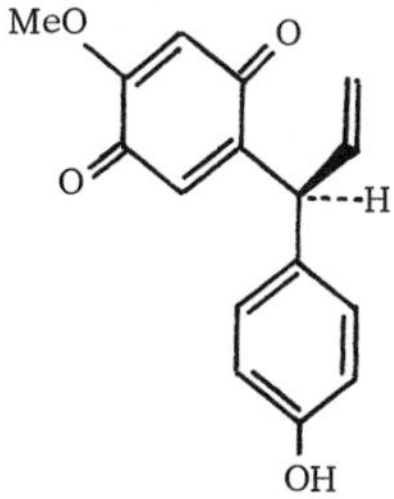

CIII. S-4'-Hydroxy-4'-methoxydalbergione

Dalbergia miscolobium [d,g]

(syn. D. violacea)

Dalbergia nigra [c,d]

Fig. 9. The natural occurrence of the dalbergiones. [a]Bhatia et al. (1965); [b]Donnelly et al. (1965b); [c]Eyton et al. (1962); [d]Eyton et al. (1965a); [e]Donnelly et al. (1965a); [f]Cook (1965); [g]Gregson (1965); [h]Roberts (1966); [i]Eyton et al. (1965b).

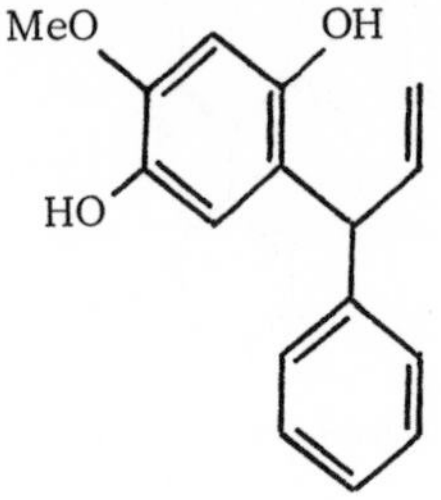

CIV — Dalbergia obtusa [a]

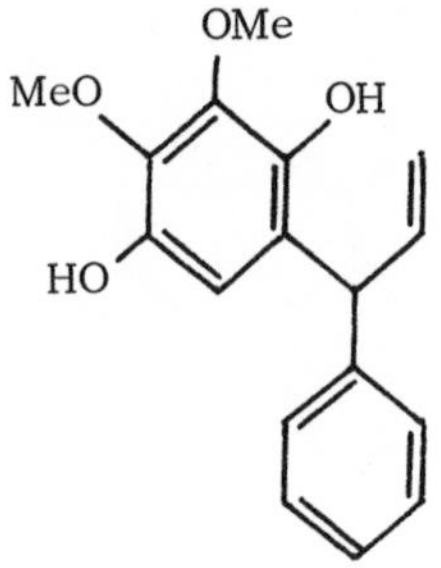

CV — Machaerium kuhlmannii [b], Machaerium scleroxylon [c]

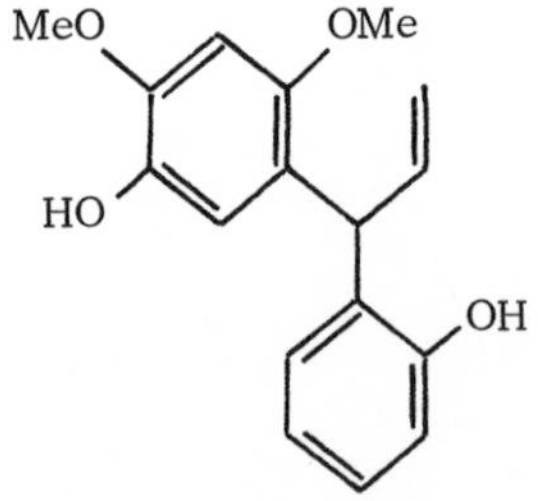

CVI. Latifolin — Dalbergia latifolia [d, e, f]

Fig. 10. Natural dalbergiquinols. [a]Gregson (1966); [b]Roberts (1966); [c]Eyton et al. (1965b); [d]Darshan et al. (1965); [e]Bhatia et al. (1965); [f]D. M. X. Donnelly et al. (1965).

B. Dalbergiquinols

The quinols or quinol methyl ethers related to the dalbergiones have been isolated from various species and in order to match the IUPAC proposals (IUPAC Information Bulletin, 1966) for the naming of natural isoprenoid quinones and related compounds (Isler, Langemann, Meyer, Rüegg and Schudel, 1965), these 3,3-diarylprop-1-enes have been called dalbergiquinols (see Fig. 10).

C. Neoflavenes

The neoflavenes (see Fig. 8) might well be expected to occur with neoflavanoids in that they could be predicted as intermediates in the transformation of dalbergiquinols or dalbergiones into dalbergins. Biogenetic precedents for these types of reaction are provided by (a) the postulated oxidative transformation of *o*-dimethylallylphenols to 2,2-dimethylchromenes (Ollis and Sutherland, 1961) or (b) the base- or acid-catalyzed transformation of quinones to chromenes. Reactions of type (b) have been frequently observed in the ubiquinone and tocoquinone series (Isler, Langemann, Meyer, Rüegg and Schudel, 1965):

(a) OH Oxidation O O

(b) R_3 R_2 O R O R_1 R_3 R_2 O R HO R_1

The reaction (b) is, of course, a model for the transformation of a dalbergione to a neoflavene. So far, only one neoflavene, 6-hydroxy-7,8-dimethoxy-4-phenylchromene (CVII), from *Machaerium kuhlmannii* (Roberts, 1966), has been recognised as a natural product, but the existence of other neoflavenes is to be expected.

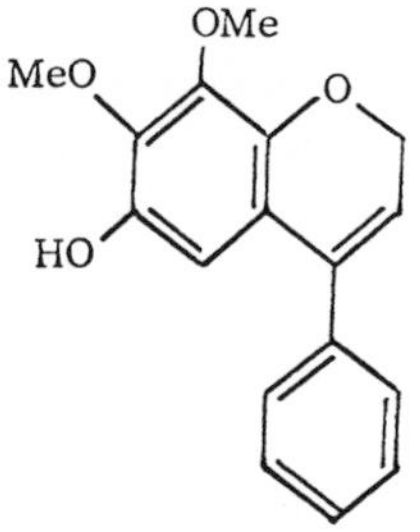

OMe
MeO O
HO

CVII

D. Dalbergins

The 4-arylcoumarins CVIII and CIX (Fig. 11) were first isolated from *Dalbergia sissoo* (Ahluwalia and Seshadri, 1957) and then later from *Dalbergia latifolia* (Balakrishna, Rao, and Seshadri, 1962; Rao and Seshadri, 1963) and *Machaerium scleroxylon* (Eyton, Ollis, Fineberg, Gottlieb, Salignac, de Souza Guimaraẽs, and Taveira Magalhaẽs, 1965b). When first isolated, the 4-arylcoumarins (CVIII) and (CIX) did not obviously fit in with other biogenetically defined groups of natural phenols, but the subsequent discovery of the dalbergiones led to our proposal that the 4-arylcoumarins belonged to the neoflavanoid class.

The name dalbergin was originally used for the one compound CVIII, but it is now proposed that the term dalbergin should be used to describe a class of natural products (see Fig. 11) and that the members of this class should be named as coumarin derivatives.

VII. PROPOSALS FOR THE BIOSYNTHESIS OF NEOFLAVANOIDS

Two routes have been considered for the biosynthesis of the neoflavanoid skeleton (Eyton, Ollis, Fineberg, Gottlieb, Salignac, de Souza Guimaraẽs, and Taveira Magalhaẽs, 1965b; Ollis, 1966) One route could be regarded as an extension of the established biosynthetic programs leading to flavonoids and isoflavonoids (Grisebach, Chap. 11 of this volume), in that as it is known that one 1,2-aryl migration is involved in the biosynthesis of isoflavonoids, so a second aryl migration could lead to the neoflavanoids. The second hypothetical route that has been proposed suggests that the neoflavanoid skeleton could be formed by alkylation, by a cinnamyl phosphate for example, of a phenolic or polyketide precursor.

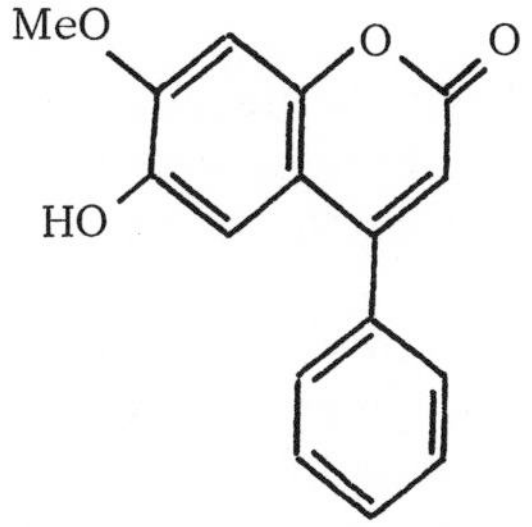

CVIII. 6-Hydroxy-7-methoxy-4-phenylcoumarin

Dalbergia cearensis [a]
Dalbergia latifolia [b, c, d]
Dalbergia miscolobium [f]
(syn. D. violacea)
Dalbergia sissoo [e]
Dalbergia spruceana [a]
Machaerium kuhlmannii [g]
Machaerium scleroxylon [h]

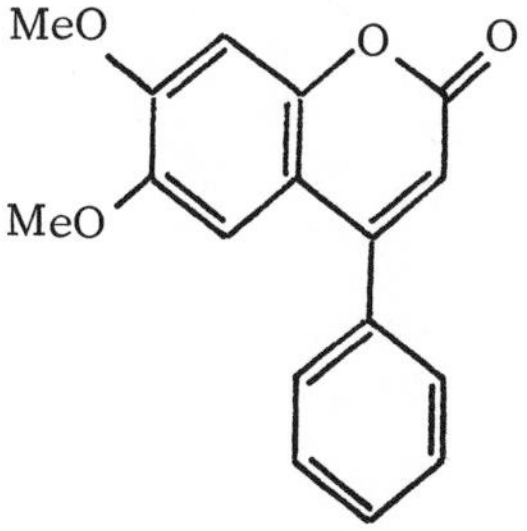

CIX. 6, 7-Dimethoxy-4-phenylcoumarin

Dalbergia sissoo [e]
Machaerium scleroxylon [h]

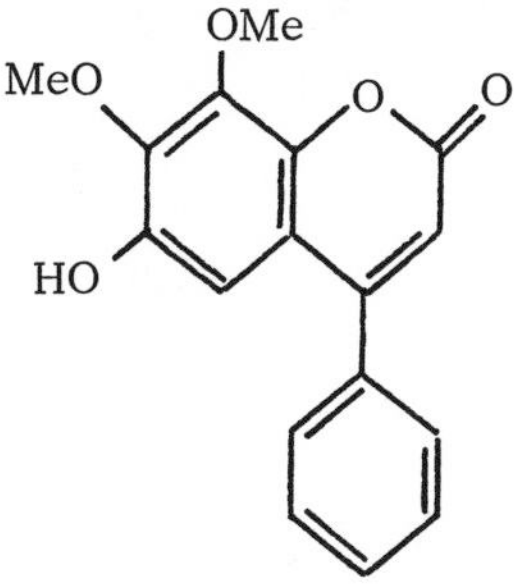

CX. 6-Hydroxy-7, 8-dimethoxy-4-phenylcoumarin

Machaerium kuhlmannii [g]

Fig. 11. Natural dalbergins. [a]Cook (1965); [b]Balakrishna et al. (1962); [c]Rao and Seshadri (1963); [d]B. J. Donnelly et al. (1965b); [e]Ahluwalia and Seshadri (1957); [f]Gregson (1965); [g]Roberts (1966); [h]Eyton et al. (1965b).

A definite decision regarding the biosynthesis of neoflavanoids cannot be taken until the appropriate feeding experiments have been done, but we feel that, on balance, the scheme indicated in Figure 12 is favoured.

The well-known reactions and reactivities of allylic derivatives (De Wolfe and Young, 1956, 1964) suggested that if the biosynthesis of neoflavanoids took the course indicated in Figure 12, then the duality of substitution characteristic of allylic compounds might also result in the appearance of cinnamyl-phenols as natural products. It was therefore particularly gratifying

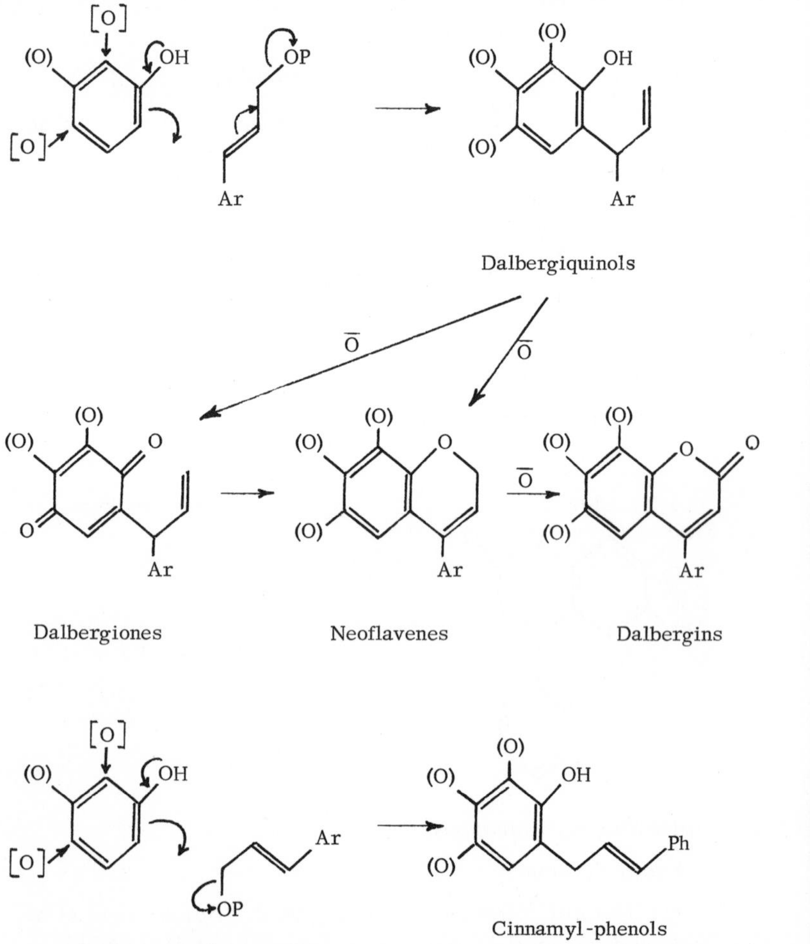

Dalbergiquinols

Dalbergiones

Neoflavenes

Dalbergins

Cinnamyl-phenols

Fig. 12. Proposed biosynthetic pathways leading to neoflavanoids and cinnamyl-phenols.

when a number of cinnamyl-phenols were indeed recognised as natural products in association with neoflavanoids. The cinnamyl-phenols and associated neoflavanoids show a striking structural correspondence in good support of the proposals indicated in Figure 12.

In connection with the mechanism proposed (Fig. 12) for the biosynthesis of dalbergiquinols and related neoflavanoids, a similar mechanism (CXI) has already been suggested (Birch, 1963) for the formation of natural allyl phenols. The isolation of elemicin (CXII) and 3,4,5-trimethoxycinnamic aldehyde (CXIII) from *Dalbergia spruceana* (Cook, 1965) fits in well, therefore, with the general biosynthetic patterns that are under consideration. A structural relative of these two compounds is the cinnamic ester (CXIV), which has been isolated from *Machaerium kuhlmannii* (Roberts, 1966).

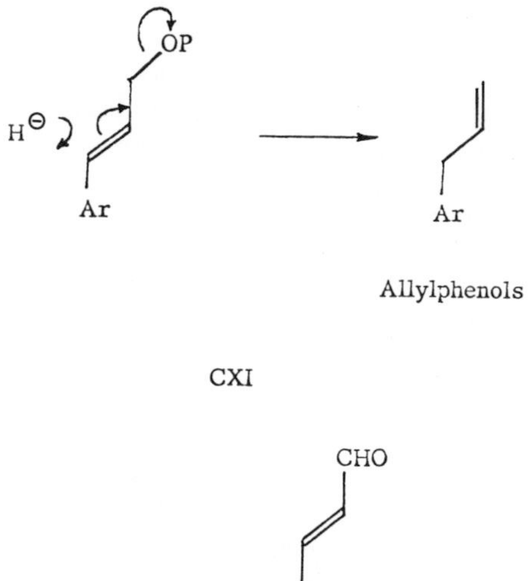

$H^{\ominus}$ OP Ar → Ar

Allylphenols

CXI

MeO OMe OMe CHO MeO OMe OMe CO_2Me OH OMe

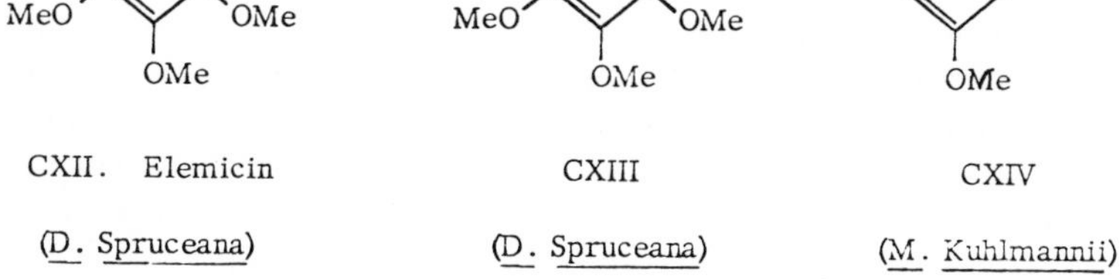

CXII. Elemicin (D. Spruceana)

CXIII (D. Spruceana)

CXIV (M. Kuhlmannii)

In some cases some simple phenolic derivatives (CXV, CXVI, CXVII, CXVIII) have been found in association with neoflavanoids.

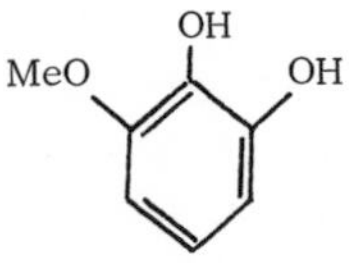

CXV (*M. kuhlmannii*)

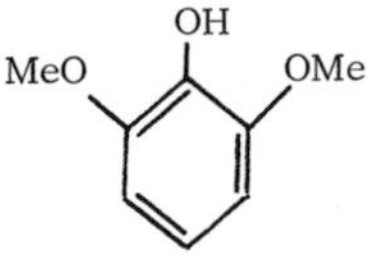

CXVI (*M. kuhlmannii*) (*M. mucronulatum*)

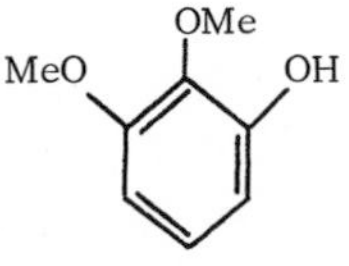

CXVII (*M. kuhlmannii*) (*M. opacum*)

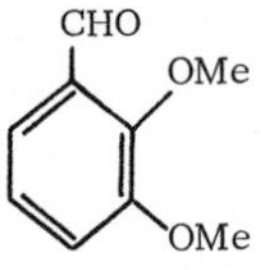

CXVIII (*M. kuhlmannii*)

VIII. THE NATURAL CINNAMYL-PHENOLS

Encouraged by the biogenetic speculation summarized in Section VII, we have examined a number of *Dalbergia* and *Machaerium* species for cinnamyl-phenols; the substances that have so far been isolated are listed in Fig. 13. The constitutions and stereochemistry of these compounds have been established by standard methods and confirmed by synthesis of either the compounds themselves or suitable derivatives (Gregson, 1965, 1966; Kurosawa, 1965).

IX. THE POSSIBLE ROLE OF NEOFLAVANOIDS AS PRECURSORS OF OTHER NATURAL PRODUCTS

During our examination of *Dalbergia* and *Machaerium* species we have isolated several other types of natural products in addition to the dalbergiones (Fig. 9), the dalbergiquinols (Fig. 10), the neoflavene (CVII), and the dalbergins (Fig. 11). Some of these are plant steroids and terpenoids, which are not considered in this review. However, a number of phenolic benzophenones has also been isolated (see Fig. 14) and proposals regarding their biogenetic relationship to the neoflavanoids (Fig. 8) are possible.

The natural benzophenones have a rather restricted phytochemical distribution (Geissman and Hinreiner, 1952; Karrer, 1958), and it is established that some natural benzophenones are formed by the polyketide route. However, the close similarity of the constitutions of cearoin (CXXIV) and scleroin (CXXV) to

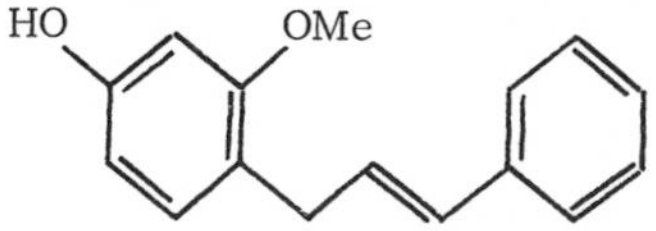

HO OMe

CXIX. Obtustyrene

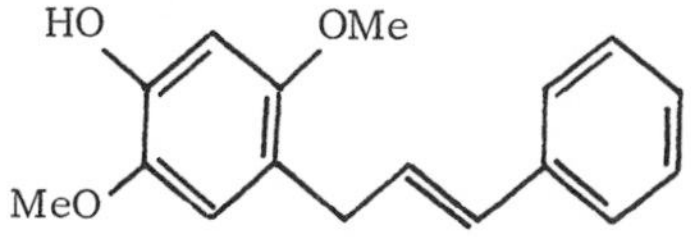

Dalbergia obtusa [a]

HO OMe MeO

CXX. Violastyrene

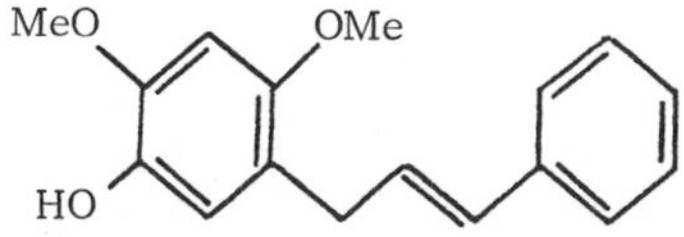

Dalbergia miscolobium [b] (syn. *D. violacea*)

MeO OMe HO

CXXI. Isoviolastyrene

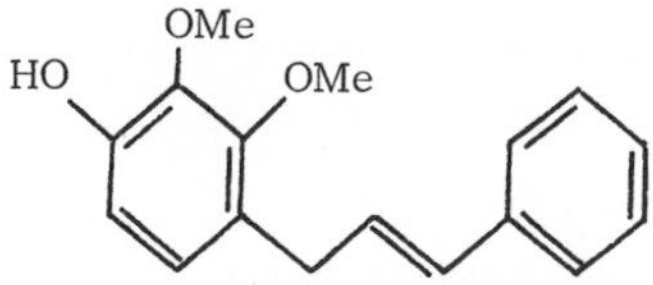

Dalbergia miscolobium [b] (syn. *D. violacea*)

OMe HO OMe

CXXII. Mucronustyrene

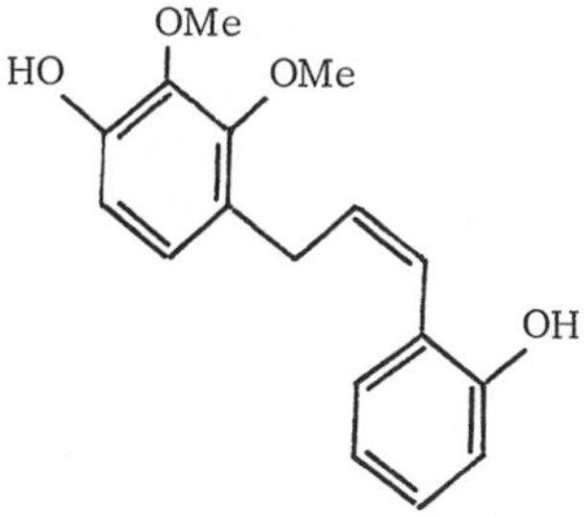

Machaerium mucronulatum [c]

OMe HO OMe OH

CXXIII. Mucronulastyrene

Machaerium mucronulatum [c]

Fig. 13. Natural cinnamyl-phenols. [a]Gregson (1966); [b]Gregson (1965); [c]Kurosawa (1965).

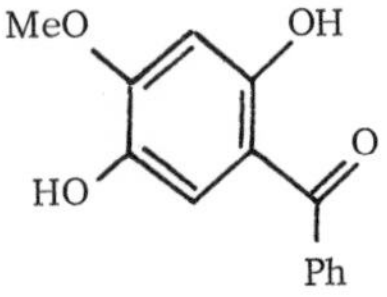

MeO OH HO O Ph

CXXIV. Cearoin

Dalbergia cearensis [a]
Dalbergia miscolobium [b]
(syn. D. violacea)

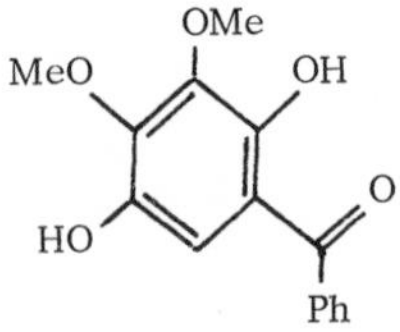

OMe MeO OH HO O Ph

CXXV. Scleroin

Machaerium scleroxylon [c]

Fig. 14. Natural benzophenones from *Dalbergia* and *Machaerium* species. [a]Cook (1965); [b]Gregson (1965); [c]Eyton et al. (1965).

that of the neoflavanoids (XCIX, C, CI, CV, CVIII, CIX) certainly suggests that these benzophenones represent an extension of the biodegradative pathway: dalbergiquinols ⟶ dalbergiones ⟶ neoflavenes ⟶ dalbergins (see Fig. 12). In this connection one can inquire; does the co-occurrence of *S*-4-methoxydalbergione (C), 6-hydroxy-7-methoxy-4-phenylcoumarin (CVIII), and benzoic acid in *Dalbergia spruceana* justify the extension of the above scheme to the following: dalbergiquinols ⟶ dalbergiones ⟶ neoflavenes ⟶ dalbergins ⟶ benzophenones ⟶ aromatic acids?

Thus the neoflavanoids provide a pattern of structural correlation that persuades us that they should be regarded as a new general class of natural phenolic compounds. Although they are structural analogues of the flavonoids and isoflavonoids (Fig. 7), we nevertheless feel that this similarity is superficial so far as the biosynthesis of neoflavanoids is concerned. On the evidence summarized in this review, we should like to propose (Ollis, 1966), as a basis for further experimental inquiry, the biogenetic relationships summarised in Figure 15.

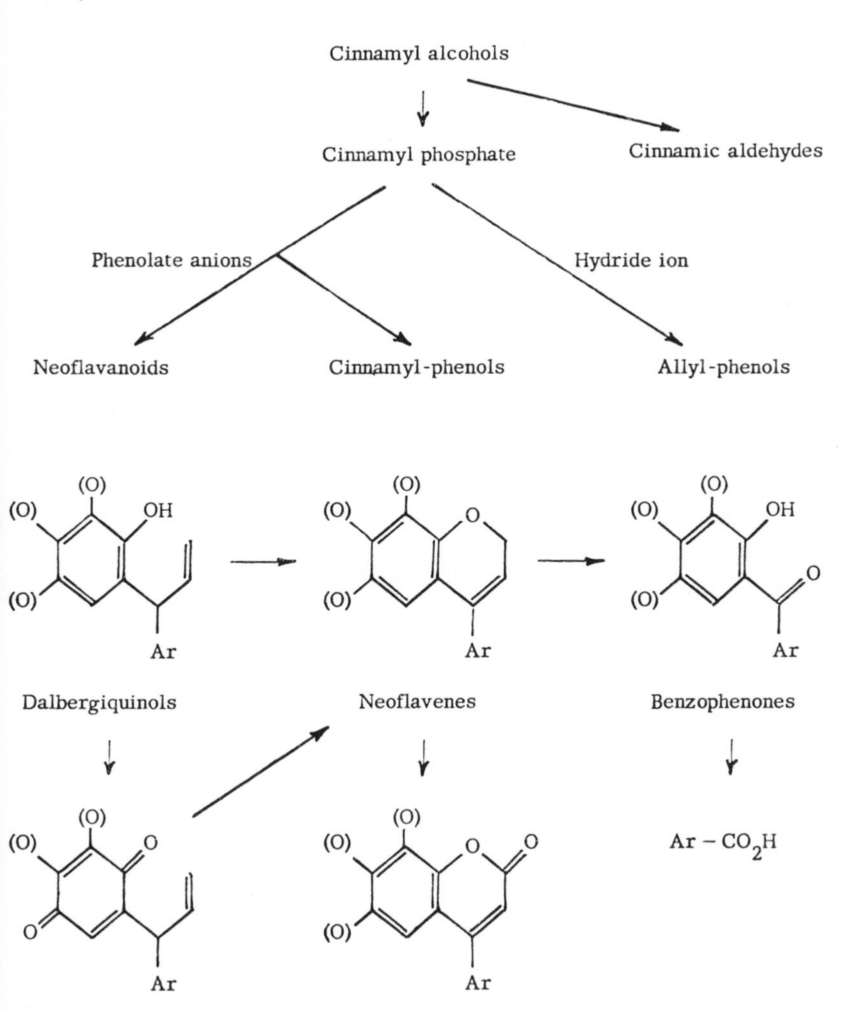

Fig. 15. A biogenetic proposal to interrelate the neoflavanoids with their natural congeners.

ACKNOWLEDGMENTS

The results described in this review refer mainly to the researches that we have carried out in collaboration with our Brazilian colleagues, and I should like to refer particularly to the benefits of our close association with Professor H. Magalhaẽs Alves

(Universidade de Minas Gerais, Belo Horizonte, Brasil), Dr. W. B. Eyton (Universidade Rural do Brasil, Rio de Janeiro), Professor O. R. Gottlieb (Universidade Rural do Brasil, Rio de Janeiro), and Professor M. Taveira Magalhaẽs (Universidade de Brasilia).

I must also express my warmest thanks to my colleagues of the University of Sheffield. Their studies have demanded extremely careful work and their specific contributions are indicated in the text.

This review was written during the tenure of a Visiting Professorship at the University of Texas. I should like to thank Dr. W. G. Whaley, Dr. H. C. Bold, the late Dr. R. E. Alston, and Dr. T. J. Mabry for their excellent hospitality.

REFERENCES

Ahluwalia, V. K., and T. R. Seshadri. 1957. J. Chem. Soc.: 970.

Akisanya, A., C. W. L. Bevan, and J. Hirst. 1959. J. Chem. Soc.: 2679.

Alves, Magalhaẽs, H. 1965. University of Minas Gerais and University of Sheffield, forthcoming publication.

Anderson, E. L., and G. F. Marrian. 1939. J. Biol. Chem., 127: 649.

Balakrishna, S., M. M. Rao, and T. R. Seshadri. 1962. Tetrahedron, 18: 1503.

Bevan, C. W. L., A. J. Birch, B. Moore, and S. K. Mukerjee. 1964. J. Chem. Soc.: 5991.

Bhatia, G. D., S. K. Mukerjee, and T. R. Seshadri. 1965. Indian J. Chem., 3: 422.

Birch, A. J. 1963. *In* Swain, T., ed., Chemical Plant Taxonomy, 141, London and New York, Academic Press.

_____, and M. Salahuddin. 1964. Tetrahedron Lett.: 2211.

Bowyer, W. J., J. N. Chatterjea, S. P. Dhoubadel, B. O. Handford and W. B. Whalley. 1964. J. Chem. Soc.: 4212.

Bredenberg, J. B., and P. K. Hietala. 1961. Acta Chem. Scand., 15: 936.

_____, and J. N. Shoolery. 1961. Tetrahedron Lett.: 285.

Brooks, B. T. 1910. Philippine J. Sci., 5A: 439 (1911, Chem. Zentr., 2: 649).

Burrows, B. F., W. D. Ollis, and L. M. Jackman. 1960. Proc. Chem. Soc.: 177.

Carter, A. L., and R. H. Common. 1964. Biochim. Biophys. Acta, 86: 56.
Clark, E. P. 1943. J. Org. Chem., 8: 489.
Clark-Lewis, J. W. 1962. Rev. Pure Appl. Chem., 12: 96.
_____, I. Dainis, and G. C. Ramsay. 1965. Aust. J. Chem., 18: 1035.
Cocker, W., T. Dahl, C. Demsey, and T. B. H. McMurry. 1962. Chem. Industr.: 216; J. Chem. Soc.: 4906.
_____, T. B. H. McMurry, and P. A. Staniland. 1965. J. Chem. Soc.: 1034.
Cook, Mrs. J. T. 1965. University of Sheffield, forthcoming publication.
Cooke, R. G., and I. D. Rae. 1964. Aust. J. Chem., 17: 379.
Condon, P., and J. Kuc. 1962. Phytopathology, 52: 182.
Crombie, L. 1963. Fortschr. Chem. Org. Naturst., 21: 275.
_____, and D. A. Whiting. 1963. J. Chem. Soc.: 1569.
Cruickshank, I. A. M., and D. R. Perrin. 1964. *In* Harborne, J. B., ed., Biochemistry of Phenolic Compounds, 528, London and New York, Academic Press.
Darshan, K., S. K. Mukerjee, and T. R. Seshadri. 1965. Tetrahedron, 21: 1495.
Dean, F. M. 1963. Naturally Occurring Oxygen Ring Compounds, 366, London, Butterworths.
De Wolfe, R. H., and W. G. Young. 1956. Chem. Rev., 56: 753.
_____, and W. G. Young. 1964. *In* Patai, S., ed., Chemistry of Alkenes, 681, New York, Interscience.
Dirscherl, W., and K. Schodder. 1940. Z. Physiol. Chem., 264: 57.
Donnelly, B. J., D. M. X. Donnelly, and C. B. Sharkey. 1965a. Phytochemistry, 4: 337.
Donnelly, D. M. X., M. R. Geoghegan, B. J. Nangle, and R. A. Laidlaw. 1965b. Tetrahedron Lett.: 4451.
Dyke, S. F., W. D. Ollis, M. Sainsbury, and J. S. P. Schwarz. 1964. Tetrahedron, 20: 1331.
East, A. J. 1963. Ph.D. thesis, Bristol, forthcoming publication.
Eyton, W. B., W. D. Ollis, I. O. Sutherland, L. M. Jackman, O. R. Gottlieb, and M. Taveira Magalhães. 1962. Proc. Chem. Soc.: 301.
_____, W. D. Ollis, I. O. Sutherland, O. R. Gottlieb, M. Taveira Magalhaẽs, and L. M. Jackman. 1965a. Tetrahedron, 21: 2683.
_____, W. D. Ollis, M. Fineberg, O. R. Gottlieb, Salignac de Souza Guimarães, and M. Taveira Magalhães. 1965b. Tetrahedron, 21: 2697.

Falshaw, C. P. 1965. University of Sheffield, forthcoming publication.
_____, and W. D. Ollis. 1966. Chem. Commun.: 305.
_____, W. D. Ollis, J. A. Moore, and K. Magnus. 1966. Tetrahedron. Supplement No 7:333.
Geissman, T. A., and Hinreiner, E. H. 1952. Boran Rev., 18:77.
Gilbert, A. H., A. McGookin, and A. Robertson. 1957. J. Chem. Soc.: 3740.
Gottlieb, O. R., and A. de Silva Braga. 1965. University of Brasilia. (Present address: Universidade Rural do Brasil, Rio de Janeiro), forthcoming publication.
Gregson, M. 1965. Ph.D. thesis, Sheffield, forthcoming publication.
_____. 1966. University of Sheffield, forthcoming publication.
Hardegger, E., M. Schellenbaum, and H. Corvodi. 1963. Helv. Chim. Acta, 46: 1171.
Harmer, R. A. 1965. Ph.D. thesis, Sheffield, forthcoming publication.
Harper, S. H. 1942. J. Chem. Soc.: 181.
_____, A. D. Kemp, and W. G. E. Underwood. 1965a. Chem. Industr.: 562.
_____, A. D. Kemp, and W. G. E. Underwood. 1965b. Chem. Commun.: 309.
Hertelendy, F., and R. H. Common. 1964. J. Chromatogr., 13: 570.
Hinman, J. W., E. L. Caron, and Hocksema. 1957. J. Amer. Chem. Soc., 79: 3789.
Isler, O., A. Langemann, H. Mayer, R. Rüegg, and P. Schudel. 1965. Bull. Nat. Inst. Sci. India (No. 28): 132.
Itô, S., Y. Fujise, and A. Mori. 1965. Chem. Commun.: 595.
IUPAC Information Bulletin. 1966. Number 25: 24, Report on Nomenclature of Quinones with Isoprenoid Side-Chains.
Iyers, R. N., K. H. Shah, and K. Venkataraman. 1951. Proc. Indian Acad. Sci., 33A: 228.
Johnson, A. P., A. Pelter, and P. Stainton. 1966. J. Chem. Soc. (C): 192. See also 1964, Tetrahedron Lett.: 1209, 2817.
Jones, H. A. 1934. J. Amer. Chem. Soc., 56: 1247.
_____, and H. L. Haller. 1943. J. Org. Chem., 8: 493.
Jurd, L. 1962. *In* Geissman, T. A., ed., The Chemistry of Flavonoid Compounds, 107, Oxford, Pergamon.
Karrer, W. 1958. Konstitution und Vorkommen der organischen Pflanzenstoffe, Basel, Birkhäuser.
King, F. E., C. B. Cotterill, D. H. Godson, L. Gurd, and T. J. King. 1953. J. Chem. Soc.: 3693.

Klyne, W., and A. A. Wright. 1957. Biochem. J., 66: 92.

_____, and A. A. Wright. 1959. J. Endocr., 18: 32.

Krishna, S., and T. P. Ghose. 1942. Indian Forest Leaflet No. 2.

Kubota, T., and T. Matsuura. 1953. J. Chem. Soc. Jap., 74: 248.

Kurosawa, K. 1965. University of Sheffield, forthcoming publication.

Livingston, A. L., E. M. Bickoff, R. E. Lundin, and L. Jurd. 1964. Tetrahedron, 20: 1963.

_____, S. C. Witt, R. E. Lundin, and E. M. Bickoff. 1965. J. Org. Chem., 30: 2353.

MacRae, H. F., D. G. Dale, and R. H. Common. 1960. Canad. J. Biochem. Physiol., 38: 523.

Marrian, G. F., and D. Beall. 1935. Biochem. J., 29: 1586.

McGookin, A., A. Robertson, and W. B. Whalley. 1940. J. Chem. Soc.: 787.

Ollis, W. D. 1961. Proceedings of the Symposium on Phytochemistry, 128, Hong Kong Univ. Press.

_____. 1962. *In* Geissman, T. A., ed., The Chemistry of Flavonoid Compounds, 353, Oxford, Pergamon.

_____. 1966. Experientia. 22: 777.

_____, and I. O. Sutherland. 1961. *In* Ollis, W. D., ed., Recent Developments in the Chemistry of Natural Phenolic Compounds, 74, Oxford, Pergamon.

_____, M. V. J. Ramsay, and I. O. Sutherland. 1965a. Aust. J. Chem., 18: 1787, and references there cited.

_____, M. V. J. Ramsay, I. O. Sutherland, and S. Mongkolsuk. 1965b. Tetrahedron, 21: 1453.

Perrin, D. D., and D. R. Perrin. 1962. J. Amer. Chem. Soc., 84: 1922.

Perrin, D. R. 1964. Tetrahedron Lett.: 29.

_____, and W. Bottomley. 1962. J. Amer. Chem. Soc., 84: 1919.

Rao, M. M., and T. R. Seshadri. 1963. Tetrahedron Lett.: 211.

Redman, B. T. 1966. University of Sheffield, forthcoming publication.

Roberts, R. J. 1966. University of Sheffield, forthcoming publication.

Robertson, A., and W. B. Whalley. 1954. J. Chem. Soc.: 1440.

Ryan, H., and R. Fitzgerald. 1913. Proc. Roy. Irish Acad. [B], 30: 106 (1913, Chem. Zentr., 2: 2048).

Schwarz, J. S. P., A. I. Cohen, W. D. Ollis, E. A. Kaczka, and L. M. Jackman. 1964. Tetrahedron, 20: 1317.

Seshadri, T. R. 1963. J. Indian Chem. Soc., 40: 497.

Shibata, S., and Y. Nishikawa. 1963. Chem. Pharm. Bull. (Tokyo), 11: 167.

Simonitsch, E., H. Frei, and H. Schmid. 1957. Mh. Chem., 88: 541.
Späth, E., and J. Schläger. 1940. Chem. Ber., 73: 1.
Sreerama Murti, V. V., N. V. Subba Rao, and T. R. Seshadri. 1948. Proc. Indian Acad. Sci., 27A: 111.
Stamm, O. A., H. Schmid, and J. Büchi. 1958. Helv. Chim. Acta, 41: 2006.
Subba Rao, N. V., and W. A. Khan. 1961. J. Sci. Indian Res., 20B: 87.
_____, and W. A. Khan. 1963. Indian J. Chem., 1: 74, 295.
_____, and T. R. Seshadri. 1946a. Proc. Indian Acad. Sci., 24A: 365.
_____, and T. R. Seshadri. 1946b. Proc. Indian Acad. Sci., 24A: 465.
Suemitsu, R., M. Huira, and M. Makajima. 1955. J. Agric. Chem. Soc. Japan, 29: 591 (1958, Chem. Abst., 52: 20511).
Suginome, H., and T. Iwadare. 1962. Experientia, 18: 163.
Urech, J., B. Fechtig, J. Nuesch, and E. Vischer. 1963. Helv. Chim. Acta, 46: 2759.
Uritani, I., T. Akazawa, and M. Uritani. 1954. Nature (London), 174: 1060.
Van Duuren, B. L. 1961. J. Org. Chem., 26: 5013.
Weinges, K. 1964. Proc. Chem. Soc.: 138.
_____, and E. Paulus. 1965. Liebig Ann. Chem., 681: 154.
Wessely, F., and F. Prillinger. 1939. Ber., 72B: 629.
_____, H. Hirschel, G. Schögl-Petziwal, and F. Prillinger. 1938. Mh. Chem., 71: 215.
Wheeler, R. E. 1963. Ph.D. thesis, Bristol, forthcoming publication.
Wolfrom, M. L., and B. S. Wildi. 1951. J. Amer. Chem. Soc., 73: 235, and references there cited.

11

RECENT INVESTIGATIONS ON THE BIOSYNTHESIS OF FLAVONOIDS

HANS GRISEBACH
Chemical Laboratory, University of
Freiburg, Germany

I. INTRODUCTION

In recent years considerable progress has been made in the elucidation of the biosynthesis of flavonoids, mainly by studies with labeled precursors in vivo. The recent review by Grisebach (1965) covers the work to the end of 1963, and in this chapter the older work will be mentioned only if it is necessary for the understanding of the newer results.

II. ACTIVATION OF CINNAMIC ACIDS IN HIGHER PLANTS

As an extension of Birch's hypothesis (Birch and Donovan, 1953) we had postulated (Grisebach, 1962) that the first reaction in flavonoid biosynthesis is the condensation of an activated cinnamic acid (probably the CoA ester) with three molecules of malonyl-CoA. The first stable intermediate formed from such a condensation would be a chalcone, which is in an enzyme-catalyzed equilibrium with the corresponding flavanone (Fig. 1).

If the reaction formulated in Fig. 1, or a similar reaction, is the first step in the biosynthesis of flavonoids, the plant must be able to activate cinnamic acid (or acids). We have carried out experiments on the activation of cinnamic acids with protein preparations from *Cicer arietinum* L. and parsley (*Apium petroselinum*) (Hahlbrock, 1965). Buckwheat and red cabbage seedlings were found to give very poor yields of protein and were, therefore, not used in these investigations.

Protein preparations from leaves of *Cicer* or parsley were prepared according to the following scheme (Fig. 2). An important purification step was the treatment of the protein from the ammonium sulfate precipitation with Dowex 1. Only after this treatment could a sharp protein peak be obtained upon elution from the Sephadex G-50 column. Several acids were tested with the hydroxamic acid test (Millerd and Bonner, 1954) for their activation by the protein extract from parsley. A typical result is shown in Table 1. The activation of cinnamic and *p*-coumaric

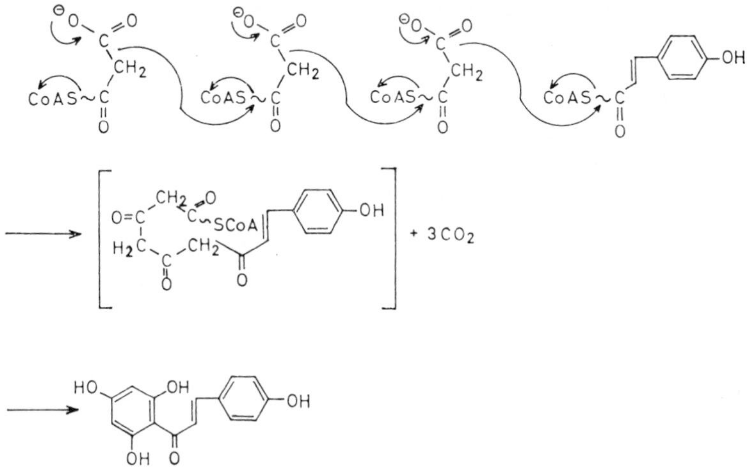

Fig. 1. Hypothetical formation of a chalcone from *p*-coumaryl-CoA and malonyl-CoA.

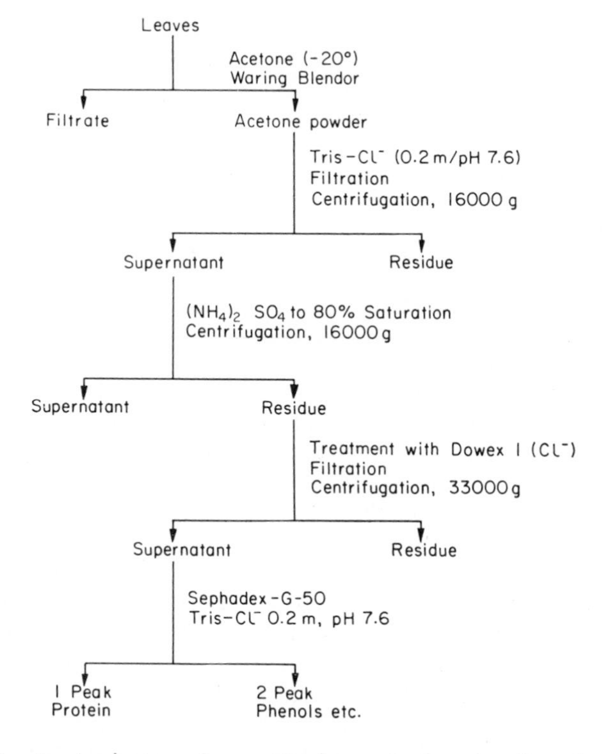

Fig. 2. Isolation of a purified protein fraction from leaves of *Cicer arietinum* or parsley.

TABLE 1

Comparison of the Activation of Several Acids with the Protein from Parsley by the Hydroxamic Acid Test[a]

Compound	Hydroxamic Acid, mμmoles
Monocarboxylic acids	
Acetic	1040
Propionic	150
Butyric	60
Capronic	80
Palmitic	30
Stearic	(5)
Dicarboxylic acids	
Oxalic	810
Malonic	540
Succinic	1290
Glutaric	25
Tricarboxylic acids	
Citric	50
Phenylpropane acids	
Cinnamic	15
p-Coumaric	15
Phenyllactic	15
Phenylpyruvic	20

[a]The reaction mixture contained 100 μmoles Tris-Cl^- (pH 7.6), 5 μmoles of ATP, 5 μmoles of $MgSO_4$, 5 μmoles of glutathione, 0.3 μmoles of CoA, 5 μmoles of the potassium salt of the acid, 200 μmoles of NH_2OH, and 5.5 mg protein, and was incubated at 37° for 2 hr in an atmosphere of N_2. At the end of the incubation, 0.25 ml of the trichloroacetic acid—$FeCl_3$ solution was added and the extinction at 546 μm was measured. Controls contained no CoASH or acid.

acid, although very low, is outside of the limits of experimental error. Whether this low activation is due to an unspecific reaction, or whether it is brought about by an enzyme that is responsible for the activation of cinnamic acids in plants, cannot be decided from these results. No evidence was found for a transacylation reaction between acetyl-CoA or succinyl-CoA and cinnamic or *p*-coumaric acid.

III. INTACT INCORPORATION OF CHALCONES (FLAVANONES) INTO FLAVONOIDS

In a number of earlier experiments, we had demonstrated the incorporation of suitably substituted ^{14}C-labeled chalcones (flavanones) into cyanidin and quercetin (Grisebach and Patschke, 1961), into the isoflavones biochanin A and formononetin (Grisebach and Patschke, 1960; Grisebach and Brandner, 1961, 1962), and into the coumaranocoumarin coumestrol (Grisebach and Barz, 1964b). In each case it was proved by degradation experiments that over 95 percent of the labeled atoms were in the expected position (Fig. 3). In recent experiments, it has also been

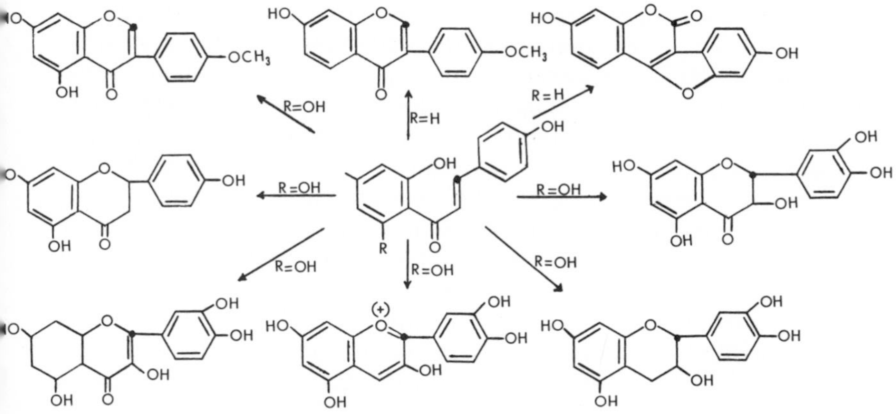

Fig. 3. Incorporation of chalcones into various flavanoids.

demonstrated that the 4,2′,4′,6′-tetrahydroxychalcone 2′-glucoside-[β-^{14}C] (I) is incorporated into catechins in tea leaves (Patschke and Grisebach, 1965a) and into the flavone apigenin in parsley (Grisebach and Billhuber, 1967) with no randomization of activity (Fig. 3).

In *Cicer arietinum*, the chalcone, I, is incorporated only into an isoflavone with phloroglucinol type of substitution in ring A (biochanin A), and the 4,4′,6′-trihydroxychalcone 4′-glucoside-[β-^{14}C] (II) served only as a precursor for an isoflavone with a resorcinol type of ring A (formononetin). These results are consistent with an intact incorporation of the chalcones. However,

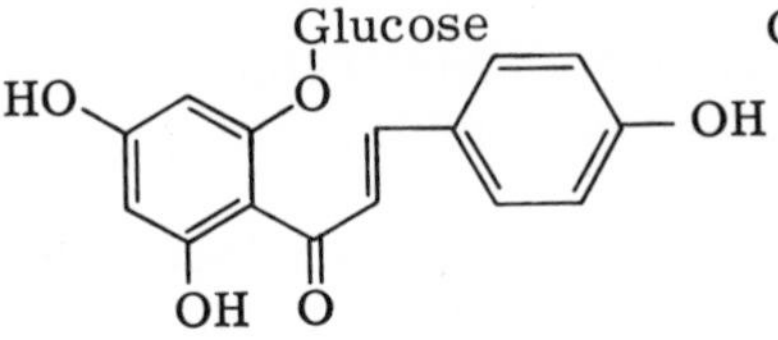

Fig. 4. Degradation of chalcones (flavanones) in the plant, R = OH or H.

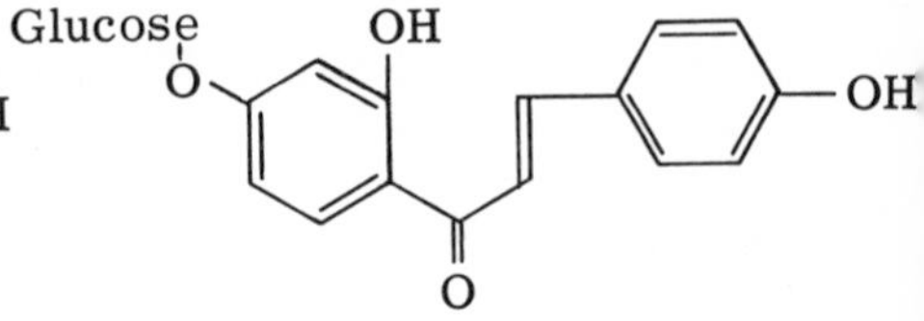

it remained uncertain if the chalcone, I, is incorporated intact into both cyanidin and quercetin, because the results discussed above would also have been obtained if the chalcone had been degraded to a phenylpropane compound prior to incorporation into the flavonoids. By analogy to the experiments with *Cicer arietinum*, we first compared the incorporation of chalcones I and II into cyanidin in red cabbage seedlings, since it was expected from the results with isoflavones that only chalcone I would serve as a precursor for cyanidin. Contrary to expectation, both chalcones were incorporated into cyanidin, and in both cases the radioactivity was located at C-2 in cyanidin. An explanation for these results could be either that a hydroxyl group is introduced at C-5 at some stage of the biosynthesis, or that a degradation of the chalcone-[β-^{14}C] to *p*-coumaric acid-[β-^{14}C] takes place in the seedlings. The incorporation of this acid into cyanidin would then lead to the same labeling pattern as that expected from intact incorporation of the chalcone. By dilution analysis it could indeed be shown that *p*-coumaric acid-[β-^{14}C] is formed from chalcone II in red cabbage seedlings (Patschke et al., 1964b). The incorporation rate into this acid was about 0.09 percent. In further experiments with *Petunia hybrida*, it was shown that chalcone I could also be degraded by the plant to *p*-coumaric acid. Radioactive ferulic acid containing less activity than the *p*-coumaric acid was also isolated from the same plant (Fig. 4).

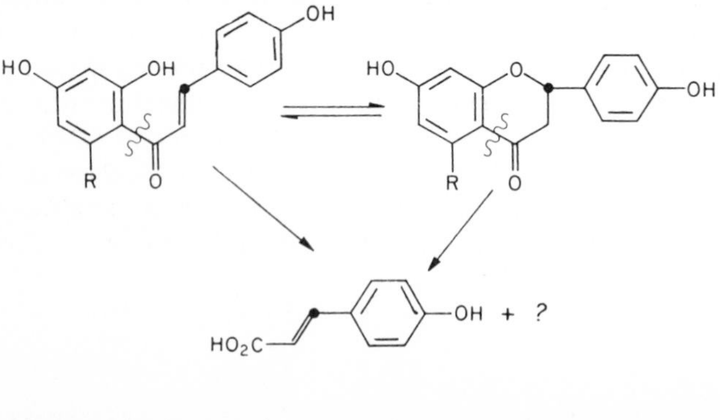

What happens to ring A during this degradation is as yet unknown. The fact that this ring can be further degraded by the plant is demonstrated by the observation that radioactive CO_2 is evolved when chalcone I, labeled only in ring A, is fed to *Cicer arietinum* (Patschke et al., 1964a).

The discovery of this unexpected degradation of chalcones in plants made it necessary to reinvestigate some of our earlier experiments, which were carried out with chalcones labeled only in the β-position with ^{14}C (C-2 of the corresponding flavanone). The best way to prove the intact incorporation of chalcones into flavonoids seemed to be the administration of a chalcone labeled both in ring A and in the β-position. We therefore synthesized 5,7,4′-trihydroxyflavanone-[2,6,8,10-$^{14}C_4$] (III)[1] (Fig. 5). This

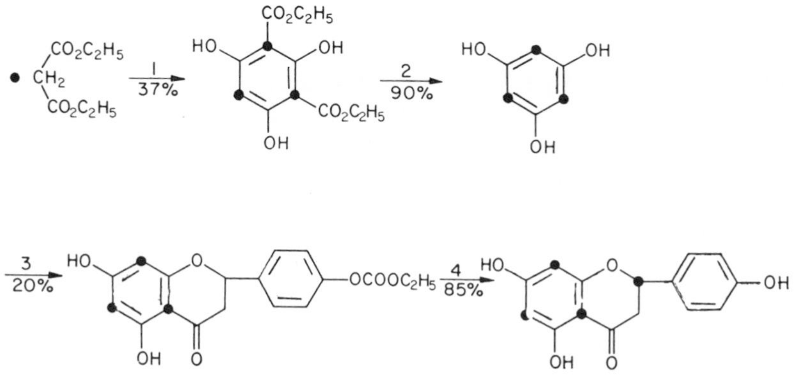

$CO_2C_2H_5$–CH_2–$CO_2C_2H_5$ $\xrightarrow[37\%]{1}$ (HO, OH, OH; $CO_2C_2H_5$, $CO_2C_2H_5$) $\xrightarrow[90\%]{2}$ (HO, OH, OH) $\xrightarrow[20\%]{3}$ (HO, OH, O; $OCOOC_2H_5$) $\xrightarrow[85\%]{4}$ (HO, OH, O; OH)

Fig. 5. Synthesis of 5,7,4′-trihydroxyflavonone-[2,6,8,10-$^{14}C_4$]. (1) $NaOC_2H_5$, 135°; (2) 60 percent KOH, N_2; (3) *p*-Carboethoxycinnamyl chloride, $AlCl_3$ in nitrobenzene; (4) 1, 2 *N* NH_3 and mixture with flavanone-[2-^{14}C].

compound was administered to red cabbage seedlings whose roots had been removed, and the ratio of radioactivity of ring A/C-2 in the isolated cyanidin was determined by degradation of this compound to phloroglucinol and protocatechuic acid. The results are summarized in Table 2. It can be seen that the ratio of radioactivity of ring A/C-2 remains unchanged upon incorporation of the flavanone III into cyanidin, thereby proving the

[1] 4,2′,4′,6′-Tetrahydroxychalcone-[1′,3′,5′-β-$^{14}C_4$]. Since the C-2′ hydroxyl group is free, the equilibrium in aqueous solution is entirely on the side of the flavanone.

TABLE 2

Incorporation of Multiple-Labeled Flavanone into Cyanidin

Compound	cpm/mmoles	Activity Ratio Ring A/C-2
5,7,4′-Trihydroxyflavanone-[2, 6, 8, 10-^{14}C] (III)	—	1:2.04
Cyanidin	—	1:2
Protocatechuic acid	11,920	
Phloroglucinol	5,915	

intact conversion of III into cyanidin (Patschke et al., 1964a).

The multiple-labeled flavanone III was also fed to shoots of young chana seedlings (*Cicer arietinum*). In agreement with our earlier results (Grisebach and Brandner, 1962), the incorporation of III into the isoflavone with the same hydroxylation pattern (biochanin A) was much higher than into the 7,4′-dihydroxyisoflavone (formononetin; Table 3). Degradation of the isoflavones showed that there was only a slight change in the ratio of radioactivity of ring A/C-2 in biochanin A, whereas in formononetin this ratio fell to 1 : 7 (Table 4). These results prove the intact

TABLE 3

Comparison of the Activities of the Isoflavones with the Flavanone III as Precursor

Compound	Dilution[a]	Incorporation Rate, %
Biochanin A	46	0.05
Formononetin	1180	0.004

[a]Specific activity of III to specific activity of compound.

TABLE 4

Activity Ratios in the Isoflavones

Compound	Activity Ratio Ring A/C-2
Trihydroxyflavanone (III)	1:2.02
Biochanin A	1:2.24
Formononetin	1:7

and specific incorporation of III into biochanin A. The slight drop in the activity ratio in biochanin A could be attributed to the concomitant incorporation of a small amount of *p*-coumaric acid-[β-^{14}C] (formed by degradation of the flavanone). The minor activity in formononetin apparently resulted mainly from an incorporation of *p*-coumaric acid-[β-^{14}C], since the relative activity in C-2 of the isoflavone increased considerably. This final proof of the intact incorporation of chalcones into cyanidin and into isoflavones strongly supports our earlier conclusions that these compounds are important intermediates in flavonoid biosynthesis (Grisebach, 1965).

IV. STEREOSPECIFIC INCORPORATION OF (−)-(2*S*)-FLAVANONE INTO FLAVONOIDS

Earlier investigations with labeled flavanones were all carried out with racemic compounds. However, the naturally occurring flavanones are all optically active (Clark-Lewis, 1962), and, so far, only (−)-flavanones with the (2*S*) configuration (Arakawa and Nakazaki, 1960) have been found in nature. To investigate the question of whether the transformation of flavanones to other flavonoids is a stereospecific process we have synthesized (+)- and (−)-5,7,4′-trihydroxyflavanone 5-β-D-glucoside-[2-^{14}C] (naringenin 5-glucoside). (±)-Naringenin 5-glucoside-[2-^{14}C] was obtained by cyclization (Zemplén et al., 1943) of the corresponding chalcone glucoside-[β-^{14}C] (Grisebach and Patschke, 1961) and purification on a polyamide column (Neu, 1958, 1960). The (±)-naringenin glucoside is a mixture of C-2 configurational isomers and can be separated into the diastereomers by paper chromatography with methanol–water (30 : 70; (−), $R_f = 0.5$; (+), $R_f = 0.61$) (Hänsel and Heise, 1959; Hänsel et al., 1963).

A. Experiments with Buckwheat Seedlings

In two parallel experiments, 5 μC of the (+) and the (−) compound, respectively, was administered to eight-day-old, etiolated buckwheat seedlings, without roots, for 24 hours in the presence of light. In both experiments about 60 percent of the radioactivity was taken up by the seedlings. Quercetin and cyanidin were then isolated in the usual way (Patschke et al., 1964a) and purified to constant specific activity. The results are shown in Table 5. The incorporation rate of the (−)-flavanone glucoside into quercetin was 12 times higher but into cyanidin only 1.7 times higher than

TABLE 5

Incorporation of (–)–and (+)–Naringenin 5–Glucoside–[2–^{14}C] into Flavonoids in Buckwheat Seedlings[a]

Precursor or Compound	Incorporation Rate, %	Dilution
(–)-Naringenin 5-glucoside-[2-^{14}C]		
Quercetin	3.7	114
Cyanidin	1.27	26
(+)-Naringenin 5-glucoside-[2-^{14}C]		
Quercetin	0.3	1348
Cyanidin	0.71	39

[a]The seedlings were grown in the dark before initiation of the experiments.

that of the (+) compound. Since the formation of cyanidin does not begin until 12 hours after the period of illumination commences, the small difference in the incorporation rate of the (+)- and the (–)-flavanone into cyanidin was probably the result of racemization of this precursor prior to its incorporation into cyanidin. Therefore, in a second experiment, the seedlings were exposed to light for 15 hours before the radioactive flavanone was made available. It can be seen from Table 6 that the difference in the incorporation of the (+) and (–) compound into cyanidin was now more significant.

It can be concluded from the above results that the (–)-flavanone is the natural intermediate for the biosynthesis of quercetin and cyanidin in buckwheat seedlings. The incorporation of the (+) compound is probably the result of racemization (via the chalcone) of this compound prior to its incorporation into the other flavonoids. The degree of racemization could be different in different cells of the plant, and thus explain the higher incorporation of the (+)-flavanone into cyanidin in comparison with the incorporation into quercetin. For example, cyanidin is formed in the epidermis and subepidermis of the hypocotyl, whereas quercetin is synthesized only in the cotyledons of buckwheat seedlings (Mohr and van Nes, 1963; Harraschain and Mohr, 1963).

TABLE 6

Incorporation of (–)–and (+)–Naringenin 5–Glucoside into Flavonoids in Buckwheat Seedlings[a]

Precursor or Compound	Incorporation Rate, %	Dilution
(–)-Naringenin 5-glucoside-[2-^{14}C]		
Quercetin	3.07	146
Cyanidin	2.36	28
(+)-Naringenin 5-glucoside-[2-^{14}C]		
Quercetin	0.19	2514
Cyanidin	0.66	99

[a]The seedlings were exposed to light 15 hours before initiation of the experiments.

B. Experiments with Chana Seedlings

Experiments similar to those described above were carried out with chana seedlings (*Cicer arietinum* L.), and the incorporation of the (+)- and (–)-flavanone into biochanin A and formononetin was determined. Table 7 shows that the (–) compound was

TABLE 7

Incorporation of (–)–and (+)–Naringenin 5–Glucoside into Isoflavones in Chana Seedlings

Precursor or Compound	Incorporation Rate, %	Dilution
(–)-Naringenin 5-glucoside-[2-^{14}C]		
Biochanin A	0.21[a]	163
Formononetin	0.0054	27000
(+)-Naringenin 5-glucoside-[2-^{14}C]		
Biochanin A	0.015	5440
Formononetin	0.0023	46000

[a]Part of the biochanin A was lost during the extraction procedure.

incorporated into biochanin A 14 times better than was the (+) compound, whereas, in agreement with earlier results (Grisebach, 1965; Patschke et al., 1964a), the incorporation of both enantiomers into formononetin was very low. These results prove the stereospecific incorporation of the naturally occurring (–)-(2*S*)-5,7,4′-trihydroxyflavanone into other flavonoids including isoflavones (Patschke et al., 1966a) and strongly corroborate the assumption that flavanones (chalcones) are normal intermediates in flavonoid biosynthesis.

V. INVESTIGATIONS ON THE MECHANISM OF THE CONVERSION OF FLAVANONES TO 3-HYDROXYFLAVANONES

Wong (1964) has recently shown that 4,2′,4′-trihydroxychalcone is also incorporated into 3,7,4′-trihydroxyflavanone (garbanzol) in cell-free extracts of *Cicer arietinum*. In an independent investigation we have studied the mechanism of the conversion of 4,2′,4′,6′-tetrahydroxychalcone to taxifolin in *Chamaecyparis obtusa* (Grisebach and Kellner, 1965). For the formation of dihydroflavonols (3-hydroxyflavanones) from flavanones (chalcones), several pathways can be formulated (Fig. 6):

1. Direct oxidation of a flavanone at C-3.
2. Formation via a chalcone-epoxide.
3. The pathway flavanone ⟶ flavone ⟶ flavonol ⟶ dihydroflavonol.
4. Addition of water to the 2,3 double bond of a flavone.

It should be possible to eliminate either pathways 1 and 2 or pathways 3 and 4 by using, as a precursor, a chalcone labeled in the β-position with tritium and ^{14}C. This tritium should be lost if the dihydroflavonol is formed via pathway 3 or 4, whereas if pathway 1 or 2 is utilized, tritium should appear in the dihydroflavonol. The tritium-labeled chalcone was synthesized as indicated in Fig. 7. This tritium-labeled compound was then mixed with the corresponding ^{14}C-labeled chalcone to produce 4,2′,4′,6′-tetrahydroxychalcone 2′-glucoside-[β-^{14}C-β-T] (IV) with a $^{14}C/T$ ratio of 1:2.1. This material was fed to leaves of *Chamaecyparis obtusa* from which we had previously isolated taxifolin 3-xyloside (3,5,7,3′,4′-pentahydroxyflavanone 3-xyloside). After several purification steps, the radioactive taxifolin xyloside (V) had a $^{14}C/T$ ratio of 1:2.3. When V was oxidized with sodium

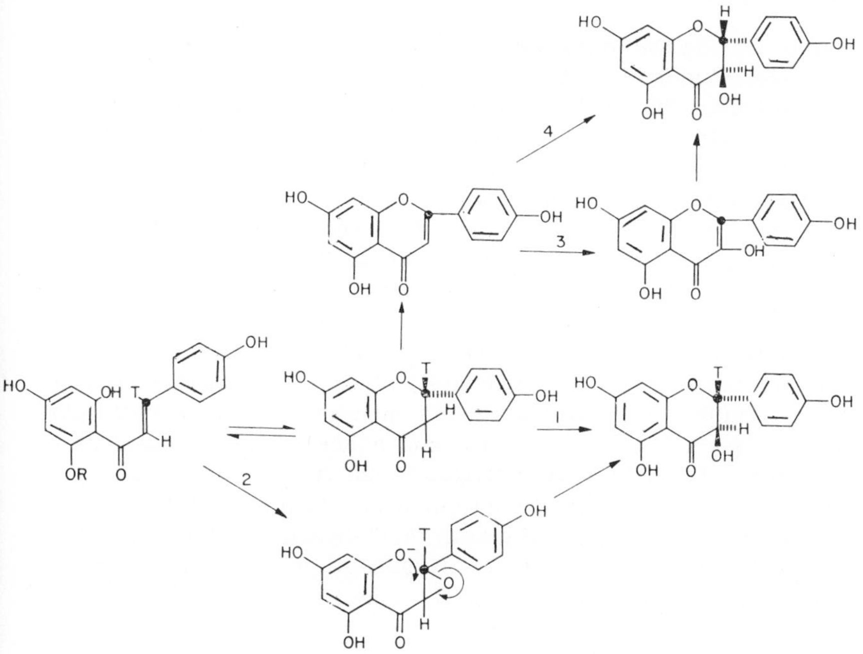

Fig. 6. Possible pathways for the formation of 3-hydroxyflavanones from chalcones (flavanones).

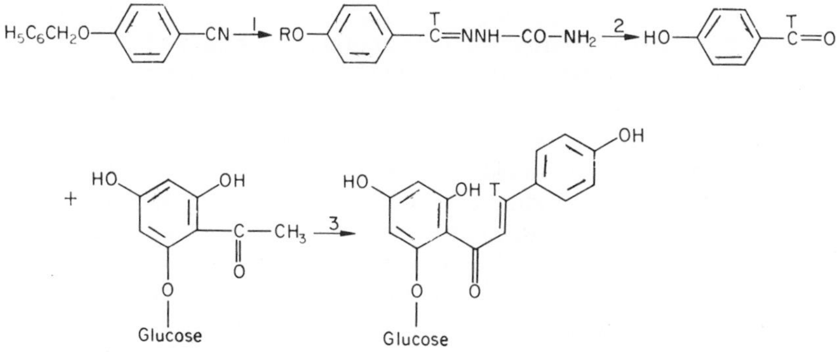

Fig. 7. Synthesis of tritium-labeled chalcone glucoside. (1) HT, Raney-Ni, semicarbazide, 65°; (2) formaldehyde, then HCl; (3) KOH, N_2.

bisulfite to quercetin xyloside, this ratio increased to 1: 0.13, proving that the bulk of the tritium in V was localized at C-2 or C-3 or both. These results demonstrate that the tritium in the β-position of the chalcone (2-position of the flavanone) is retained during the biosynthesis of taxifolin. Pathways 3 and 4 can, therefore, be excluded.

In connection with pathways 1 and 2, it is necessary to discuss briefly some stereochemical aspects. All dihydroflavonols, that have so far been found to occur in nature have the 2-aryl and 3-hydroxyl group in a *trans* position (equatorial, equatorial). The same is true of the compounds in the catechin series. Furthermore, (–)-flavanones and (+)-dihydroflavonols have the same absolute stereochemistry at C-2 (Whalley, 1962). The in vitro synthesis of dihydroflavonols via chalcone epoxides also leads to *trans* products (Fischer and Arlt, 1964). On the other hand, the direct oxidation of flavanones at C-3, with Fenton's reagent, for example, seems to lead to the unnatural *cis*-3-hydroxyflavanones (Dean, 1963). However, oxygenases are known that can substitute an equatorial hydrogen atom with a hydroxyl group in vivo without change of configuration (Hayano, 1962). It is therefore possible that in plants the 3-hydroxyflavanones are formed from flavanones by an oxygenase, which replaces the equatorial hydrogen atom at C-3 of the flavanone with a hydroxyl group.

VI. THE ROLE OF 3-HYDROXYFLAVANONES (DIHYDROFLAVONOLS) IN FLAVONOID BIOSYNTHESIS

The 3-hydroxyflavanones (dihydroflavonols) have been postulated as intermediates in the biosynthesis of flavonols, anthocyanins, and isoflavones (Grisebach, 1965). To investigate the role of dihydroflavonols in flavonoid biosynthesis, 3,5,7,4′-tetrahydroxyflavanone (dihydrokaempferol) (VI) was labeled with tritium by a modified Wilzbach technique (Wollenberg and Wenzel, 1963) and purified to radio purity by paper chromatography and recrystallization. The radio purity was checked by dilution analysis. When compound VI, having a specific activity of 11.7 mC/mmole, was oxidized with sodium bisulfite (Pachéco, 1960), kaempferol-[T] (VII) with a specific activity of 6.84 mC/mmole was obtained. This result proves that 41.5 percent of the tritium in VI was located at carbon atoms 2 and 3.

A. Incorporation into Quercetin and Cyanidin

In two parallel experiments dihydrokaempferol-[T] (VI) or kaempferol-[T] (VII) together with phenylalanine-[1-^{14}C] as an internal standard was fed in light to etiolated seven-day-old buckwheat seedlings without roots over a period of 48 hours. Since dihydrokaempferol might be dehydrogenated to kaempferol in the plant, the experiment with kaempferol was necessary as a control. Phenylalanine is a good biosynthetic precursor for flavonoids, and the $T/^{14}C$ ratio found in the flavonoids measures the incorporation of dihydrokaempferol or kaempferol in comparison with phenylalanine. The results shown in Table 8 prove

TABLE 8

Incorporation of Dihydrokaempferol-[T], Kaempferol-[T], and Phenylalanine-[1-^{14}C] into Quercetin and Cyanidin[a]

Precursor or Compound	Dilution		Incorporation, %		
	^{14}C	T	^{14}C	T	$T/^{14}C$
Dihydrokaempferol-[T], Phenylalanine-[1-^{14}C]					9.5
Quercetin	2330	315	0.42	0.58	12.9
Cyanidin	227	48	0.43	0.38	8.3
Kaempferol-[T], Phenylalanine-[1-^{14}C]					7.0
Quercetin	4890	4690	0.22	0.075	1.8
Cyanidin	219	539	0.44	0.045	0.7

[a]Incorporations and dilutions are corrected for loss of tritium at C-2 and C-3 of dihydrokaempferol.

that dihydrokaempferol, but not kaempferol, is an efficient precursor for quercetin and cyanidin. Dihydrokaempferol (VI) is a better precursor for quercetin than is phenylalanine, as is shown by the increase of the $T/^{14}C$ ratio in quercetin. During the conversion of VI to quercetin or cyanidin the tritium at C-2 and C-3 is lost. If an isotope effect occurs during the dehydrogenation reaction, the actual incorporation rate of VI into these flavonoids should be higher. The much lower incorporation of kaempferol into quercetin and cyanidin could occur via dihydrokaempferol if the dehydrogenation step were reversible. The introduction of the 3′-hydroxyl group into ring B (see Section VII) was apparently possible only in the case of dihydrokaempferol and not in the case of kaempferol, since, otherwise, an efficient transformation of kaempferol to quercetin should be observed. According to these results and earlier investigations, the biosynthetic pathway

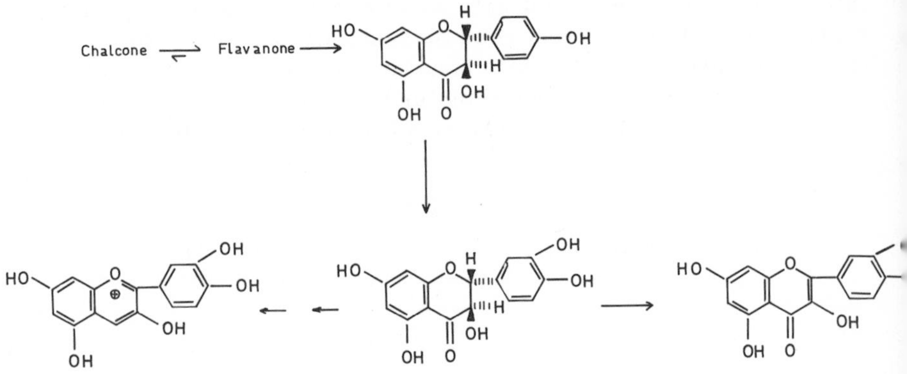

Chalcone ⇌ Flavanone → [dihydrokaempferol structure] ↓ [dihydroquercetin structure] → → cyanidin; → quercetin

Fig. 8. Biosynthesis of the flavonoids in buckwheat. Glycosylation is not indicated.

to quercetin and cyanidin in buckwheat can be formulated as shown in Figure 8 (Barz et al., 1965; Patschke et al., 1966b).

B. Incorporation into Isoflavones

The 3-hydroxyflavanones have been postulated as potential intermediates for the biosynthesis of isoflavones (Grisebach, 1961), since the 2(axial)H: 3(axial)H *trans* structure of the naturally occurring dihydroflavonols (Whalley, 1962) would make the 1,2-aryl shift via a phenonium cation intermediate possible. Attempts to bring about such a rearrangement in vitro have not met with success. Only flavones (formed by dehydration of dihydroflavonols) could be isolated from the reaction mixture (Mehta and Whalley, 1963).

In experiments similar to those carried out with buckwheat seedlings, dihydrokaempferol-[T] or kaempferol-[T] together with phenylalanine-[1-^{14}C] was fed to chana seedlings (*Cicer arietinum*), and the incorporation of both tritium and ^{14}C activity into the isoflavones was determined (Table 9). Whereas the incorporation of phenylalanine into the isoflavones was good, the

TABLE 9

Incorporation of Dihydrokaempferol–[T], Kaempferol–[T], and Phenylalanine–[1–^{14}C] into the Isoflavones of Chana Seedlings[a]

Precursor or Compound	Dilution		Incorporation, %		$T/^{14}C$
	^{14}C	T	^{14}C	T	
Dihydrokaempferol-[T]					
Biochanin A	—	1.48×10^5	—	1.57×10^{-3}	
Formononetin	—	2.98×10^4	—	3.04×10^{-3}	
Dihydrokaempferol [T], Phenylalanine-[1-^{14}C]					2.6
Biochanin A	2957	3.47×10^5	0.25	1.45×10^{-3}	1.5×10^{-2}
Formononetin	230	1.09×10^5	1.3	3.86×10^{-3}	7.4×10^{-3}
Kaempferol-[T], Phenylalanine-[1-^{14}C]					2.87
Biochanin A	3135	1.5×10^5	0.21	2.8×10^{-3}	3.6×10^{-2}
Formononetin	219	2.0×10^5	1.1	7.8×10^{-4}	1.9×10^{-5}

[a]Incorporations and dilutions are corrected for loss of tritium at C-2 and C-3 of dihydrokaempferol.

incorporation of dihydrokaempferol and kaempferol was very low and unspecific; that is, both biochanin A and formononetin were labeled. Furthermore, no significant difference in the incorporation rate of dihydrokaempferol or kaempferol could be found.

In addition, a competition experiment between dihydrokaempferol and phenylalanine-[1-^{14}C] was carried out. When dihydrokaempferol and phenylalanine-[1-^{14}C] in a molar ratio of 2 : 1 were fed to chana seedlings, the incorporation rate into biochanin A was not significantly different from that obtained after administration of phenylalanine-[1-^{14}C] alone (Table 10). The difference in the incorporation rate of phenylalanine into biochanin A and formononetin can be explained by the fact that formononetin biosynthesis proceeds at a faster rate.

From these results it must be concluded that the incorporation of dihydrokaempferol and kaempferol into the isoflavones is unspecific and occurs only via an indirect route or via breakdown products (Barz et al., 1965; Barz and Grisebach, 1966).

The dihydroflavonol garbanzol (3,7,4′-trihydroxyflavanone), which corresponds in its oxygenation pattern in ring A and ring B to formononetin, has recently been found in chana seedlings[2]

[2] The seedlings also contain 4,2′,4′-trihydroxychalcone, the corresponding flavanone (7,4′-dihydroxyflavanone), and 7,4′-dihydroxyflavonol.

TABLE 10

Competition Experiment between DL–Phenylalanine–[1–^{14}C] and Dihydrokaempferol in Chana Seedlings

Compound	Yield, μg	Specific Activity, cpm/mmole	Dilution	Incorporation Rate, %
6.35 μmole DL-Phenylalanine-[1-^{14}C]				
Biochanin A	2.85	1.18×10^7	744	0.21
Formononetin	1.10	4.32×10^7	202	0.32
6.35 μmole DL-Phenylalanine-[1-^{14}C] plus 12.8 μmole Dihydrokaempferol				
Biochanin A	4.24	6.97×10^6	1255	0.19
Formononetin	1.69	6.08×10^7	144	0.69

(Wong et al., 1965). Wong (1965), therefore, investigated the transformation of this compound into formononetin. In agreement with our results, no incorporation of the ^{14}C-labeled garbanzol into formononetin took place in vivo or in experiments with cell-free extracts from chana seedlings.

In contrast to the above results, Imaseki et al. (1965) reported the specific incorporation of radioactivity from garbanzol-[4-^{14}C] into formononetin in chana seedlings (Table 11). It can be seen from Table 11 that the incorporation of radioactivity from garbanzol into formononetin was 10 to 30 times as high as in-

TABLE 11

Incorporation of Garbanzol–[4–^{14}C] into Isoflavones in Chana Seedlings

Seedlings	Etiolated	Green	Green
Conditions of feeding	Cut end dark, 2 days	Cut end light, 2 days	Root light, 5 days
Formononetin (cpm/mmole)	5074	6317	2060
Biochanin A (cpm/mmole)	50	229	164

TABLE 12

Dilution Factors in Isoflavones with Garbanzol-[4-^{14}C] as Precursor

Seedlings	Etiolated Rootless, 5 Days		Green Whole Seedlings, 2 Days	
	Free[a]	Bound[b]	Free[a]	Bound[b]
Formononetin	5.1×10^4	7×10^4	2.5×10^5	2.1×10^4
Biochanin A	1.9×10^6	—	1.6×10^6	1.3×10^6

[a]aglycone
[b]glycoride

corporation into biochanin A. It was further shown that about 90 percent of the radioactivity in formononetin was located at C-2 (Geissman, 1965). However, the dilution of the radioactivity in formononetin was very high (Geissman, 1965; Table 12), and it must therefore be concluded that in this instance, also, the incorporation of radioactivity from garbanzol into formononetin occurred by an indirect route.

On the basis of these and earlier results, dihydroflavonols, flavonols, and flavones (Brandner, 1962) can be excluded as intermediates in isoflavone biosynthesis. Because of the mechanism of their rearrangement, chalcone epoxides are also unlikely intermediates for isoflavones (Grisebach and Barz, 1964a). Flavanones, however, are very efficiently and specifically incorporated into isoflavones (Patschke et al., 1964a, 1966a). One can therefore postulate that the aryl migration leading to isoflavones takes place at the flavanone stage (Birch, 1963). A reasonable mechanism for the rearrangement of a flavanone would be the oxidation of an enolic tautomer (Geissman, 1965; Fig. 9).

VII. DETERMINATION OF THE OXYGENATION PATTERN OF RING B OF FLAVONOIDS

The question of whether the oxygenation pattern of ring B of flavonoids is determined at the cinnamic acid stage or at the

H O Ar H H O ⇌ H O Ar OH → OX$^{\oplus}$ H O Ar O$^{\oplus}$ $-H^{\oplus}$ → isoflavone

Fig. 9. Rearrangement of a flavanone to an isoflavone.

stage of a C_{15} intermediate remains to be solved.

We have recently compared the incorporation of 4,2′,4′,6′-tetrahydroxychalcone 2′-glucoside-[β-^{14}C] (I) and 3,4,2′,4′,6′-pentahydroxychalcone 2′-glucoside-[β-^{14}C] (VIII) into cyanidin and quercetin in buckwheat seedlings (Patschke and Grisebach, 1965b). If the 3′-hydroxyl group[3] of quercetin or cyanidin is introduced before or at the chalcone stage, then VIII would be expected to be a better precursor than I for both flavonoids. However, the experiment demonstrated that both chalcones are incorporated to the same extent (Table 13). This result can be

TABLE 13

Incorporation of Tetra– and Pentahydroxychalcone into the Flavonoids of Buckwheat

Precursor or Compound	Incorporation Rate, %	Dilution
Tetrahydroxychalcone-[β-^{14}C] (I)		
Quercetin	0.11	3040
Cyanidin	2.63	14
Pentahydroxychalcone-[β-^{14}C] (VIII)		
Quercetin	0.07	3930
Cyanidin	2.26	15

explained by assuming that the introduction of the 3′-hydroxyl group takes place after the chalcone stage and that the hydroxylation reaction is not rate limiting. This assumption is in agreement with the following results:

1. With cinnamic, *p*-coumaric, and caffeic acid as precursors, the dilution values in quercetin were 148, 185, and 1268, respectively (Underhill et al., 1957). Though caffeic acid has the same hydroxylation pattern as ring B of quercetin, it is the poorest precursor. However, the significance of this result has recently been questioned by Hess (1966), who found an inhibition of flavonoid biosynthesis by caffeic acid in flowers of *Petunia hybrida*.

2. In a yellow aster mutant, 4,2′,4′,6′-tetrahydroxychalcone 4′-glucoside and the corresponding flavanone accumulate, where-

[3] Corresponds to the 3-hydroxyl group of the chalcone VIII.

as, in the red aster, cyanidin 3-glucoside but no chalcone is found (Seyffert, 1963). Since the tetra- and not the pentahydroxychalcone is accumulated when the biosynthesis of cyanidin is blocked, it must be assumed that the 3′-hydroxyl group is introduced after the chalcone–flavanone stage.

3. In the experiments with dihydrokaempferol and kaempferol, it was shown that the former but not the latter compound is an efficient precursor for quercetin and cyanidin (see p. 393). It follows that dihydrokaempferol is a good substrate for the hydroxylation at C-3′.

On the other hand, when ferulic acid-[methyl-^{14}C] or sinapic acid-[methyl-^{14}C] is fed to flowers of *Petunia hybrida*, the methylated anthocyanins, which have a substitution pattern in ring B corresponding to that of the precursors, possess the highest radioactivity (Hess, 1964). Furthermore, the best substrate for the methylating enzyme from *Petunia hybrida* petals is caffeic acid, not cyanidin (Hess, 1966). It seems, therefore, that in this material the substitution pattern of the anthocyanins is already determined at the cinnamic acid stage (Hess, 1965).

As in other cases of secondary plant metabolism (Bu'Lock, 1965), it may well be that there is not just one pathway leading to flavonoids with different substitution patterns in ring B. Perhaps determination of the substitution pattern of ring B at the cinnamic acid stage, as well as at the stage of a C_{15} intermediate, is possible.

VIII. BIOSYNTHESIS OF ROTENOIDS

Theoretical biosynthetic relationships between rotenoids and isoflavones have been discussed in detail by Grisebach and Ollis (1961). The first experimental proof of the similarity between isoflavonoid and rotenoid biosynthesis was reported by Crombie and Thomas (1965). Feeding of phenylalanine-[2-^{14}C] to one-year-year-old *Derris elliptica* plants resulted in a labeling pattern in rotenone that would be expected for a 6a ⟶ 12a aryl migration in rotenoid biosynthesis (Fig. 10).

Experimental evidence for the origin of the methylene bridge is still lacking, but it can be assumed that this carbon atom is derived from a C_1 precursor. The cooccurrence of the rotenoid dolineone (X) and the isoflavanone (neotenone; XI) in *Neorautanenia pseudopachyrrhiza* and *Pachyrrhizus erosus* (Crombie and

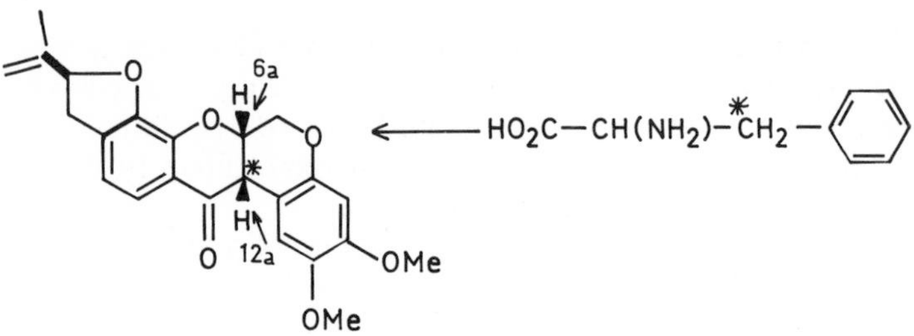

Fig. 10. Incorporation of phenylalanine-[2-^{14}C] into rotenone.

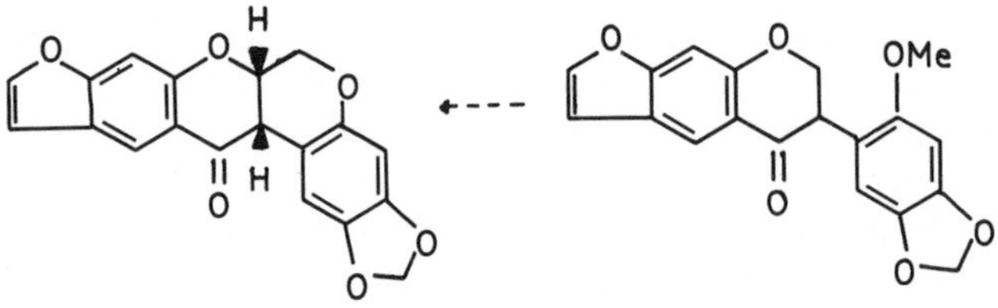

Fig. 11. Hypothetical formation of a rotenoid (dolineone) from a 2′-methoxyisoflavanone (neotenone).

Whiting, 1963) suggests that the biosynthesis of rotenoids could proceed by oxidation of 2′-methoxyisoflavanones (Fig. 11).

IX. BIOGENETIC RELATIONSHIPS OF FLAVONOIDS

From earlier investigation (Grisebach, 1965) and the experimental results discussed in this chapter, the biogenetic relationships among flavonoids shown in Fig. 12 can be postulated. The broken arrows indicate pathways that have not been investigated experimentally.

ADDENDUM

The transformation of daidzein-[T] (7,4′-dihydroxyisoflavone-[T]) into coumestrol (7,12-dihydroxycoumestan) in lucerne was investigated. D,L-Phenylalanine-[carboxyl-^{14}C] was used as an internal standard. Daidzein is a better precursor for coumestrol than phenylalanine because the T/^{14}C ratio in coumestrol increases by a factor of 2.94. These results are compatible with the assumption that coumestrol is an oxidation product of daidzein (Barz and Grisebach, 1966).

The aglycone of a second flavone glycoside present in parsley plants and fruits has been identified as chrysoeriol (5,7,4′-trihydroxy-3′-methoxyflavone). According to its hydrolytic and spectral properties the glycoside is the 7-apiosyl-glucoside of chrysoeriol. 4,2′,4′,6′-Tetrahydroxychalcone-2′-glucoside-[β-^{14}C] is incorporated into apigenin (5,7,4′-trihydroxyflavone) in parsley without randomization of the label. When a mixture of 5,7,4′-trihydroxyflavanone-[2-^{14}C] with either *trans*-3, 5, 7, 4′-tetrahydroxyflavanone-[T] or 3,5,7,4′-tetrahydroxyflavone-[T] was fed to 2-month-old parsley shoots, the incorporation of the trihydroxyflavanone into apigenin or chrysoeriol was higher by a factor of from 67 to 400 than that of the tetrahydroxyflavanone or tetrahydroxyflavone. These results prove that the flavones are formed by dehydrogenation of flavanones and not by dehydration of dihydroflavonols (Grisebach and Bilhuber, 1967).

In order to determine at what stage in the biosynthesis of biochanin A (5,7-dihydroxy-4′-methoxyisoflavone) and formononetin (7-hydroxy-4′-methoxyisoflavone) the methylation of the 4′-hydroxy group occurs, the incorporations of *p*-methoxycinnamic acid-[3-^{14}C-methyl-T], 4-methoxy-2′,4′-dihydroxychalcone-[β-^{14}C-methyl-T], and a number of unmethylated precursors into the isoflavones of chana seedlings (*Cicer arietinum* L.) were compared. p-Methoxycinnamic acid and 4,2′,4′-trihydroxychalcone were found to be the best precursors. Although the incorporation of *p*-methoxycinnamic acid was very good (incorporation rate 6.2%) the T/^{14}C ratio dropped by 95% upon incorporation into the isoflavones, a fact that showed that a rapid demethylation takes place in the plants. In addition, kinetic experiments were carried out with *p*-methoxycinnamic acid-[3-^{14}C-methyl-^{14}C-methyl-T] and 4-methoxy-2′,4′-dihydroxychalcone-[β-^{14}C-methyl-^{14}C-methyl-T]. But in this case also the rapid demethylation did not permit a definite conclusion to be drawn as to the stage in the biosynthesis at which the methylation step occurs. The very good incorporation of 4,2′,4′-trihydroxychalcone-[β-^{14}C] into the isoflavones (5.3% incorporation, dilution 61), however, shows that methylation can occur at a late stage in the biosynthesis (Barz and Grisebach, 1967).

Wong and Moustafa (1966) have partially purified an enzyme from soybean seedlings which catalyzes the conversion of 4,2′,4′-trihydroxychalcone to (–)-7,4′-dihydroxyflavanone.

p-Coumaric acid + 3 malonate
(acetate)

HO
OH
(OH)
O
CH
Aurones

HO
OH
(OH)
O
Flavones

HO
OH
(OH)
O
Chalcones

HO
OH
(OH)
O
H
Flavanones

HO
OH
(OH)
O
Dihydrochalcones

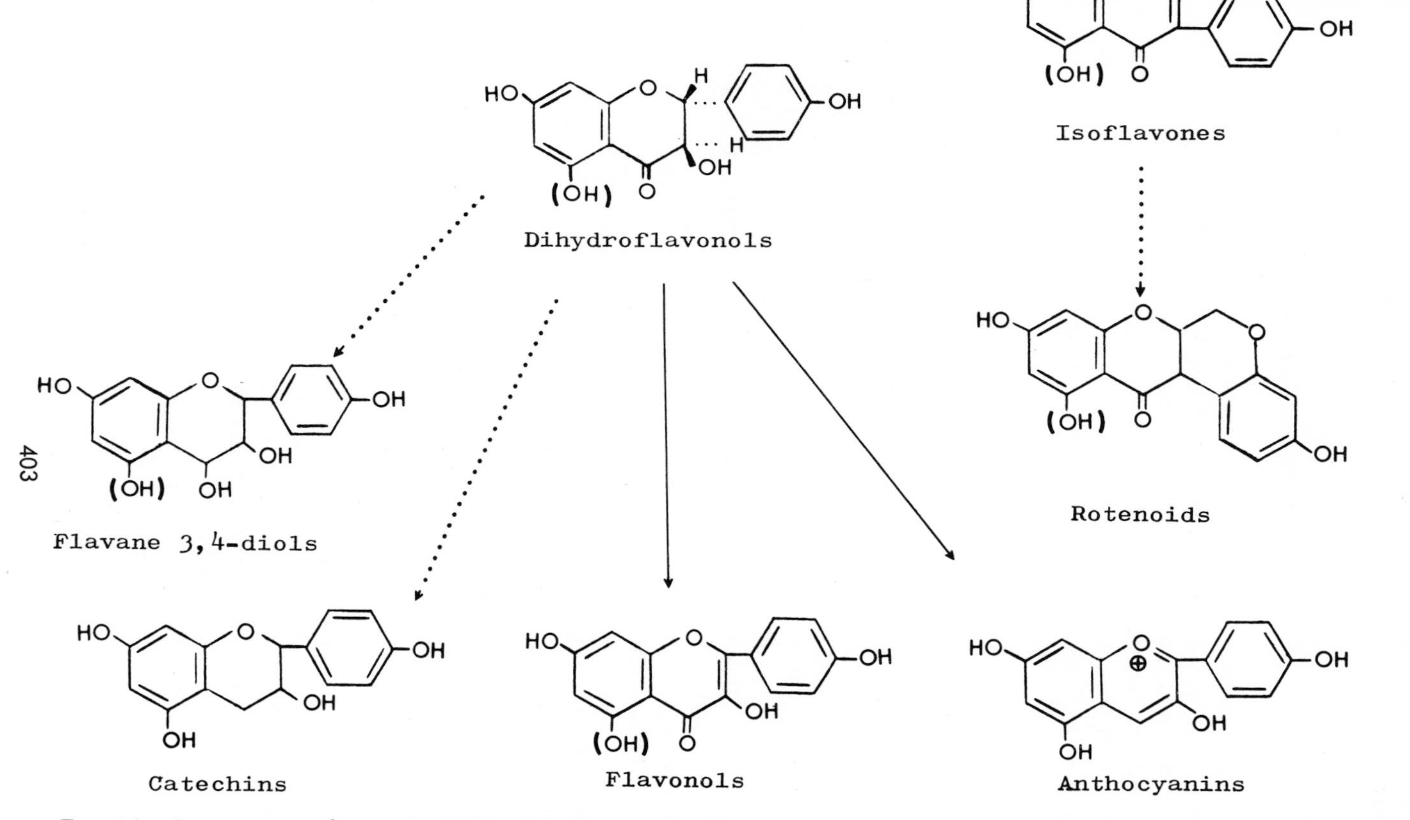

Fig. 12. Biogenetic relationships between flavonoids. The substitution pattern in ring B has not been specified. Broken arrows indicate pathways that have not been investigated experimentally.

ACKNOWLEDGMENTS

Our results, which are discussed in this chapter, have not always been obtained in such a straightforward manner as might be assumed from this short review. I owe these results to the excellent and patient work of my co-workers, who are mentioned in the appropriate references. The support of our work by the Deutsche Forschungsgemeinschaft and by the Fonds der Chemie is gratefully acknowledged.

REFERENCES

Arakawa, H., and M. Nakazaki. 1960. Chem. Industr. 73. Liebig Ann. Chem., 636: 111.
Barz, W., and H. Grisebach. 1966a. Z. Naturforsch. [B], 21: 47.
_____, and H. Grisebach. 1966b. Z. Naturforsch. [B], 21: 1113.
_____, and H. Grisebach, 1967. Z. Naturforsch. [B], 22: 627.
_____, L. Patschke, and H. Grisebach. 1965. Chem. Commun., 7: 400.
Birch, A. J. 1963. *In* Swain, T., ed., Chemical Plant Taxonomy, 149, London, Academic Press.
_____, and F. W. Donovan. 1953. Aust. J. Chem., 6: 360.
Brandner, G. 1962. Doctoral Thesis, Freiburg i. Br.
Bu'Lock, J. D. 1965. The Biosynthesis of Natural Products, 82, New York, McGraw-Hill.
Clark-Lewis, J. W. 1962. Pure Appl. Chem., 12: 96.
Crombie, L., and M. B. Thomas. 1965. Chem. Commun., 8: 155.
_____, and D. A. Whiting. 1963. J. Chem. Soc.: 1569.
Dean, F. M. 1963. Naturally Occurring Oxygen Ring Compounds, 350, London, Butterworths.
Fischer, F., and W. Arlt. 1964. Chem. Ber., 97: 1910.
Geissman, T. A. 1965. Personal Communication.
Grisebach, H. 1961. *In* Ollis, W. D., ed., Recent Developments in the Chemistry of Natural Phenolic Compounds, 69, London, Pergamon.
_____. 1962. Planta Medica, 10: 385.
_____. 1965. *In* Goodwin, T. W., ed., Chemistry and Biochemistry of Plant Pigments, 279, New York, Academic Press.
_____, and W. Barz. 1964a. Chem. Ber., 97: 1688.
_____, and W. Barz. 1964b. Z. Naturforsch. [B], 19: 569.
_____, and W. Billhuber, 1967. Z. Naturforsch. [B], 22: 746.

_____, and G. Brandner. 1961. Z. Naturforsch. [B], 16: 2.
_____, and G. Brandner. 1962. Experientia, 18: 400.
_____, and S. Kellner. 1965. Z. Naturforsch. [B], 20: 446.
_____, and W. D. Ollis. 1961. Experientia, 11: 1.
_____, and L. Patschke. 1960. Chem. Ber., 93: 2326.
_____, and L. Patschke. 1961. Z. Naturforsch. [B], 16: 645.
Hahlbrock, K. 1965. Doctoral thesis, Freiburg i. Br.
Hänsel, R., and D. Heise. 1959. Arch. Pharm. (Weinheim), 292: 398.
_____, D. Heise, H. Rimpler, and G. Pinkewitz. 1963. Arch. Pharm. (Weinheim), 296: 468.
Harraschain, H., and H. Mohr. 1963. Z. Bot., 51: 277.
Hayano, M. 1962. *In* Hayaishi, O., ed., Oxygenases, 181, New York, Academic Press.
Hess, D. 1964. Planta, 60: 568.
_____. 1965. Umschau in Wissenschaft und Technik, 65: 49.
_____. 1966. Pflanzenphysiol., 55: 374.
Imaseki, H., R. E. Wheeler, and T. A. Geissman. 1965. Tetrahedron Lett., 23: 1785.
Mehta, P. P., and W. B. Whalley. 1963. J. Chem. Soc.: 5327.
Millerd, A., and J. Bonner. 1954. Arch. Biochem. Biophys., 49: 343.
Mohr, H., and E. van Nes. 1963. Z. Bot., 51: 1.
Neu, R. 1958. Nature, 182: 660.
_____. 1960. Arch. Pharm. (Weinheim), 293: 169.
Pachéco, H. 1960. C. R. Acad. Sci. (Paris), 251: 1077.
Patschke, L., and H. Grisebach. 1965a. Z. Naturforsch. [B], 20: 399.
_____, and H. Grisebach. 1965b. Z. Naturforsch. [B], 20: 1039.
_____, W. Barz, and H. Grisebach. 1964a. Z. Naturforsch. [B], 19: 1110.
_____, D. Hess, and H. Grisebach. 1964b. Z. Naturforsch. [B], 19: 1114.
_____, W. Barz, and H. Grisebach. 1966a. Z. Naturforsch. [B], 21: 201.
_____, W. Barz, and H. Grisebach. 1966b. Z. Naturforsch. [B], 21: 45.
Seyffert, W. 1963. Personal communication.
Underhill, E. W., J. E. Watkin, and A. C. Neish. 1957. Canad. J. Biochem., 35: 219.
Whalley, W. B. 1962. *In* Geissman, T. A., ed., The Chemistry of Flavonoid Compounds, 455, Oxford, Pergamon.

Wollenberg, H., and M. Wenzel. 1963. Z. Naturforsch. [B], 18: 8.
Wong, E. 1964. Chem. Industr.: 1985.
_____. 1965. Biochem. Biophys. Acta, 111: 358.
_____, and E. Moustafa. 1966. Tetrahedron Lett., 26: 3021.
_____, P. J. Mortimer, and T. A. Geissman. 1965. Phytochemistry, 4: 89.
Zemplén, G., R. Bognár, and J. Székely. 1943. Ber. Dtsch. Chem. Ges., 76: 386.

SUBJECT INDEX

SYSTEMATIC INDEX

Recent advances in phytochemistry. v. 1–
1968–
New York, Appleton-Century-Crofts.

v. illus. 24 cm. annual.

Vol. for 1968 is the Proceedings of the 6th annual symposium of the Plant Phenolics Group of North America, 1966.

1. Botanical chemistry—Collected works. I. Plant Phenolics Group of North America. Proceedings of the annual symposium.

QK861.R38 581.1'92 67-26242

Library of Congress [5]